Advances in Industrial Control

Springer
London
Berlin
Heidelberg
New York
Barcelona
Budapest
Hong Kong
Milan
Paris
Santa Clara
Singapore
Tokyo

Other titles published in this Series:

Microcomputer-Based Adaptive Control Applied to Thyristor-Driven D-C Motors
Ulrich Keuchel and Richard M. Stephan

Expert Aided Control System Design
Colin Tebbutt

Modeling and Advanced Control for Process Industries, Applications to Paper Making Processes
Ming Rao, Qijun Xia and Yiquan Ying

Modelling and Simulation of Power Generation Plants
A.W. Ordys, A.W. Pike, M.A. Johnson, R.M. Katebi and M.J. Grimble

Model Predictive Control in the Process Industry
E.F. Camacho and C. Bordons

H_∞ Aerospace Control Design: A VSTOL Flight Application
R.A. Hyde

Neural Network Engineering in Dynamic Control Systems
Edited by Kenneth Hunt, George Irwin and Kevin Warwick

Neuro-Control and its Applications
Sigeru Omatu, Marzuki Khalid and Rubiyah Yusof

Energy Efficient Train Control
P.G. Howlett and P.J. Pudney

Hierarchical Power Systems Control: Its Value in a Changing Industry
Marija D. Ilic and Shell Liu

System Identification and Robust Control
Steen Tøffner-Clausen

Genetic Algorithms for Control and Signal Processing
K.F. Man, K.S. Tang, S. Kwong and W.A. Halang

Advanced Control of Solar Plants
E.F. Camacho, M. Berenguel and F.R. Rubio

Control of Modern Integrated Power Systems
E. Mariani and S.S. Murthy

Advanced Load Dispatch for Power Systems: Principles, Practices and Economies
E. Mariani and S.S. Murthy

Supervision and Control for Industrial Processes
Björn Sohlberg

Modelling and Simulation of Human Behaviour in System Control
Pietro Carlo Cacciabue

Krzysztof Kozlowski

Modelling and Identification in Robotics

With 50 Figures

Springer

Professor Krzyzstof Kozlowski
Faculty of Electrical Engineering, Technical University of Poznan,
ul. Piotrowo 3a, 60 -965 Poznan, Poland

ISBN 3-540-76240-X Springer-Verlag Berlin Heidelberg New York

British Library Cataloguing in Publication Data
Kozlowski, Krzysztof
Modelling and identification in robotics. - (Advances in industrial control)
1. Robotics 2. Robots - Dynamics - Computer simulation
3. Robots - Programming
1. Title
629.8'92'0113
ISBN 354076240X

Library of Congress Cataloging-in-Publication Data
Kozlowski, Krzysztof, 1951-
Modelling and identification in robotics / Krzysztof Kozlowski.
p. cm. -- (Advances in industrial control)
Includes bibliographical references (p.).
ISBN 3-540-76240-X (hardcover : alk. paper)
1. Robotics--Mathematical models. 2. Robots--Dynamics--Mathematical models. 3. System identification. I. Title.
II. Series.
TJ211.K69 1998 97-51769
629.8'92--dc21 CIP

Typesetting: Camera ready by author
Printed and bound at the Athenæum Press Ltd., Gateshead, Tyne and Wear
69/3830-543210 Printed on acid-free paper

Advances in Industrial Control

Series Editors

Professor Michael J. Grimble, Professor of Industrial Systems and Director
Dr. Michael A. Johnson, Reader in Control Systems and Deputy Director

Industrial Control Centre
Department of Electronic and Electrical Engineering
University of Strathclyde
Graham Hills Building
50 George Street
Glasgow G1 1QE
United Kingdom

Series Advisory Board

Professor Dr-Ing J. Ackermann
DLR Institut für Robotik und Systemdynamik
Postfach 1116
D82230 Weßling
Germany

Professor I.D. Landau
Laboratoire d'Automatique de Grenoble
ENSIEG, BP 46
38402 Saint Martin d'Heres
France

Dr D.C. McFarlane
Department of Engineering
University of Cambridge
Cambridge CB2 1QJ
United Kingdom

Professor B. Wittenmark
Department of Automatic Control
Lund Institute of Technology
PO Box 118
S-221 00 Lund
Sweden

Professor D.W. Clarke
Department of Engineering Science
University of Oxford
Parks Road
Oxford OX1 3PJ
United Kingdom

Professor Dr -Ing M. Thoma
Westermannweg 7
D-30419 Hannover
Germany

Professor H. Kimura
Department of Mathematical Engineering and Information Physics
Faculty of Engineering
The University of Tokyo
7-3-1 Hongo
Bunkyo Ku
Tokyo 113
Japan

Professor A.J. Laub
College of Engineering - Dean's Office
University of California
One Shields Avenue
Davis
California 95616-5294
United States of America

Professor J.B. Moore
Department of Systems Engineering
The Australian National University
Research School of Physical Sciences
GPO Box 4
Canberra
ACT 2601
Australia

Dr M.K. Masten
Texas Instruments
2309 Northcrest
Plano
TX 75075
United States of America

Professor Ton Backx
AspenTech Europe B.V.
De Waal 32
NL-5684 PH Best
The Netherlands

To my wife,
Wanda,
and children,
Stanisław and Marta

SERIES EDITORS' FOREWORD

The series *Advances in Industrial Control* aims to report and encourage technology transfer in control engineering. The rapid development of control technology impacts all areas of the control discipline. New theory, new controllers, actuators, sensors, new industrial processes, computing methods, new applications, new philosophies..., new challenges. Much of the development work resides in industrial reports, feasibility study papers and the reports of advanced collaborative projects. The series offers an opportunity for researchers to present an extended exposition of such new work in all aspects of industrial control for wider and rapid dissemination.

The robotics control field readily exemplifies the multi-disciplinary nature of control engineering. In this monograph by Krzysztof Kozlowski are the science of mechanics, and computer science and the engineering disciplines of control, electronics and mechanical engineering. The work has an objective of deriving and validating robotic arm models for different types of industrial robots. Consequently the methods of model identification play a significant role in the research reported. How to do practical parameter identification is always a mixture of theory, computational consideration, insight and art. The way in which this combination has come together to yield consistent and repeatable results is well demonstrated in this monograph. The work concludes with a chapter on approaches to position control for different robot types.

The systematic approach to robots engineering, model derivation, parameter estimation and robot control should be useful for the student and the expert engineer alike. However, the practical insights reported should be especially valuable to all engineers working with robotic systems.

M.J. Grimble and M.A. Johnson
Industrial Control Centre
University of Strathclyde,
Glasgow, G1 1QE.
Scotland, U.K.

PREFACE

The idea of this monograph was conceived when I wrote my Habilitation Thesis entitled "Mathematical dynamic robot models and identification of their parameters" [122] concerning theoretical aspects of robot dynamics modelling and identification of their parameters. I have been interested in robotics for the last ten years. My first project in robotics entitled "Optimisation methods for nonlinear robot control", was for the Institute of Fundamental Problems in Technology, Polish Academy of Sciences. The results of this work are presented in four reports [97, 98, 112, 119]. In these reports the theoretical aspects of robot modelling and identification of dynamic parameters of robot models are presented. Results of this work were published later in the Habilitation Thesis [122]. In the thesis dynamic equations of motion for an open kinematic chain consisting of n rigid bodies are formulated. The equations of motion are formulated based on the Lagrangian and Newton-Euler formalism. These equations are presented in a form which is suitable for the purpose of identification of their parameters. The parameters which appear in these equations are: mass, centre of mass (moment of the first order) and elements of the inertia tensor of each individual rigid body. As a result of this formulation, a dynamic model of a robot, which consists of dynamic parameters which are linear combinations of the link dynamic parameters, is obtained. These models are know as canonical models which are expressed in terms of a minimum number of dynamic parameters. In the Habilitation Thesis, possible ways to obtain the canonical models are described. In addition, the theoretical results presented in [122] are illustrated by simulation considerations which were satisfactory, but only from a theoretical point of view.

Robotics is an interdisciplinary field and involves such disciplines as control theory, mechanics, computer science and electronics. My background is control engineering and I have been working in this practical field for many years. I was interested in obtaining experimental results which would verify robotic theories concerning the dynamics of robots and the identification of their parameters. I started to build an experimental set–up around the industrial IRp-6 robot in our Robot Control Laboratory at the Chair of Control, Robotics, and Computer Science, Poznan University of Technology, Poland. The IRp-6 robot at that time was available on the Polish market and we decided to buy it both for teaching and research purposes. Unfortunately, this robot has a closed architecture and the manufacturer, Factory for Automation, Ostrów Wielkopolski, Poland did not want to provide us with any detailed technical information about the robot. The Institute of Control Engineering, Technical University of Warsaw built a bus access interface [144] between the robot controller and the host computer and also a data acquisition system which collects motor positions, velocities, and currents of the IRp-6 robot [143]. Our experimental set–up is equipped with these two devices. In addition we bought a force/torque sensor JR3 [93] which allows us to collect forces and torques for the purpose of load identification and hybrid control.

A supporting programming system which performs various functions was written for the experimental set–up. Several experiments were carried out, most of them by Dr. P. Dutkiewicz while writing his Ph.D. Thesis, under my supervision [50], entitled "Robot and load identification of the IRp-6 robot with hybrid control and programming elements". Parts of my Habilitation Thesis and Dr. P.Dutkiewicz's Thesis are in the book entitled "Modelling and identification in robotics" published in Polish by the Poznan University of Technology Press [139]. In both references many experimental results concerning the practical identification of robot and load dynamics and hybrid control are included. A lot of attention was put on optimal trajectory design for the purpose of identification in robotics and planning the experiments. These problems are very interesting in the robotics field since, in general, the identification of dynamic parameters of the mechanical systems is not a trivial problem. The last two references are a logical consequence of the theoretical considerations presented in my Habilitation Thesis. I strongly believe that the experimental results constitute a good illustrations of the theory. It is hard to consider theoretical results in such a practical and interdisciplinary field as robotics. It was my main motivation in writing this monograph, while at the same time I believe that I have maintained a reasonable balance between theory and practice throughout.

Most of the robots in industrial practice are geared robots. Therefore the results which are included in the monograph are devoted to this type of robot. Identification of their dynamic parameters is not a trivial problem due to friction phenomena. I believe that friction torques versus velocity have to be precompensated and these torques cannot be neglected. Friction depends on many factors which are sometimes difficult to describe analytically. This problem is described in the monograph. Due to the fact that geared robots are equipped with gears, the dynamic effects are to some extent decoupled because the inertia of the rigid body, seen from the motor side is divided by the square of the gear ratio. At the same time, friction phenomena within the transmission mechanisms cannot be neglected. In general, geared robots do not move very fast and this further complicates the practical identification of their dynamic parameters

Another type of robot used in industry is the direct drive robot (DDA) which is not as popular as the geared robot. These robots are usually faster than geared robots and their dynamics cannot be neglected. Although the friction effects are not as dominant as for geared robots they cannot be neglected. As an example of a robot of the first type we took the IRp-6 robot and of the second type EDDA (Experimental Direct Drive Robot) [124] built for research purposes. These two robots were used to carry out experimental work concerning robot and load dynamics identification. The experimental results were compared to those existing in robotics literature. In order to carry out the experiments on an open architecture, an experimental set–up of a one link geared robot was constructed in the Robot Control Laboratory. The identification experiments performed on it and this set-up was designed for teaching and research purposes. These results are also compared with the simulation results concerning these robots. I believe that the comparison is wide and applies to recently obtained results in the field of robotics.

The experimental work was made possible due to the kind support of the State Committee for Scientific Research (SCSR). Two research projects strictly connected to the contents of the monograph were carried out under my supervision. The first project entitled "Programming system of the IRp-6 ASEA robot, its model and force and torque sensor feedback", No 1096/S5/92/06 [34] was carried out in 1992. The second project supported by the SCSR was entitled "Acquisition, knowledge processing, and parallel computations of the optimal trajectory for robots", No 8S505 009.06 [133] was carried out in 1994-95. Besides that, a small part of the research work was supported by the Poznan University of Technology under the project "Implementation of a force and torque sensor in hybrid controller of the IRp-6 industrial robot" [127] in 1994. Under these three projects it was possible to buy specialised and expensive equipment which was necessary to perform the experiments described in the monograph. Further support came from the for Foundation for Promotion and Advanced Automation Technology, a Japanese organisation which sponsored the project entitled "Development and experimental validation of real–time inverse and forward dynamics algorithms for an industrial robot" [126] and also assisted with equipment and travelling expenses for me to attend the IEEE Robotics and Automation Conference, of one the most important conferences in the field of robotics. These scientific contacts played a very important role in the research carried out at our laboratory.

I would like to state clearly that without the support of the above institutions and Japanese organisation the experimental part of the research would be not possible. I should like to express my gratitude and thanks for their support.

The experimental work concerning the EDDA robot was performed in the Institut für Robotik und Prozessiformatik, Technische Universität Braunschweig, Germany. I was there under the DAAD (Deutscher Akademischer

Austauschdienst) in TU Braunschweig from 1st October, 1991 to 28th February, 1992. The experimental set–up was built by the German side and the experiments were carried out together with Dr. M. Prüfer. I should like to express my gratitude and thanks to, the DADD organisation and to Dr. M. Prüfer for his joint research effort.

Finally, it gives me genuine pleasure to acknowledge some people who have made this monograph possible. I express my thanks to my colleague Dr. P. Dutkiewicz for carrying out most of the experimental work presented in the monograph. His comments on the draft version of the monograph were very valuable. I should also like to thank many of my graduate students who wrote parts of the programming system and took part in experimental work too. Now they work in different fields in industry and factory automation.

I am very grateful to Mr. K. Romanowski, M.S. for his suggestions concerning the merits of the monograph and improving my English. I am also grateful to Mr. G. Niwczyk, M.S. for his help and patience and the typeset of the monograph.

Prof. M. J. Grimble and Dr. M. J. Johnson editors of the Advances in Industrial Control Monograph Series are gratefully acknowledged for their encouragement in pursuing this project.

Last but not least, thanks are due to Mr. G. Cash for reading the monograph and polishing the English that had initially been "Polished" by me in the early version of the monograph.

Poznań, December, 1997 Krzysztof R. Kozłowski

CONTENTS

LIST OF FIGURES

LIST OF TABLES

NOMENCLATURE

$a_i, d_i, \alpha_i, \theta_i$ – parameters of the modified Denavit-Hartenberg notation,
$\boldsymbol{A}(\boldsymbol{q})$ – moment of inertia matrix with dimensions $n \times n$, symmetric and nonsingular, with elements A_{ij},
$\boldsymbol{B}(\boldsymbol{q})$ – $n \times (n(n-1)/2)$ Coriolis coefficient matrix with elements $B_{ijk} = \frac{\partial A_{ij}}{\partial q_k} + \frac{\partial A_{ik}}{\partial q_j} - \frac{\partial A_{jk}}{\partial q_i}$,
$\bar{\boldsymbol{C}}(\boldsymbol{q})$ – $n \times n$ centrifugal coefficient matrix with elements $\bar{C}_{ij} = \frac{\partial A_{ij}}{\partial q_j} - \frac{1}{2}\frac{\partial A_{jj}}{\partial q_i}$,
$\boldsymbol{C}(\boldsymbol{q}, \dot{\boldsymbol{q}})$ – $n \times n$ Coriolis matrix with elements C_{ij},
${}^i\boldsymbol{c}_i$ – vector connecting the origin of the coordinate frame i with the centre of mass of rigid body i, left superscript denotes that this vector is expressed in the coordinate frame i, therefore when the manipulator moves the three components of this vector are constant,
c_i – stiffness constant of the drive system i,
C_i – centre of mass of the rigid body i,
dd_{ir} – differential of the function d_{ir},
$\boldsymbol{d}^T(\boldsymbol{q}, \dot{\boldsymbol{q}})$ – vector function with dimensions $1 \times 12n$ which appears in the integral model, for the canonical integral model its dimensions are $1 \times m_c$; this function is also named as the regression function for of integral model,
$\boldsymbol{D}(\boldsymbol{q}, \dot{\boldsymbol{q}}, \ddot{\boldsymbol{q}})$ – matrix function with dimensions $n \times 10n$ which appears in the differential model, it is also called as the input regression function of the differential model; for the differential canonical model this matrix has dimensions $n \times m_c$,
$\boldsymbol{dl}^T(\boldsymbol{q}, \dot{\boldsymbol{q}})$ – vector function with dimensions $1 \times 10n$ which appears in the integral model without viscous and Coulomb friction coefficients,
$\boldsymbol{df}_s^T = \boldsymbol{df}_s^T = [\int_{t_1}^{t_2} |\dot{q}_1| dt, \cdots, \int_{t_1}^{t_2} |\dot{q}_n| dt]$ – vector of the function associated with vector $\boldsymbol{f}_s$ in the integral model,
$\boldsymbol{df}_v^T = \boldsymbol{df}_v^T = [\int_{t_1}^{t_2} \dot{q}_1^2 dt, \cdots, \int_{t_1}^{t_2} \dot{q}_n^2 dt]$ – vector of the function associated with vector $\boldsymbol{f}_v$ in the integral model,
DDA – Direct Drive Arm,
ΔT – time increment,
E – expectation operator,
E_{ki} – kinetic energy rigid body i of the manipulator,
E_{kc} – total kinetic energy of the manipulator,

$E_{kc}(t_i)$ – total kinetic energy of the manipulator at time instant t_i
E_{pi} – potential energy of rigid body i of the manipulator,
E_{pc} – total potential energy of the manipulator,
$E_{pc}(t_i)$ – total potential energy at time instant t_i,
$\boldsymbol{C}_v$ – $diag[\dot{q}_1, \dot{q}_2 \dots \dot{q}_n]$ diagonal $n \times n$ matrix of the functions associated with the vector of velocity friction,
$\boldsymbol{C}_s$ – $diag[sgn(\dot{q}_1), sgn(\dot{q}_2) \dots sgn(\dot{q_n})]$ diagonal $n \times n$ matrix of the functions associated with the Coulomb friction coefficients,
$f(k\Delta T) = f[\boldsymbol{s}(k\Delta T), \ddot{\boldsymbol{q}}(k\Delta T)]$ – function which describes the dynamical behaviour of the system for the optimisation procedure of the input trajectory for the differential model. For the integral model, this function is a function of generalised positions and velocities,
F_{id} – coefficient which is a difference between the break away friction coefficient and the Coulomb friction coefficient for joint i,
F_{is} – viscous friction coefficient for joint i,
F_{iv} – velocity friction coefficient (or viscous friction coefficient) for joint i,
$\boldsymbol{f}_s$ – $n \times 1$ vector of the Coulomb friction coefficients for all manipulator joints,
$\boldsymbol{f}_v$ – $n \times 1$ vector of the velocity friction coefficients for all manipulator joints,
$\frac{\partial F_2}{\partial \phi_2}, \frac{\partial F_3}{\partial \phi_3}$ – nonlinear functions which describe gear ratios between joint and motor velocities of the second and third links of the IRp-6 robot. These functions depend on the geometry of the parallel structure of the IRp-6 robot and joint displacement of the second and third link, respectively,
$\boldsymbol{f}$ – spatial force exerted by a force/torque sensor on the body in point P,
FFT – Fast Fourier Transform,
$\boldsymbol{\Phi}(\boldsymbol{q}, \dot{\boldsymbol{q}}, \ddot{\boldsymbol{q}})$ – input regression matrix in the canonical differential model written as the upper triangular matrix with elements $\boldsymbol{\Phi}_{ij}^T$,
$\phi_1, \phi_2, \phi_3, \phi_4, \phi_5$ – generalised joint coordinates of the IRp-6 robot (see Fig 3.6b) and Table 3.2,
$\dot{\phi}_1, \dot{\phi}_2, \dot{\phi}_3, \dot{\phi}_4, \dot{\phi}_5$ – generalised joint velocities of the IRp-6 robot,
G_i – component i of the vector of the gravitational forces,
$\boldsymbol{\Gamma}_i$ – an update gain matrix (also called the adaptation gain matrix).
$\boldsymbol{g} = [0, 0, g_0]^T$ – gravity acceleration vector, where $g_0 = -9,81$ m/s^2,
$\boldsymbol{G}(\boldsymbol{q})$ – $n \times 1$ vector of gravitational forces,
$H(t_i) = E_{kc}(t_i) + E_{pc}(t_i)$ – sum of the total kinetic and potential energy of the manipulator at time instant t_i,
$H(k\Delta T)$ – Hamiltonian at instant time k,
${}^i\boldsymbol{I}_i$ – inertia tensor of link i (or second order moment) expressed in local coordinate frame i, therefore it is constant when the manipulator moves,
$\boldsymbol{I}_c$ – inertia tensor of the load with respect to the centre of mass of the load expressed in the coordinate frame with origin at the centre of mass,
I_{ia} – inertia moment of the actuator i,
$\boldsymbol{J}$ – Jacobian of a manipulator,

${}^{i}\boldsymbol{J}_i$ – pseudo inertia matrix of link i expressed in the local coordinate frame, and therefore is constant when the manipulator moves,
$\boldsymbol{K}_p$ – global compliance matrix of the manipulator,
$\boldsymbol{K}_p^{-1}$ – global stiffness matrix of the manipulator,
${}^{i}\boldsymbol{l}_{i+1}$ – vector which connects the origin of coordinate frame i and $(i+1)$ expressed in coordinate frame i,
$L(\cdot)$ – the Laplace transformation of a function $(\cdot)$,
$L^{-1}(\cdot)$ – inverse of the Laplace transformation,
$\boldsymbol{\lambda}$ – vector of Lagrange's multipliers,
$\mathcal{L}$ – Lagriangian,
m_i – mass of rigid body i,
$m_i^i\boldsymbol{c}_i = [m_i c_{ix}, m_i c_{iy}, m_i c_{iz}]^T$ – first order moment of rigid body i expressed in the local coordinate frame i, note that components of this vector are calculated as a result of the multiplication of the mass m_i by the vector of centre mass of body i,
m_c – number of parameters in the canonical model,
n – number of degrees of freedom of the manipulator,
$\boldsymbol{m}$ – torque vector exerted by the force/torque sensor on the rigid body,
$\boldsymbol{n}, \boldsymbol{o}, \boldsymbol{a}$ – three orthonormal vectors associated with the tool centre point, $\boldsymbol{n}$ - normal vector, $\boldsymbol{o}$ - orientation vector, and $\boldsymbol{a}$ - approach vector,
O_i – origin of coordinate frame i,
$\boldsymbol{P}$ – input correlation matrix,
$\Psi_1, \Psi_2, \Psi_3, \Psi_4, \Psi_5$ – motor joint variables for the IRp-6 robot,
$\dot{\Psi}_1, \dot{\Psi}_2, \dot{\Psi}_3, \dot{\Psi}_4, \dot{\Psi}_5$ – motor joint velocities of the IRp-6 robot,
q_i – joint i generalised position,
$\dot{q}_i$ – joint i generalised velocity,
$\ddot{q}_i$ – joint i generalised acceleration,
$\boldsymbol{q}$ – $n \times 1$ vector of generalised positions,
$\dot{\boldsymbol{q}}$ – $n \times 1$ vector of generalised velocities,
$\ddot{\boldsymbol{q}}$ – $n \times 1$ vector of generalised accelerations,
$\boldsymbol{Q}_i(q_i)$ – friction force acting at joint i,
r – current number of the basis function,
${}^{i-1}_{\ i}\boldsymbol{R}$ – direction cosine matrix between coordinate frames i and $(i-1)$,
$\boldsymbol{s} = [\boldsymbol{q}, \dot{\boldsymbol{q}}]$ – state of the system for the differential model; in the case of the integral model $\boldsymbol{s} = \boldsymbol{q}$,
$sgn(\cdot)$ – sign function of the argument $(\cdot)$,
$Tr(\cdot)$ – trace of the matrix $(\cdot)$,
$\sigma_i = 0$ – denotes that the joint i is rotational,
$\sigma_i = 1$ – denotes that the joint i is translational,
$\bar{\sigma}_i = 1 - \sigma_i$,
$(\cdot)^T$ – transpose operation,
τ_i – joint i generalised force,
$\boldsymbol{\tau}$ – $n \times 1$ vector of generalised forces,

τ_e – $n \times 1$ vector of non-potential forces (or equivalently non-dissipative forces at the system),

τ_{ifs} – Coulomb friction force or torque at joint i,

$\boldsymbol{\tau}_{fs}$ – $n \times 1$ vector of the Coulomb friction forces or torques for all the joints of the manipulator,

τ_{ifv} – viscous friction force or torque at joint i,

$\boldsymbol{\tau}_{fv}$ – $n \times 1$ vector of viscous forces or torques for all the joints of the manipulator,

${}^i\boldsymbol{v}_i = [v_{ix}, v_{iy}, v_{iz}]^T$ – 3×1 vector of linear velocity of the origin of the coordinate frame i,

$\boldsymbol{w}_s$ – 6×1 vector of 3 forces and 3 torques; (or equivalently wrench),

$\boldsymbol{x}$ – $12n \times 1$ vector consisting of dynamic parameters of all links and Coulomb and velocity friction coefficients for all joints of the manipulator,

$\boldsymbol{x}'$ – $10n \times 1$ vector of dynamic parameters of the manipulator,

$\boldsymbol{x}'_i$ – 10×1 vector consisting of 10 individual link i dynamic parameters namely; mass, three components of the first order moment and six components of the inertia tensor,

$\boldsymbol{x}'_{ic}$ – vector consisting of agregated parameters of link i which has number less or equal to 10,

$\boldsymbol{x}'_c$ – $m_c \times 1$ vector of the aggregated parameters of the canonical model (vector with a minimum number of dynamic parameters),

$\hat{\boldsymbol{x}}$ – 12×1 vector of the estimates of the parameter vector $\boldsymbol{x}$,

$\boldsymbol{X}_i$ – dynamic parameter i which appears in the vector of dynamic parameters,

$\hat{\boldsymbol{X}}_i$ – estimate of parameter $\boldsymbol{X}_i$,

${}^i\boldsymbol{\omega}_i = [\omega_{ix}, \omega_{iy}, \omega_{iz}]^T$ – angular velocity of link i,

$y = \int \boldsymbol{\tau}^T \dot{\boldsymbol{q}} dt$ – integral of the mechanical energy,

$y = F(\boldsymbol{P}_K)$ – cost function in the optimisation procedure, F can be either the condition number or the reciprocal of the smallest singular value or the determinant of the input correlation matrix,

$\boldsymbol{Y}(\boldsymbol{q}, \dot{\boldsymbol{q}}, \ddot{\boldsymbol{q}})$ – input regression matrix in the differential model written as the upper triangular matrix with elements $\boldsymbol{Y}_{ij}^T$,

CHAPTER 1
INTRODUCTION

This monograph is devoted to the research issues concerning a type of geared robots which are the most common in industrial practice. These robots are used, for example, on car industry assembly lines and in a variety of industrial applications. By a geared robot we mean a robot with drive systems which are connected with links via gear mechanisms. Therefore each motor drives a link not directly but via a gear [168]. These types of robots are described in detail in this monograph. We present the programming system, problems of formulating equations of motion describing their dynamical behaviour in time and identification of dynamic parameters which appear in the equations of motion.

A second class of robot which appears in industry, but rather seldom, is the direct drive robot (DDA). This type of robot, is not equipped with gears. As a consequence a drive system for each link and the link itself form one rigid body. Obviously methods which can be applied to formulate the equations of motion are exactly the same as used for the first type of robot. However, their dynamical behaviour is different. Direct drive robots are usually faster than geared robots due to the lack of a transmissions system. The dynamical behaviour of geared robots is different because the gears partly decouple the dynamic coupling between the links of the manipulator. This can be explained easily. An effective inertia along the axis of rotation when transformed to the motor side is divided by the square of the gear ratio which is in the range of 140–200. Therefore we can say that the dynamical behaviour is attenuated. Nevertheless, due to the gear mechanism, the friction phenomenon becomes dominant and definitely cannot be neglected. The friction torque versus velocity depends on temperature, direction of rotation, generalised positions, velocities, and other factors which are very difficult to describe in an analytical form. On the contrary, for direct drive robots friction torques can be neglected, however it is only an approximation which has to be verified experimentally. As will be shown in the monograph the friction phenomenon for direct drive robots sometimes cannot be neglected.

We want to keep the considerations presented in this monograph general and to introduce to the reader problems which can be encountered in both types of robots. The research problems discussed are basic and experimental. Basic results are concentrated on planning the identification experiment for finding an unknown set of dynamic parameters which is not trivial from a theoretical and practical point of view. The identification techniques which are used are known and come from the estimation of linear systems. However, the input correlation matrix proved to be ill–conditioned and the identification failed. Therefore, it is necessary to design a special trajectory for the purpose of identification. Engineering intuition can be wrong and the problem of finding on optimal trajectory is widely discussed in this monograph. Besides that we address some aspects of building robot dynamics models in combination with building a friction model for both types of robots. Theoretical considerations also concern methods of finding a minimum set of dynamic parameters.

The experimental part of the monograph is devoted to the identification experiments in combination with the identification of friction parameters and its modelling, in general, and to the measurements of the local and global stiffness of a manipulator. At the same time many practical problems were solved during the realisation of the main parts of this work. Experimental results were carried out using an experiment set–up built around an industrial geared IRp-6 robot. The IRp-6 robot in its original form, is not suitable for experimental work because of its closed architecture and inaccessible torque loop. However, after a lot of hard and tedious work, we eventually managed to build hardware and software tools which enabled us to carry out many experiments while also being able to use the robot for teaching purposes.

In the monograph we also present experimental results concerning the identification of dynamic parameters of a direct drive robot. This robot is called EDDA (Experimental Direct Drive Robot) [124, 181, 185, 193] and was constructed at the Institut für Robotik und Prozessinformatik of the Technische Universität Braunschweig, Germany. Part of the experimental work is a joint effort between this Institute in Germany and Chair of Control, Robotics, and Computers Science of the Poznań University of Technology, Poland.

Due to the limitations of the set–up described above we built our own experimental set–up which has direct access to the motor current loop. This set-up is a one link geared robot and any signal which is important for identification and control can be measured. This is an alternative solution to overcome the difficulties with the closed architecture of the typical industrial robot controller. Unfortunately in this case we have a robot with only one degree of freedom. All experimental work described in this monograph was carried out on the above set-ups and the results were compared to those existing in robotics literature. We wanted to show the difficulties and the practical problems which have to be solved in order to perform advanced experiments. Comparison of the results shows the potential of the approach and the advantages and disadvantages of the considerations given in this monograph.

Now we describe in a systematic way problems addressed in the monograph. The first problem is devoted to robot programming. Nowadays there are many programming languages for robots [80, 153, 226, 232, 238, 239]. There have been many attempts to find a universal language, non of which have been successful. A possible standardisation is described in references [162, 170, 171]. Unfortunately these attempts were not successful. It is mainly due to the fact that robot manufacturers are not interested in it, they design their own programming tools which result from their experience and their product is rather hermetically closed to other users. The quality of the software is sometimes quite good but unfortunately manufacturer's software is not suitable for robotics research. Roughly 300 - 400 programming languages exist, but almost none of these can be used for our research goals. A survey of robot programming languages can be found in the Habilitation Thesis written by C. Zieliński [239]. One of the aims of this monograph is to show the reader which basic structures constitute a modern programming language and cite several references by way of illustration. Since most of the experiments described in this work concern the IRp-6 robot, we present here features for this robot, but at the same time, we want to discuss these considerations in a universal way. As a consequence some considerations are specific to this particular robot, but most of them can be implemented to any robot in industrial practice. We have to state clearly that there is a substantial difference between a robot programming language and an algorithmic programming language such as, for example, Pascal or C++. The industrial robot works in a real environment which is difficult to predict and any programme working correctly and executing a particular task in a given area, cannot be repeated even if the changes in the environment are small. Because of that, robot programming, in general, is a difficult task.

The main part of this monograph is devoted to the identification of robot and load dynamic parameters with regard to geared robots and direct drive robots. The class of geared robots is represented by two robots, the IRp-6 robot which is equipped with position joint controllers and an experimental one link geared robot designed and built in our laboratory. The second class of robots is represented by the EDDA robot which has two degrees of freedom. In addition, we show how to implement position controllers for the purpose of hybrid control of robot manipulators and how to measure local and global stiffness of the IRp-6 robot. These are experiments which can be carried out on the IRp-6 robot without having direct access to the torque loop. We say that the controller has a closed structure.

The robot is a complicated mechanical structure which consists of several links (treated as rigid bodies), motors, gears and transmissions. Usually, the information about its kinematical structure is available. The robot is designed in such a way that the distances and angles between the joints are known and can be treated as data collected at a data plate of the robot. The kinematical structure of a robot is very important because most robots are kinematically controlled. This structure is assigned according to the notation used, for example, the modified Denavit-Hartenberg notation [31]. In the monograph we assume that the kinematical structure is known. If it is not the case the kinematical parameters can be identified using, for example, techniques described by Khalil and Gautier [101].

Besides the kinematical parameters which characterise a robot there are also dynamic parameters. Each link considered as a rigid body, has ten dynamic parameters: mass, centre of mass, of which there are three (which multiplied by mass defines first order moment), and six elements of the inertia tensor. All these parameters (except mass) are expressed in a local coordinate frame which is assigned according the modified Denavit-Hartenberg notation. Therefore, they are constant regardless of the movement of the robot. This observation is very important and will be used later for the purpose of identification.

Note that the manufacturers are not usually concerned about the dynamical structure of a robot because it moves rather slowly and dynamics do not play an important role. Each link is driven by a motor which has its own dynamics compare, for example, references [3, 92, 129, 166, 174, 176, 214, 215, 218]. Link can be driven directly by the motor when it is mounted at its shaft. In this particular case we say that the link and motor form one rigid body or a one degree of freedom direct drive robot. Asada and Youcef–Toumi [9] carried out a very interesting survey about direct drive robots and their construction. Most of the industrial robots are not direct drive robots. As an example consider the PUMA 560 [3, 218] or the IRp-6 [89] which are very popular in industrial applications.

The geared robots are equipped with different transmissions, for example, harmonic, planetary, hypoid, worm gears [168] which connect links with motors. Each motor and gear system has its own dynamics [3, 174, 215, 218]. The geared mechanism is usually a source of the energy dissipation due to friction phenomena which cannot neglected. Taking into account the dynamic parameters of the links and motors and the friction effects which are associated with transmissions (we usually neglected the dynamics of the transmission, but we took into account the dynamics of the spinning parts of the gear transmission), the robot from a control point of view can be treated as a black box with unknown dynamic parameters. These parameters are not listed on the data plate. Although the main part of the monograph is devoted to the identification of dynamic parameters of robots, we also include friction effects which are very important for geared robots. In order to have a greater insight into friction and gear transmission modelling we built a one degree of freedom robot. It consisted of a motor, harmonic drive and a link and was designed at the Factory for Automation, Ostrów Wielkopolski, Poland, especially for research purposes. The identification experiments presented in this work are devoted to both classes of robots and are illustrated by the IRp-6 robot, one link geared robot and the EDDA robot.

First we have to build a dynamic model of a robot. Robot dynamics modelling is the subject of many monographs compare, for example, the following references [17, 31, 58, 95, 188, 210]. There are also many papers on this subject, for example, [4, 12, 13, 20, 103, 106, 122, 125, 146, 150, 151, 233]. Usually authors of these references are not interested in robot dynamics identification, they prefer to derive equations of motion of robots because practical implementation of the identification scheme is very expensive. In order to carry out identification experiments, the robot has to be equipped with many sensors: position, velocity, acceleration and force/torque sensors. Particularly the last one is very expensive especially when we measure all three forces and torques at the local coordinate frame associated with the sensor. Industrial robots are not equipped with force/torque sensors, seldom with velocity sensors and rarely with acceleration sensors. Of course, each sensor has an individual single board computer which allows data acquisition. From the above description we see that an experimental set-up has to be equipped with all the sensors and is therefore expensive. The Chair of Control, Robotics, and Computer Science has all these sensors which makes the experimental identification possible.

From a mathematical point of view, robot dynamic models can be devided into two groups. To the first group belongs so-called differential models. For this kind of model in order to calculate the generalised forces acting at the joints of the manipulator we need to measure joint generalised positions, velocities and accelerations and we need information about the numerical values of link kinematical and dynamical parameters. Differential models are described in the following references [22, 31, 35, 59, 172, 176, 213, 224, 228, 233, 235]. Usually we derive these models based on the Lagrange equations of the second kind. Alternatively the differential model can be derived from the d'Alembert principle, however this model is less suitable for the purposes of identification because the dynamic parameters are not readily available from the components of the model derived from this principle. It is clear that to apply any identification scheme to the differential model we need to measure joint generalised torques. Gautier named [69] the differential models as the explicit dynamic models.

The second group of models is represented by integral models. By integral model we mean a model which is described in terms of generalised positions, velocities and torques. Here we consider two cases. In the first case the integral model is derived from the energy theorem [81] hence many authors named it the energy model [21, 36, 60] while others refer to it as the integral model [124, 183, 185]. The integral model derived from the energy theorem is sometimes called as the energy difference model [203, 205] since in derivation we calculate the difference between the total energy of the system between two instants of time [81]. The energy model does not need either acceleration or implicit velocity derivation. In the second case we define the so-called implicit dynamic model [69] which does not need explicit accelerations but needs the derivation of a function of the velocity (derivation of the velocity to get the joint acceleration has bee replaced by the derivation of the function $\boldsymbol{A}(\boldsymbol{q})\dot{\boldsymbol{q}}$ of the velocity, for notation see Nomenclature). Schaefers and co-authors [191] also use the implicit dynamic model in the integral form. Many authors propose integrating directly the differential model in order to avoid acceleration signals in the final equations of motion (here by equations of motion we mean the differential model defined above). This integral form of the differential model can be found, for example, in the following references [82, 154]. The integration operation is performed in the finite period of time. Lu and co-authors [154] claim that this method is not preferred because an integrator is an infinite-gain filter at zero frequency. This means that large errors can result from small low-frequency errors such as offsets. Therefore another approach is to apply a low pass filter operation to both sides of the differential model. This method was implemented by Szynkiewicz and co-authors [216] and Schaefers and co-authors [191]. Some people named such a model as the filtered integral model. All models described above belong to the second group of models, namely, to the integral models and the only difference between them is how they have been derived. From the identification point of view there is no difference between them because they are represented by the same set of dynamic parameters.

The integral model derived from the energy theorem is represented only by one equation, while other forms of the integral model and the differential model itself are represented by n equations, where n denotes a number of degrees of freedom. This difference is essential from the identification point of view. One equation collects less information than n equations of motion and we can say that the differential and integral models are richer in information than the energy model. This observation is intuitively obvious but will be discussed in the monograph.

From the identification point of view we are interested in so-called canonical models. These models are described in terms of a minimum set of dynamic parameters. This set consists of dynamic parameters which are combinations of dynamic parameters of individual links of a manipulator. The nonlinear functions which are associated with the dynamic parameters are formally linearly independent, therefore a necessary condition for successful identification is fulfilled. However, at a certain point of the input trajectory the input correlation matrix can be poorly defined. Several methods which lead to the canonical model are described in the monograph. These methods are borrowed from references [14, 62, 63, 64, 105, 108]. We have also formally proved that the differential and integral models are described in terms of the same set of dynamic parameters. Here we extend the results presented by Sheu and Walker [203, 205] to the case when the friction torques are represented by Coulomb and velocity components.

The dynamic parameters, regardless of the model type, appear linearly and in order to identify them we can use any of the methods which are known from linear estimation theory. The simplest identification method is the least squares method which can be used in its recursive form [94, 152]. We were interested in every possible simple method which could be fast enough to allow us to implement the identification scheme on-line. In order to speed up and stabilise the calculations we applied the Agee Turner factorisation to the input correlation matrix. In the monograph we discuss other identification schemes, for example, Canudas de Wit and Aubin [26] exploit the structure of the input regression matrix which appears in the differential model and they suggest using the sequential least squares method which drastically reduces the number of operations and allows on–line identification. The batch least squares processing is suitable for identification off–line and is used, for example, by Armstrong [4, 5] or Craig [30]. In the frequency domain, authors [95, 192] suggest using the covariance method or correlation techniques and Vandanjon and co–authors [222] the Fast Fourier Transform in the frequency domain which also leads to the least squares method. Regardless of the type of the identification method used, the identification of robot dynamic parameters is a difficult problem, particularly for the integral model. This is mainly due to the fact that the input correlation matrix depends on the input trajectory described in terms of generalised positions, velocities and accelerations. As mentioned above the input correlation matrix can be poorly defined depending on the chosen input trajectory and is one of the problems associated with the identification of robot dynamics.

The problem of input trajectory design is solved in the monograph. Here we extend the results discussed by Armstrong [4, 5] which were obtained originally for the differential model to the integral model. We optimised the condition number of the input correlation matrix. We carried out numerical calculations concerning the optimisation algorithm of the condition number for differential and integral models. Some numerical and experimental results can be found in references [48, 51, 130, 132]. The optimisation algorithm was extended to the differential model of the load dynamic parameters identification. It is shown that using optimal trajectories we are able to identify the Coulomb and viscous coefficients of the integral model of the IRp-6 robot, which is not possible at all using non-optimal trajectories. We show many experiments concerning the integral model of the IRp-6 robot which illustrate the usefulness of the integral model for the purposes of robot dynamic parameters identification. Since both differential and integral models are described in terms of exactly the same dynamic parameters, those defined for the integral model can be used in the differential model for the purposes of control. Note that Prüfer and co-authors [183] and other researchers prefer to use the differential model instead of the integral model. It is one of the reasons why we discuss the integral model widely in this work and shows that it can be used successfully for the purpose of identification of robot dynamics. The results presented here are extensions to those known for the differential model in engineering practice. Compare, for example, the following references which are devoted to the investigations of the identification of the dynamic parameters of the differential models [1, 22, 68, 70, 150, 151, 196, 197, 220, 240].

We also propose another method which allows identification of complicated friction characteristics (here friction torque versus joint velocity). Functions associated with velocity and Coulomb friction coefficients are prone to be linearly dependent particularly for the integral model. Seeger and Leonhard [196] and later Seeger [197] suggested precomputing the friction torque characteristics from the generalised forces for the differential model. Here we extend these results to the integral model regardless of the robot type. It is shown that these characteristics can be measured in a systematic way and that the dissipative energy can be precomputed from the total energy of the mechanical systems. This approach seems to be interesting and was verified for the IRp-6 robot and for the one link geared robot. When we precompute the friction characteristics from the integral model the dynamic parameters can easily be identified.

Another important problem which is discussed in this monograph is a verification problem of the identified model. The IRp-6 robot is equipped with position controllers and it is not possible to implement the computed torque method [1, 28, 31, 195] to verify the accuracy of the identified model. The computed torque method is a model based control method, however it requires direct access to the torque loop. Other model based control schemes (see [30]) are not suitable because they do not require the model to be known precisely. Besides that, the linear relationship between the functions associated with the parameters is not a necessary condition for the successful implementation of model based schemes. Even if the linear relationship exists, a robot trajectory can be recovered accurately even though the knowledge about the model may be poor. In order to overcome these difficulties we propose three methods for the verification of the identified dynamic model. Note that in the monograph we are interested in the accuracy of the identified model because we want to learn about the quality of the particular identification scheme. Here the art of the identification scheme is of concern and its quality.

The first method calculates the integral of a product of generalised torques and velocities and compares it with the energy difference calculated at two instants of time based on the identified model. Obviously this method is applicable to the energy model only. The second method compares generalised torques or filtered torques with the appropriate torques calculated based on the identified model. This method is applicable to the differential and integral models (except the energy model). The third method, and in our opinion the most reliable, compares the input trajectory with the trajectory calculated based on the identified model. This comparison can be applied to the differential models and integral models (except the energy model) and seems to be the most difficult way of discussing the identification results.

Most of the experimental work in this monograph is concentrated around the IRp-6 robot which has a closed architecture and is position controlled. We needed to open its architecture to some extent (namely allowing direct access to the IRp-6 bus controller) and to construct the software tools to realise this aim. Therefore we built the necessary hardware and software tools for the experimental set–up around the IRp-6 robot. As a consequence we were able to carry out many experiments which otherwise would be not have been possible at all. We also performed special measurements on the experimental set-up. They constitute the second part of the monograph.

First we measured the local stiffness of each joint of the IRp-6 robot. Here by local stiffness we mean the stiffness of each drive mechanism which actuates a particular joint. Based on local stiffness measurements we calculated the so-called global stiffness of the IRp-6 robot which depends on its configuration. The local stiffness measurements are not easy because the IRp-6 robot has parallelogram kinematical structure, nevertheless we propose a method which overcomes this difficulty. We use a force/torque sensor in these measurements, which makes the experimental set-up more attractive and universal for other possible applications.

Secondly, we implemented a hybrid control scheme for robots having position controllers. The hybrid controller consisted of two independent controllers: position and force [107, 157, 187, 223, 235]. Due to the fact that the IRp-6 robot does not have direct access to the torque loop we cannot apply a classical hybrid controller. We model contact phenomena between the robot and environment as spring. It is known that the external forces acting on the manipulator (which can be measured by the force/torque sensor), assuming that the gravitational and frictional forces are precompensated, are related to the joint forces by means of the Jacobian transpose [9]. The increments of the generalised forces can be recalculated using the local stiffness measurements, to the increments of the joint generalised positions. This is consistent with the assumption that the local behaviour at the joint can be modelled as a local spring (compare, for example, [9, 166, 209]) and we can use the local stiffness measurements here. The local stiffness is constant and depends on the gear mechanisms between the motors and links. Based the global stiffness measurements we are able to find directions in Cartesian space in which regulation can be performed easily. This information is helpful in designing the hybrid controller for the IRp-6 robot. Examples of the hybrid controller known in literature for robots having only position controllers can be found in [33] for the PUMA 560 robot and in [77, 78] for the IRp-6 robot. Here we implemented the algorithm presented in references [77, 78] which we updated with information of the stiffness directions of the force controller in Cartesian space. Equivalent increments of the forces and torques are transformed to joint increments and the force controller is realised. The stiffness measurements are consistent to those presented in references [56, 57], but were published independently in references [42] and [49]. The original method for local stiffness measurements from the experimental point of view has been developed and implemented in our laboratory and is devoted to the IRp-6 robot.

The monograph consists of six main chapters. Two chapters are devoted to theoretical considerations, namely to the modelling of dynamics for the purposes of control and to different identification schemes used to estimate unknown robot dynamic parameters. These chapters are illustrated by analytical examples as far as dynamics is concerned and by simulation results of the least squares method used to

identify the unknown dynamic parameters. The four other chapters describe the experimental results for different robots. The obtained results are widely discussed and compared to those existing in robotics literature because we want to keep all considerations as general as possible. The specific features of the IRp-6 robot can be considered as "local". However, many comments and discussion included in this work generalise the results obtained for the IRp-6 robot and make them applicable to classes of geared and non–geared robots.

Now we summarise briefly the contents of all chapters. The second Chapter describes a concept of the experimental set-up around the IRp-6 robot. Here we discuss both hardware and software issues concerning the building of a possible universal and open experimental set–up for further different experiments. A general concept of a robot programming language is presented which allows high level programming of the IRp-6 robot. Specific features of the IRp-6 robot are included in this language. A frame concept and basic data structures are introduced which allow advanced trajectory planning in both joint and Cartesian space. These general considerations are illustrated by a simple example of palletising and by a virtual programming panel. We also present a hardware description of a one link geared robot which shows that it is possible to develop a system which has completely open architecture and is ideally suited for experiments.

The third Chapter is devoted to theoretical considerations of robot dynamics modelling for the purposes of identification of their parameters. Both types of models are considered, differential and integral. In these models we incorporate friction characteristics expressed in terms of the velocity and Coulomb friction coefficients. Different types of models are compared against each other and their usefulness for the purposes of identification is discussed. It is shown in a rigorously mathematical way that both models contain the same minimum set of parameters. A formal proof which includes both friction coefficients (velocity and Coulomb) is given. Different techniques in obtaining canonical models are discussed and compared. Finally, two analytical examples of robot dynamics are considered. The first one is devoted to the derivation of the differential and integral models of the EDDA robot. The second example shows a complete derivation of the differential and integral models of the IRp-6 robot.

The next chapter is devoted to various identification schemes which can be used for the purposes of identification of robot dynamics. Off-line and on-line identification is of interest. Identification schemes are considered in time and frequency domain. We consider the least squares method, ridge regression, sequential least squares, correlation techniques, covariance and FFT methods. Both deterministic and stochastic cases are presented. In the first case all measurement signals are assumed to be deterministic signals. In the second case the measurement signals are corrupted by additive noise with known characteristics. The estimates are evaluated and different statistical features of the estimates are calculated. The considerations in this chapter are illustrated by simulation results for the PUMA 560 robot in the case of the differential model (both deterministic and stochastic case are considered) and for the integral model of the IRp-6 robot.

The fifth Chapter describes the experimental results for three different robots: the IRp-6, one linked geared robot and the EDDA. At the beginning of the chapter optimisation results of the input trajectory design are presented. The numerical results are presented for three degrees of freedom of the IRp-6 robot for both its differential and integral models. Both results are compared and discussed with those known in robotics literature. Next, we present a new experimental method of measuring the friction characteristics for the integral model of the IRp-6 robot. The experimental results of the identification of dynamic parameters of the IRp-6 robot are discussed. Here we consider mainly the integral model, however, some comments on the differential model identification are presented too. The identification is performed by using optimal and non-optimal trajectories and these results are compared. It is demonstrated that by using optimal trajectories it is possible to identify the friction coefficients of the integral model which is not usually the case. The next experimental results concern the identification of dynamic parameters of the one link geared robot. Here we also discuss the identification of the gear transmission and electrical constants of the drive system. Finally, we present identification results for the EDDA robot [124, 185] which represent the class of direct drive robots. The experimental results presented in this chapter are compared to those existing in robotics literature.

In Chapter 6, we present identification results of load dynamics. First, we derive differential and integral models for the identification of load dynamic parameters. Here we present appropriate equations for the IRp-6 robot. Then, we discuss numerical results of the optimisation procedure of optimal trajectory design for the load identification for both types of model. The next two sections describe static (when manipulator does not move) and dynamic (when manipulator follows a prescribed trajectory in joint space) measurements of load dynamic parameters identification. As before, the obtained results are discussed with those presented in robotics literature.

Chapter 7 is devoted to the hybrid control of robots equipped with position controllers. Two control schemes are presented. We present also an original experimental method of local stiffness measurements and consequently the global stiffness calculations for the IRp-6 robot. Finally, in the last chapter we present concluding remarks and at the end of this work a list of publications.

CHAPTER 2

ROBOT HARDWARE AND PROGRAMMING SOFTWARE DESCRIPTION

2.1 Introduction

In research on robot control system the main problem is that the structure of the controller is closed. Industrial robots are equipped with controllers which are not suitable for any kind of research. These controllers are designed so that the robot is able to repeat simple operations. Therefore the control structure is not universal and the controllers are not appropriate for research purposes. In order to solve this problem the best way would be to build a new controller with an open architecture, suitable for research. Such a solution, discussed at the end of this chapter, is usually expensive. A more cost-effective alternative is to build a modular open programming architecture, which uses the resources of the existing controller but is more open and useful for research purposes. This solution is discussed first in this chapter. As a representative example we take the IRp-6 robot, which belongs to the class of geared robots.

The programming system developed here is used to programme the five degrees of freedom IRp-6 robot. The hardware structure of the robot remains unaltered. In order to obtain a direct access to the controller resources a 16–bit interface which connects an IBM compatible PC computer directly to the bus of the robot controller was built [143]. This allows the user to programme the robot in an arbitrary manner using at the same time the resources of the controller with its limitations and restrictions. In addition the robot is equipped with a measuring system which makes it possible to collect measured data such as positions, velocities, and DC currents of all the motors of the IRp-6 robot. Note that the system measures directly the motor quantities (not the joint quantities). Besides that the system is equipped with a force/torque sensor which is mounted at the wrist of the robot. In literature, for example [75, 169], specialised control systems are reported which are a part of the controller; these systems cannot be applied here since we assumed that the original structure of the controller is not altered and the same industrial robot can be used both for industrial and research purposes.

In this chapter we consider two robot systems. One is the IRp-6 robot which is a typical industrial robot with five degrees of freedom, driven by DC motors, each controlled by 8-bit microprocessor board. The industrial robot controller is based on 16-bit microprocessor. Around this robot we built an experimental set-up designed to overcome the problem of the robot controller having closed and non-universal architecture and therefore not being suitable for research purposes. The IRp-6 robot was originally position controlled. Due to the lack of the feedback torque it was not possible to design any torque control scheme for this robot.

This motivated us to build a simple, one degree of freedom robot, designed for teaching and research purposes. The second part of this chapter describes the hardware of the experimental set-up involving the robot, built at the Chair of Control, Robotics, and Computer Science at the Poznań University of Technology.

These two parts, though conceptually different are complementary. They show that regardless of the type of the robot we can built around it an experimental set-up so that the robot can be used for experimental work. Software and hardware play different roles in the two cases. In the first case the main computational burden is in the software system, specially designed for the research. In the second case the main part of the design is located in the hardware system, making the software less complex.

The most important part in our design is the software system. A number of different languages for programming industrial robots have been developed since early 70's. In references [80], [153], and [238] some of the commercially available high-level programming languages developed by industrial robot builders and research laboratories are reviewed. Rembold and Dillmann [189] give an overview of some languages used in Europe. An interesting survey of robot programming systems with different approaches and current issues in this area is given in [226]. The software system described in the monograph is based on ideas presented in [39, 40, 47, 114, 116, 120, 121]. It is written in Pascal, making use of the advantages of a general-purpose programming language without requiring to build a new language from scratch. It is to some extent based on the PASRO language by Blume and Jakob in [16], though its application area is much wider (see Zieliński [239], where the C language in programming the IRp-6 robot for research purposes is used). The whole system has open architecture and incorporates all the features of the original programming system of the IRp-6 robot. This is done through an interface of direct access to the resources of the controller. As far as robot programming language classification is concerned, the software system belongs to object-level programming languages. A set of software blocks satisfies the following goals: portability, manipulator independence, world modelling, Cartesian programming, sensor integration and force control. These goals are described in this chapter.

In order to avoid the drawbacks of the hardware and software system the IRp-6 robot we built a simple one degree of freedom robot for research purposes. The experimental set-up consists of a DC-motor, a rate generator, a harmonic drive, a control system, and a load. A harmonic drive transmission is often used in industrial robots and is very interesting because of nonlinear friction effects. Tuttle [219] investigated different harmonic drive characteristics and pointed out that harmonic drive plays significant role in control. Design of a controller which incorporates the dynamic model of the harmonic drive is presented in references [18, 186, 199]. In these references the harmonic drive is a part of the drive mechanism of an industrial robot.

Our system has open architecture and can be used both for research and educational purposes. It is a one link geared robot. Part of the system, due to the mechanical precision, was built at the Industrial Automation Factory in Ostrów Wielkopolski, Poland. A standard power controller for the system was also delivered by the same vendor.

This chapter is organised as follows. In section 2.2 considerations concerning the hardware system of the IRp-6 robot are presented. Next section, which is the main section of this chapter, is dedicated to software. It introduces the frame concept and basic data types which are required for a programming language. The concepts of trajectory generation in configuration space and Cartesian space are discussed. Next, the notion of virtual programming panel is introduced. All considerations in Section 2.3 are illustrated by a simple programming example. In Section 2.4 we present the experimental set-up designed and built for research purposes. Finally in Section 2.5 we discuss further comments on the robot programming system.

2.2 General hardware description

In this section we present general hardware requirements for the experimental set-up built around the IRp-6 industrial robot.

A robot control system dedicated for research purposes, consists of a host computer, an industrial controller, a robot position manipulation module, and a sensor module [34]. According to this, the control system for the IRp-6 robot of the structure presented in Fig.2.1 was developed. As the industrial controller for this robot has closed architecture, an interface connecting an IBM compatible host computer (PC/486-50MHz in our set-up) and the controller has been introduced. This allows the host computer to directly access the resources of the controller [144, 236]. The interface consists of two boards:

- one, called MI-AT, is placed on the controller bus,
- the other, called the IRp-6 Adapter, is a part of the PC/AT host computer input-output channel.

A parallel arbitration system allows both computer systems to work simultaneously, as well as gives the possibility to stop the controller processor (Intel 8086) for one read-write cycle of the host computer or for a longer time. Consequently, the host computer has free access to the controller's AMS-bus and can work as the master block on the bus. Therefore the host computer may fully use the system resources of the controller, i.e. the RAM memory, the axis-position controllers and the input-output devices of the robot. The throughput of the interface is 800 kbaud.

As the driving units of the robot, popular DC-current servos with rate generator feedback are used. One degree-of-freedom axis-position controller, based on the Intel 8080 8-bit microprocessor, computes the required driving torque τ using the position error read from the resolver–angle position transformer. The torque is then amplified by the power controller, which has connected the feedback signal from the velocity $\dot{q}$ of the driven axis.

The host computer and the robot controller work asynchronously and simultaneously. The synchronisation of their programmes is done by commands sent between the computers through the interface or by external events (robot input-output), or by using controller signals connected to the interrupt system of the host computer.

The hardware architecture of the control system, shown in Fig.2.1, is modular and open, making it possible to connect any new device required for research. The flexibility of the hardware system, defined and connected in a programmable way, is limited only by the host computer speed. So far, there have been two modules connected to the system: a sensor module and a robot manipulation module.

The sensor module consists of an external measurement system and force/torque sensors. Two sensors are used: German DLR [76, 84, 85] and American JR3 [93]. They are of similar construction, consisting of strain gauges elements and a specialised computer which collects measurements data. The main part of the sensor is a strain gauge head, which is actually a strain gauge bridge. It consists of a sleeve and a cross which are connected by four round rods. This construction works on the basis of a measurement bridge, therefore the strain gauges are glued to the construction in eight different locations. A special glue was used by the designers in order to carry the strains of the material which result from the applied forces and torques. The head is connected directly to the analog-to-digital converter which is a part of the measurement computer (which is a single board computer - SBC). The strain gauge measurements are analog voltage signals which are very weak and have to be amplified by the amplifier and next converted by the analog-to-digital 12 bit converter.

The SBC collects both analog and digital values of the three forces and three torques measured by the bridge. These values are scaled by a special programme which calculates them directly in the local coordinate frame associated with the sensor. The SBC is connected to the host computer via a serial or parallel interface; it is equipped with quite powerful software, which makes it possible to scale the data in Newton and in Newton metre. Filtering the measurement data via different filters which can be chosen by the engineer is also possible. The local coordinate frame can be assigned to the sensor in an arbitrary way suitable for a particular measurement environment, which can vary with different applications of the sensor. We can choose only one force or torque and display its value numerically and graphically. We can also change the speed of transmission between the SBC and the host computer.

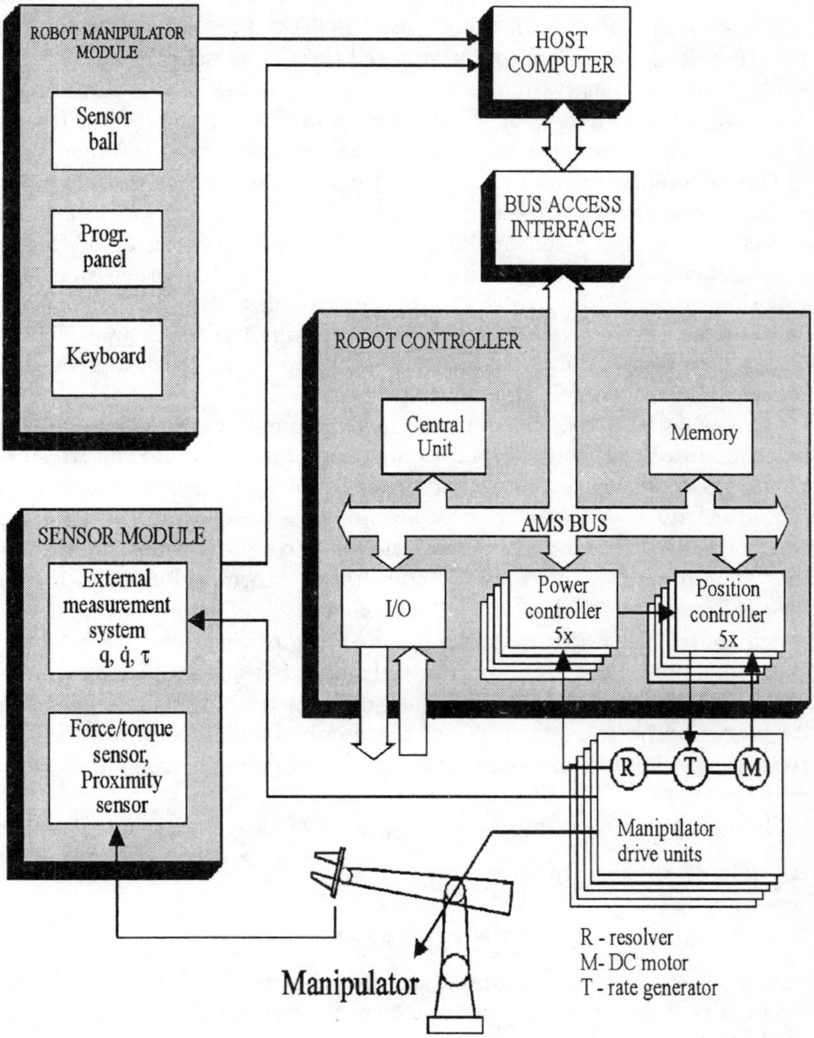

Fig. 2.1. The hardware structure of the system

The original software system of the JR3 sensor is superior to that of the DLR sensor. Details of the programming support for the sensors can be found in references [93] and [76], respectively.

In order to connect each of the two sensors to the host computer we use the interface described in reference [144], which has direct access to the resources of the IRp-6 controller (see Fig.2.1). The host computer runs a software system which collects prefiltered data from sensors.

The measurement system monitors such quantities as axis position q_i, axis velocity $\dot{q}_i$, and torque τ_i of the driving motor for all degrees of freedom of the IRp-6 robot. The JR3 force/torque sensor is mounted between the wrist and the pneumatic gripper. This sensor can measure three force components F_x, F_y, F_z and three torque components M_x, M_y, M_z with respect to the coordinate frame of the sensor. The specialised microcomputer of this sensor accomplishes such tasks as data acquisition, filtration, processing etc. and sends the processed data to the host computer through a DMA channel. Additionally, inductive proximity sensors are mounted at the gripper. They have been used to follow the path during welding operation with a given distance between the sensor itself and the path.

As mentioned before the external measurement system (see Fig.2.1) measures the motor positions, velocities and the joint torques of the IRp-6 robot. The joint torques are not measured directly; instead, the rotor currents of the DC motors are read. Note that we are not able to measure the joint torques directly due to the fact that we have no force/torque sensors mounted at the joints. The measured currents are converted to torques by use of a constant which characterises each motor. This constant is indicated on the data plate of each motor. In order to make it sure that the torques are proportional to the currents of the DC motors we dismounted the motors and measured these constants experimentally. They differed from the data given on the data plates by no more than 2% of the nominal value for all motors. Therefore it was decided to use the measurements of currents as equivalent torque measurements.

The robot manipulation module consists of three independent units:

- an industrial programming panel,
- a sensor ball,
- a keyboard connected to the host computer.

These devices are used to manipulate the robot position during the teaching process. The programming panel is connected directly to the robot controller, and the robot position can be set by a joystick mounted on the panel. The robot position can also be changed by a simple keyboard connected to the host computer. The most interesting part is the sensor ball [84] (see Fig.2.1), which allows easy, free movement of the robot arm in the workspace. This ball is a linear converter of the exerted force and torque to the displacements. These quantities are then transformed to the appropriate input values for the axis-position controllers, so one can feel as if the robot wrist is directly connected to the operator's hand. A dedicated microcomputer accomplishes the measurement, data processing, and communication with the host computer.

The control system is multiprocessor, multitasking, with hardware tasks scheduling. The demands of real-time computations make this system a distributed intelligence structure.

2.3 General software considerations

The software system is illustrated in Fig.2.2. It consists of three main parts:

- robot computer level software,
- measurement level software,
- host computer level software.

The first part was originally developed by the manufacturer in assembly language and C. It contains such blocks as the inverse and forward kinematics, which are not explicitly indicated in Fig.2.2. It also provides move statements, which execute instructions both in joint and Cartesian space. It also includes simple PD-controller programmes for one of the five degrees of freedom manipulator axes. The programming panel (shown in Fig.2.1) with joystick is used to input programmes to the 16-bit computer in the robot computer level. All other details of the original robot computer level software can be found in the technical manual. As indicated in Fig.2.2, the host computer has direct access to the resources of the robot controller through the interface [144]. This important feature allows the user to send programmes from the host computer to the memory of the controller and execute them.

The measurement level software consists of three data acquisition systems. One of them collects motor positions, velocities, and currents. The second system collects data from the force/torque sensor mounted at the wrist of the robot, and the third collects welding parameters data such as current, voltage, and velocity of the wire line feed. The host computer can read these data through serial and parallel interfaces. This part of the software is standard and we do not describe it in detail.

The most important part is the host computer level software. This part has been completely designed and developed in our Robot Control Laboratory. In its fundamental concepts it is based on the PASRO language by Blume and Jakob [16]. In addition, it incorporates some new programming elements like generation of trajectories of different shapes (both in joint and Cartesian space), procedures describing robot arm dynamics and a library of various move instructions. In the next section we describe all parts of this software according to the blocks shown in Fig.2.2.

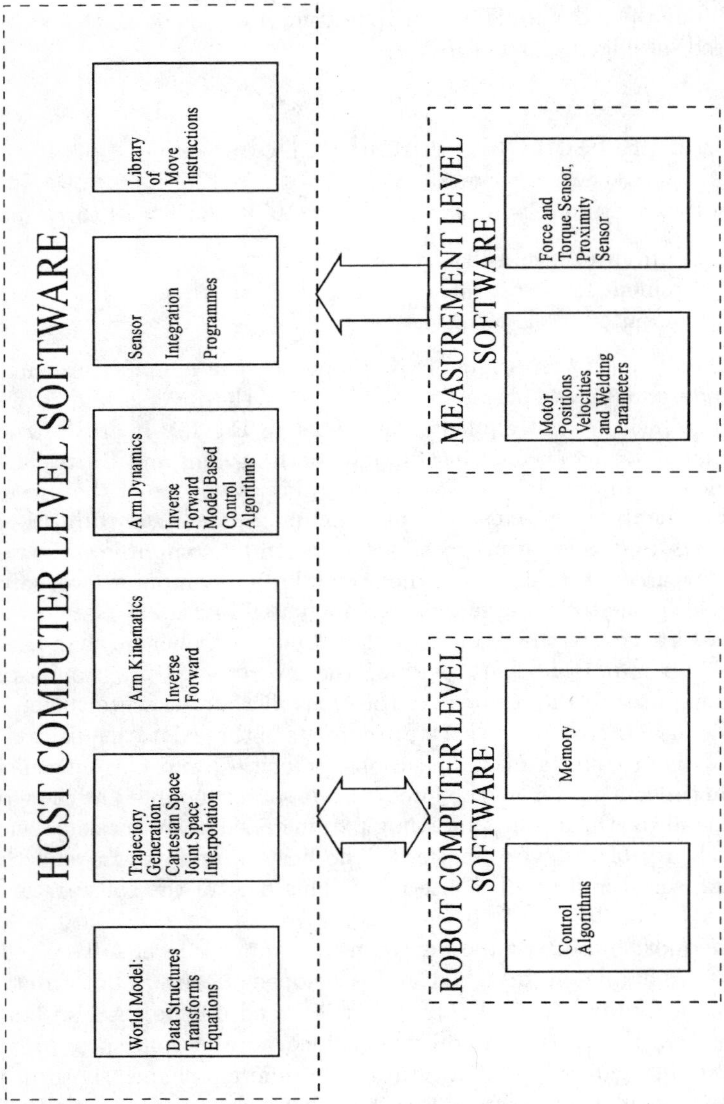

Fig. 2.2. The software system

2.3.1 Description of objects

Since any assembly process requires that objects are to be manipulated, the question arises of how to represent them in a programme. The object representation is based on the frame concept. The position of a rigid body in space can be described using six coordinates, three of them representing the position and three the orientation of the body. Such a system is called the frame [31]. The orientation of the body can be expressed in terms of the Euler angles or roll-pitch-yaw which are common in robotics literature [31].

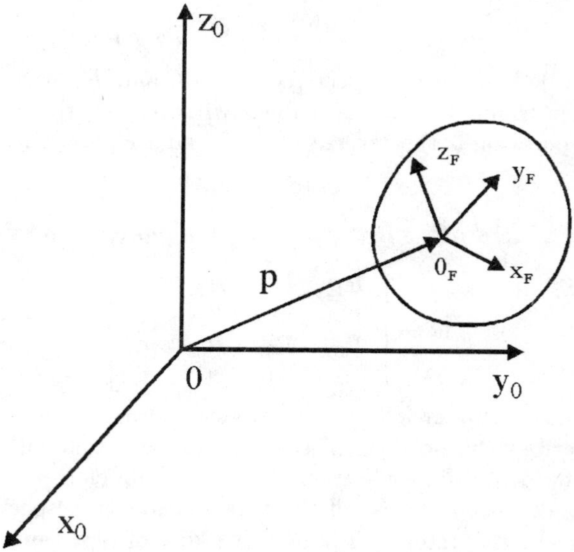

Fig. 2.3. The frame concept

The frame concept is illustrated in Fig.2.3, where the main reference coordinate system is called the base. To denote the axes of this system we use small letters with index 0. Vector $\boldsymbol{p}$ represents the distance between the origin of the base and the origin of the coordinate frame $\boldsymbol{F}$ attached to a rigid body. To represent the frame mathematically, we describe it by two components, rotation $\boldsymbol{R}$ and vector $\boldsymbol{p}$:

$$\boldsymbol{F} = (\boldsymbol{R}, \boldsymbol{p}) . \tag{2.1}$$

Rotation $\boldsymbol{R}$ can be represented as a rotation by angle α about an arbitrary axis N:

$$\boldsymbol{R} = \mathrm{Rot}(N, \alpha) . \tag{2.2}$$

Vector $\boldsymbol{p}$ has three components p_x, p_y, p_z, $\boldsymbol{p} = [p_x, p_y, p_z]^T$, where $(\cdot)^T$ means the transpose operation.

Using Euler's theorem, one can describe the equivalent angle and axis of rotation see, for example, Paul [176]. There are two equations which define very simple mathematics on rotations, namely:

$$\boldsymbol{R}^{-1} = \mathrm{Rot}(N, -\alpha) = \mathrm{Rot}(-N, \alpha) \tag{2.3}$$

and

$$\mathrm{Rot}(N, \alpha_1) \cdot \mathrm{Rot}(N, \alpha_2) = \mathrm{Rot}(N, \alpha_1 + \alpha_2) \; . \tag{2.4}$$

The multiplication of rotations or composition of rotations is not commutative so we have two alternative interpretations when we compose two rotations. The composition of two frames is a frame defined as

$$\boldsymbol{F}_1 \cdot \boldsymbol{F}_2 = (\boldsymbol{R}_1 \cdot \boldsymbol{R}_2, \boldsymbol{R}_1 \cdot \boldsymbol{p}_2 + \boldsymbol{p}_1) \; , \tag{2.5}$$

where $\boldsymbol{F}_1$ is defined with respect to the base and $\boldsymbol{F}_2$ with respect to $\boldsymbol{F}_1$. The resultant frame $\boldsymbol{F}_1 \cdot \boldsymbol{F}_2$ is defined with respect to the base. Another important operation is the inverse $\boldsymbol{F}^{-1}$ of frame $\boldsymbol{F}$, given by

$$\boldsymbol{F}^{-1} = (\boldsymbol{R}^{-1}, -\boldsymbol{R}^{-1} \cdot \boldsymbol{p}) \; . \tag{2.6}$$

In robotics literature, a frame is very often represented by a homogeneous matrix:

$$\boldsymbol{F} = \begin{bmatrix} R_{11} & R_{12} & R_{13} & p_x \\ R_{21} & R_{22} & R_{23} & p_y \\ R_{31} & R_{32} & R_{33} & p_z \\ 0 & 0 & 0 & 1 \end{bmatrix} , \tag{2.7}$$

in which the 3×3 upper-left part represents the rotation part and the last column describes the position of the origin of a frame with respect to the origin of a base coordinate system. Note that the elements of the rotation part are the direction cosines of the three axes with respect to the axes of the base coordinate system. Although this kind of representation of position of objects is redundant, it has the advantage of simplifying the computations since matrix algebra can be applied.

It is often noted [39, 46, 80, 83, 239] that modern robot programming language should be equipped with operations on frames as they play very important role in world modelling and in a programming technique called teaching by doing. Therefore a modern programming language provides the data types of scalar, vector, rotation, and frame. The standard Pascal language lacks some of these data types and special data types have to be introduced. A scalar is a real number, on which any arithmetic operation can be performed. The user can operate on objects by making use of vectors, rotations, and frames. Vectors are constructed from three scalars, and are represented as

```
type
  Vector = record
    x, y, z : real;
  end;
```

Rotation is defined as the following record:

```
type
    Rotation = record
        Axis : Vector;
        Angle : real;
        Matrix : RotMatrix;
    end;
```

where

```
type
    RotMatrix = record
        n, o, a : Vector;
    end;
```

From the above definitions it follows that rotation is represented by a 3×3 matrix, or a record of three vectors. Alternatively, rotation can be represented by the axis of rotation and the equivalent angle. Finally, rotation is represented by concatenation of these two definitions and this kind of representation is often used in robot programming.

Using previously defined vector and rotation, a frame can easily be constructed as:

```
type
    Frame = record
        Rot : Rotation;
        Transl : Vector;
    end;
```

This idea of using records for definition of new data types is well-known and has been reported by many researchers [164, 176]. It is obvious that a modern programming language has to be equipped with the data types presented above. There are several procedures which operate on vectors, rotations and frames.

First we describe the operations which can be performed on vectors. These are multiplication of a vector by a scalar, finding the length of a vector, subtraction and addition of two vectors, scalar and cross product of two vectors. Each operation is formally defined as a procedure with formal parameters. The procedures for vectors are summarised below.

- MakeVector (var Vout : Vector; x,y,z : real)
  Creates variable Vout which is of type Vector and consists of three numbers, which are the coordinates of the vector measured in metres.
- VAbs (var Sout : real; Vin : Vector)
  Calculates length Sout of vector Vin.
- VAdd (var Vout : Vector; Vin1, Vin2 : Vector)
  Calculates sum Vout of two vectors Vin1, Vin2.

- VSub (var Vout : Vector; Vin1, Vin2 : Vector)
  Calculates difference Vout of two vectors Vin1, Vin2.
- VMul (var Vout : Vector; Vin : Vector; s : real)
  Calculates product Vout of vector Vin multiplied by scalar s.
- VRot (var Vout : Vector; Rin : Rotation; Vin : Vector)
  Creates vector Vout, which results from rotating vector Vin by the angle and along the axis of rotation defined by matrix Rin.
- VDot (var Sout : real; Vin1, Vin2 : Vector)
  Calculates scalar product Sout of two vectors Vin1 and Vin2.
- VCross (var Vout : Vector; Vin1, Vin2 : Vector)
  Calculates cross product Vout of two vectors Vin1 and Vin2.

The following operations can be performed on rotations and frames.

- MakeRotation (var Rout : Rotation; Vin : Vector; s : real)
  Creates rotation Rout, which is defined by means of the axis of rotation Vin and the angle of rotation s.
- RotAngle (var Sout : real; Min : RotMatrix)
  Calculates the angle of rotation Sout from the rotation matrix Min.
- RotAxis (var Vout : Vector; Min : RotMatrix)
  Extracts the axis of rotation Vout from rotation matrix Min,
- RotRot (var Rout : Rotation; Rin1, Rin2 : Rotation)
  Calculates rotation matrix Rout, which is a superposition of two rotation matrices Rin1 and Rin2, by multiplying the rotation matrices, and then finding a new angle of rotation from procedure RotAngle and a new axis of rotation from procedure RotAxis.
- MakeFrame (var Fout : Frame; Rin : Rotation; Vin : Vector)
  Creates frame Fout from rotation matrix Rin and translational vector Vin, which has the coordinates p_x, p_y, and p_z (see Eq.(2.7)).
- SetFrame (var Fout : Frame; x,y,z : real, Vin : Vector; Sin : real)
  Creates frame Fout based on the translational vector with coordinates x, y, and z, and the rotation matrix which is formed from the axis of rotation Vin and rotation Sin (compare two alternative ways of constructing the rotation matrix).
- FrameTrans (var Fout : Frame; Fin : Frame; Vin : Vector)
  Translates frame Fin by vector Vin defined with respect to the coordinate frame associated with frame Fin.
- FrameRot (var Fout : Frame; Fin : Frame; Rin : Rotation)
  Rotates frame Fin along the axis and by the equivalent angle which result from rotation matrix Rin defined with respect to frame Fin.
- FrameInv (var Fout : Frame; Fin : Frame)
  Calculates inverse Fout of frame Fin (see Eq.(2.6)).
- FrameRel (var Fout : Frame; Fin1,Fin2 : Frame)
  Calculates the transformation which has to be applied to frame Fin1 in order to obtain frame Fin2.

Composition of frames is defined using up to six frames, which is very useful in an interactive teaching system [47, 118, 120, 121]. We can introduce the following procedures:

TransFrame2 (var Fout : Frame; Fin1,Fin2 : Frame)

...

up to

TransFrame6 (var Fout : Frame; Fin1,Fin2,Fin3,Fin4,Fin5,Fin6 : Frame).

There are also some pre-declared vectors, rotations, and frames, for instance xAxis, yAxis, zAxis, nilVector, nilRot, nilFrame, RobotFrame, EndFrame and RobotJoints. We do not describe them in detail since they are self-explanatory, for example, xAxis $= [1\ 0\ 0]^T$. All the procedures described in this section are based on the record definition from the standard Pascal language and composite library called ARYTEXP.SYS. They were tested successfully on an IBM compatible PC computer.

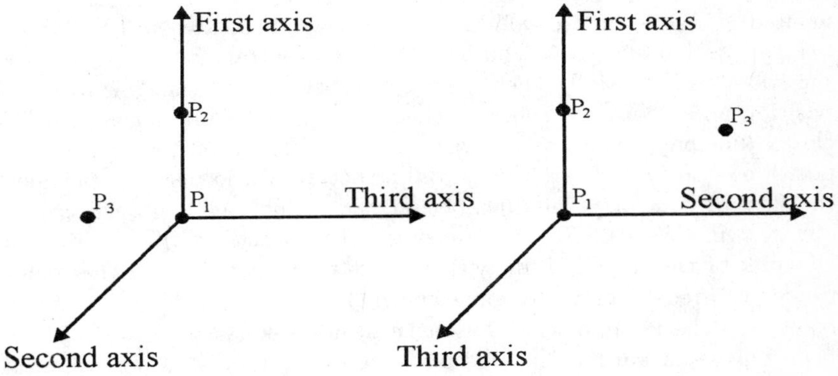

Fig. 2.4. Three pointing procedure

One of the problems with frames is in describing orientations, since it is practically quite difficult to visualise complex orientations in three-dimensional space [47, 118, 120, 217]. As mentioned before, using Euler's theorem one can define an equivalent angle and axis of rotation, and assuming a defined rotation, it is possible to extract the axis of rotation and the equivalent angle. This idea seems very simple, but it is easy only when the axis of rotation is parallel to one of the axes of the base coordinate system. With an object of an arbitrary position and orientation with respect to the base coordinate system, it is still difficult to define an unknown orientation. To overcome this we can use the robot as a measuring device in three-dimensional space. This idea is originally introduced by Grossman and Taylor [79]. They propose to use a bendy pointer which is fixed in the robot hand. First, the pointer has to be calibrated to find out the pointer transformation. A simple calibration procedure is described by Rembold and

Dillmann [189]. Constructing a frame using three pointing procedure is illustrated in Fig.2.4. Using the idea of Grossman and Taylor [79], the user has to point the robot arm at the origin of the unknown frame, then at one of the axes, and finally at the plane passing through this and another axis. Then one can easily compute:

$$\begin{aligned} \text{first axis} &= |P_2 - P_1| \\ \text{and } \{\text{third axis} &= ||P_2 - P_1| \times |P_3 - P_1||, \\ \text{second axis} &= \text{third axis} \times \text{first axis}\}, \\ \text{or } \{\text{third axis} &= ||P_3 - P_1| \times |P_2 - P_1||, \\ \text{second axis} &= \text{first axis} \times \text{third axis}\}. \end{aligned} \tag{2.8}$$

Since all industrial robots are equipped with sensors measuring the position of the robot, this idea is very useful.

In the proposed programming system, an interactive teaching system based on ideas introduced by Takase and co-authors [217] has been implemented. These ideas are well-known and several researchers have developed teaching techniques [164]. Our system only uses transformations to describe the robot workspace. In such a system, every motion clause is a request to position and orient the robot so that the position equation, representing a closed kinematic chain of homogeneous transformations, is satisfied. These position equations reflect the spatial structure of the task independently of the flow of control. A number of assigned frame names have been used in our system. We call T6 the transform that describes the position of the last link of the manipulator with respect to its base. When the manipulator is required to move to some known location POS, also described with respect to the manipulator's base, the position equation, T6=POS, is satisfied. This equation may include up to six (this number is not a limitation) transformations associated with the gripper, the objects on the scene and other specified places, which are important to execute an assembly task. A task description includes a number of these equations. Some of the transformations, which appear in these equations are known, for example, from the engineering drawings. Other transformations can be obtained by making use of the bendy pointer. A simple calibration procedure was written to compute the bendy pointer transformation. By making use of the bendy pointer and the manipulator, three different points of an object can be reached. These transformations can easily be calculated. In the teaching process, some of the frames are defined using the manipulator itself, and an equation relating any undefined transformation is obtained. This method is very systematic and allows the user to interpret the task by progressively solving for the unknown transformations. This part of the software essentially extends the original software written by the manufacturer. It makes it possible to introduce the elements of world modelling and object representation in three-dimensional space. This facilitates Cartesian programming, which is very useful to execute complex tasks, in a poorly defined environment.

Since assembly often involves attaching one object to another, our programming system has a mechanism to keep track of the location of a subsidiary piece of the assembly as the main assembly is moved; this mechanism is called affixment [164]. The programming system operates with two instructions: Affix and Unfix. They are described in more detailed in Section 2.3.6

2.3.2 Joint space trajectories

The next block of the host computer level software is the trajectory generator. The system handles two types of trajectories, those in joint space and in Cartesian space. First we describe trajectories in joint space.

Here we briefly describe joint space trajectories which are polynomials of the third and fifth order. Consider first a trajectory which is a polynomial of the third order with respect to time, namely

$$q(t) = a_0 + a_1 t + a_2 t^2 + a_3 t^3 \,, \tag{2.9}$$

where for simplicity we omitted the subscript index describing the number of the joint and t denotes the time variable. We impose the following constraints on joint variable $q(t)$:

$$\begin{aligned} &\text{– initial position } q(0) = q_0, \\ &\text{– final position } q(t_f) = q_f, \\ &\text{– initial velocity } \dot{q}(0) = \dot{q}_0, \\ &\text{– final velocity } \dot{q}(t_f) = \dot{q}_f. \end{aligned} \tag{2.10}$$

Taking into account the constraints given by Eq.(2.10) and calculating the first derivative of Eq.(2.9) we find the four unknown coefficients:

$$\begin{aligned} a_0 &= q_0 \,, \\ a_1 &= \dot{q}_0 \,, \\ a_2 &= \frac{3}{t_f^2}(q_f - q_0) - \frac{2}{t_f}\dot{q}_0 - \frac{1}{t_f}\dot{q}_f \,, \\ a_3 &= -\frac{2}{t_f^3}(q_f - q_0) + \frac{1}{t_f^2}(\dot{q}_f + \dot{q}_0) \,. \end{aligned} \tag{2.11}$$

Now we assume that the joint variable is described in terms of the fifth order polynomial trajectory with respect to time as follows:

$$q(t) = a_0 + a_1 t + a_2 t^2 + a_3 t^3 + a_4 t^4 + a_5 t^5 \,. \tag{2.12}$$

We need six initial conditions in order to calculate the polynomial coefficients given by Eq.(2.12). Therefore we impose two additional constraints to those given by Eq.(2.10), namely:

– initial acceleration $\ddot{q}(0) = \ddot{q}_0$,
– final acceleration $\ddot{q}(t_f) = \ddot{q}_f$. (2.13)

Taking into account the constraints given by Eqs.(2.10) and (2.13) and calculating the first and second derivative of function $q(t)$ given by Eq.(2.12) the six unknown coefficients can be found as follows:

$$\begin{aligned}
a_0 &= q_0\,, \\
a_1 &= \dot{q}_0\,, \\
a_2 &= \ddot{q}_0/2\,, \\
a_3 &= \tfrac{1}{2t_f^3}[20(q_f - q_0) - (8\dot{q}_f + 12\dot{q}_0)t_f - (3\ddot{q}_0 - \ddot{q}_f)t_f^2]\,, \\
a_4 &= \tfrac{1}{2t_f^4}[30(q_0 - q_f) + (14\dot{q}_f + 16\dot{q}_0)t_f + (3\ddot{q}_0 - 2\ddot{q}_f)t_f^2]\,, \\
a_5 &= \tfrac{1}{2t_f^5}[12(q_f - q_0) - 6(\dot{q}_f + \dot{q}_0)t_f - (\ddot{q}_0 - \ddot{q}_f)t_f^2]\,.
\end{aligned} \tag{2.14}$$

The coefficients given by Eqs.(2.11) and (2.14) can be found, for example, in Craig [31]. Note that the above trajectories are smooth and can be used for identification purposes.

2.3.3 Implementation of the bus access interface in trajectory planning for the IRp-6 robot

The IRp-6 robot controller has closed architecture and the user can only programme the robot using the teaching padent. This closed architecture is very inconvenient when we want to identify the robot dynamic parameters due to the limitations imposed by the standard controller. First we want to design trajectories discussed in the previous section which are suitable for the purpose of identification. In order to do this we use the bus access interface [144], which has direct access to the robot controller resources and allows the user to use them without affecting the standard controller at the same time.

Therefore a lot of effort was made to write software support for advanced programming of the IRp-6 robot. The data structures presented in Section 2.3.1 are not included in the original software of the IRp-6 robot. We built a software module named IRP6.TPU, which is essentially a library of procedures written to satisfy the requirements of the Turbo Pascal compiler. The library of procedures does not constitute a new language; it rather uses the standard Turbo Pascal compiler. As an example we show how the third-and fifth-order polynomial trajectories can be incorporated in the module. The trajectory generator uses the following procedures:

- InitIRp. This procedure assigns logically the robot controller to the host computer, which is necessary for communication between these two devices. It has to be executed at the beginning of the programme which uses the interface.

- BusLock. This procedure allows the host computer to take over the controller bus as long as it is not cancelled. As a consequence this procedure facilitates fast data transfer between the host computer and the robot controller.
- BusUnlock. This procedure releases the robot controller bus.
- wIRpIOw (Dat: word; IOAddr: word). This procedure writes a 16-bit word Dat in input device with IOAddr address in the robot controller.
- rIRpIOw (var Dat: word; IOAddr: word). This procedure reads a 16-bit word from the output device and stores it in variable Dat.

A procedure named TR_W35.EXE generates the third-and fifth-order polynomial trajectories and serves five joint controllers of type MA-70 [34], treating them as input/output ports. The I/O addresses are assigned to the joint controllers according to Table 2.1. Note that axes z_1,z_2,z_3,z_4, and z_5 are the same as in Fig.3.6a.

Table 2.1. I/O addresses assignment of the joint controllers

Joint controller axes	Port address of the position read-out and increment write-in operations (DATA)	Port address of the state read-out and command write-in operations (CTRL)
z_1	0EB00H	0EB02H
z_2	0EB04H	0EB06H
z_3	0EB08H	0EB0AH
z_4	0EB0CH	0EB0EH
z_5	0EB10H	0EB10H

Read–out operation from the DATA port is used by the TR_W35.EXE procedure in order to read the real positions of the robot joints. When the robot is synchronised (the synchronisation position is the starting position when the power electronics is turned on), the values read from the DATA port are zero. Each position is a 16-bit word coded in U2 code. Recall that in order to read correct data from this port the IRp-6 robot has to be synchronised first. Before the read-out operation of the position from the CTRL port a command with a bit not allowing write-in operation set to 1 is sent.

Write-in operation to do the DATA port is used by the procedure in order to write the increments of the each position of the joint with appropriate time. The input increment is realised in 32ms time. A 16-bit word which is sent to the DATA port has the following form:

bit	15	14	13	12	11	10	9	8	7	6	5	4	3	2	1	0
	Z	0	1	*	*	*	*	*increment*								

where: Z denotes the sign of the *increment*,
$*$ denotes that the value of a bit is irrelevant,
increment – is the absolute value of the position increment.

Before write-in operation of the new position increments the procedure waits until bit 14, named WRITTEN, in the status word of the CTRL port is equal to 1, which means that write-in operation is allowed. At the same time bit 2, of the status word named ROBSYNCHR, when set to 1 informs the programming system that the robot is in the synchronisation position.

Write-in operation to the CTRL port in the MA-70 module is executed in order to synchronise the robot. At the very beginning a status word with bit 1 set to 1 is sent out, and causes all the registers of the controllers to be set to zero. After 2ms it is necessary to sent the control word with bit STOP set to 0. Finally, in order to synchronise all position controllers it is necessary to send the status word with bit 0, named SYNCHR, set to 1. This sequence of operations makes it possible to synchronise the IRp-6 robot.

2.3.4 Description of the programme executing the polynomial trajectories.

The IRp6 robot has five degrees of freedom and constitutes of the following rigid bodies:

- the column,
- the lower arm,
- the upper arm,
- the hinge,
- the rotating end body.

The IRp-6 robot executes the following movements (see Fig.3.6b):

- rotation of the column by angle ϕ_1,
- rotation of the lower arm by angle ϕ_2,
- rotation of the upper arm by angle ϕ_3,
- rotation of the hinge by angle ϕ_4,
- rotation of the end body by angle ϕ_5.

Here we chose ϕ_1, ϕ_2, ϕ_3 ϕ_4, and ϕ_5 as the generalised coordinates. The range of each joint coordinate is [213, 237]:

- movement of the column $\phi_1 \in [-2.97\text{rad}, +2.97\text{rad}]$,
- movement of the lower arm $\phi_2 \in [-0.7\text{rad}, +0.7\text{rad}]$,
- movement of the upper arm $\phi_3 \in [-0.7\text{rad}, +0.7\text{rad}]$,
- movement of the hinge $\phi_4 \in [-2.0\text{rad}, +2.27\text{rad}]$,
- movement of the end body $\phi_{5\,\max} - \phi_{5\,\min} = 2\pi\text{rad}$.

The maximum generalised velocity of each joint is 1.57rad/s except for the velocity of the first joint which is 1.68rad/s. The maximum accelerations of the joints are the same as the magnitudes of the corresponding velocities of each joint. The generalised coordinates ϕ_1, ϕ_2, ϕ_3 ϕ_4, and ϕ_5 are associated with the links of the manipulator. Since the IRp-6 robot is a geared mechanism, the motor variables Ψ_1, Ψ_2, Ψ_3 Ψ_4, and Ψ_5 have to be calculated separately. We could use Ψ_i as generalised coordinates but from our engineering practice it is more intuitive to deal with the joint coordinates rather than with the motor coordinates. Nevertheless from a conceptual point of view both sets are equally suitable. The motor coordinates can be calculated from the joint coordinates and vice versa. This problem is discussed in detail in Section 3.7.2.
Here we summarise the results:

$$\Psi_1 = -n_1\phi_1 , \tag{2.15}$$

$$\Psi_2 = \frac{2\pi}{SK}\sqrt{B^2 - A^2\sin(\beta - \phi_{20} + \phi_2)} + C_1 , \tag{2.16}$$

$$\Psi_3 = -\frac{2\pi}{SK}\sqrt{B^2 - A^2\sin(\phi_3 + \beta)} + C_2 , \tag{2.17}$$

$$\Psi_4 = -n_4(\phi_2 + \phi_3 + \phi_4) , \tag{2.18}$$

$$\Psi_5 = -n_4(\phi_2 + \phi_3 + \phi_4) - n_4 n_5 \phi_5 , \tag{2.19}$$

where: $\Psi_1 \div \Psi_5$ – motor variables (a change in the motor variable Ψ_i by 2π is equivalent to 256 increments),
n_1, n_4, n_5 – corresponding gear ratios whose numerical values are $n_1 = 160$, $n_4 = 128$, and $n_5 = 32/19$,
SK – a slip of the screw drive, which is equal to 5 mm,
C_1, C_2 – integration constants, which are calculated from the following expressions:

$$C_1 = -\frac{2\pi}{SK}\sqrt{B^2 - A^2\sin(-\phi_{20} + \beta)} , \tag{2.20}$$

$$C_2 = \frac{2\pi}{SK}\sqrt{B^2 - A^2\sin\beta} , \tag{2.21}$$

$$A^2 = 2\sqrt{a^2 + b^2}\sqrt{d^2 + e^2} , \tag{2.22}$$

$$B^2 = a^2 + b^2 + d^2 + e^2 , \tag{2.23}$$

$$\beta = \arctan\frac{e}{d} . \tag{2.24}$$

The constants which appear in Eqs.(2.20)÷(2.24) have the following numerical values: $\phi_{20} = 0.26$rad, $e = 155$mm, $d = 183$mm, $\sqrt{a^2 + b^2} = 140$mm, $A = 259$mm, and $B = 278$mm.

In order to implement the trajectory generator we have to synchronise the robot first, using a procedure called the synchronisation of the robot, which is a part of the TR_W35.EXE programme. As a result of this procedure all the joints of the IRp-6 robot are in their initial position, which is assumed to be zero.

Trajectory planning of the IRp-6 robot is simple and can be done using the TR_W35.EXE module. In order to generate a trajectory we have to specify the order of the polynomial, the initial and final values of the joint generalised positions, and the duration of the trajectory. It is assumed that the initial and final velocities and accelerations are zero.

Program TR_W35.EXE is written in the Pascal language and is built using the Turbo Pascal 6.0 compiler. The procedures which serve the bus access interface and the trajectory generator consist of a module called SP_W35.TPU. These procedures can be used by other programmes when at the beginning of the programme we specify the statement uses st_w35. The TR_W35.EXE module consists of the following procedures.

Position (var Pos0 : Vec5) – reads the generalised positions of all the joints starting from the column of the robot and ending at the rotating end body and places these positions in vector Pos0 consisting of five elements. As mentioned before, the robot has to be synchronised and before calling the Position procedure two other procedures, InitIRp and BusLock, have to be executed.

Range (Posf : Vec5; var LeftRight : Tab; var pp: Boolean) – calculates the range of possible changes of all the generalised positions and informs the system whether the current position is within the specified range. As before the robot has to be synchronised and procedures InitIRp and BusLock have to be executed before calling the Range procedure. The arguments are:

- Posf – a five-element vector holding the input values of the generalised positions of all the joints, expressed in radians.
- LeftRight – a table holding the boundary values of the generalised positions of all the joints. The values are expressed in radians.
- pp – a logical variable which is true if all the generalised coordinates are within the working range of the manipulator and false when the positions do not satisfy this requirement.

MoveIRp6(var Pos0, Posf, Vel0, Velf, Acc0, Accf, Tim, Pol: Vec5; var pp, pr, st: Boolean) – moves the robot according to the initial conditions specified by the user when conditions pp, pr, and st are satisfied. Before executing this instruction the robot has to be synchronised. The formal arguments are:

- Pos0, Vel0, Velf – five-element vectors consisting of the initial values of the generalised coordinates, velocities, and accelerations, specified starting from the column of the robot. The values are expressed in radians, rad/s, and rad/s^2, respectively.
- Posf, Velf, Accf – five-element vectors consisting of the final values of the generalised coordinates, velocities, and accelerations, specified

starting from the column of the robot. The values are expressed in radians, rad/s, and rad/s^2, respectively.

Tim – a five-element vector consisting of the times of duration of movements of the joint variables specified starting from the column of the robot. The shortest possible time is 0.2 s.

Pol – this vector variable specifies the order of the polynomial which can be 3 or 5, for all the joints of the robot.

pp – this logical variable is exactly the same as in the Range statement above.

pr – this logical variable is true if the joint velocities are within the range of the generalised velocities specified for the robot, otherwise it has value false.

st – a logical variable which is true if the specified order of the polynomial is 3 or 5, otherwise it has value false.

Programme TR_W35.EXE allows to plan the polynomial trajectories in joint coordinates for the IRp-6 robot. Using this programme we can execute smooth trajectories with a trapezoidal velocity profile. Note that we can assign a different total time of movement to each joint.

With our programming system it is also possible to move a single joint according to a parabolic polynomial, and a cubic polynomial with via points. We used these trajectories to identify the dynamic parameters of the robot (see Chapter 5) and we are able to identify these parameters for the first three links of the manipulator. These trajectories were also used to identify the load parameters (see Chapter 6) by making use of the force/torque sensor mounted at the wrist.

Now we illustrate the considerations on the trajectory planning of the IRp-6 robot by discussing instructions MoveOnLine and MoveOnScrew, which were developed and implemented in our programming system. Both instructions use signals from the force/torque sensor. These signals are necessary to identify the dynamic parameters of the load. Recall that the force/torque sensor measures three components of the force and torque in the local coordinate frame XYZ associated with it. The axes of the coordinate frame are

$\boldsymbol{o}$ – the orientation vector,

$\boldsymbol{a}$ – the approach vector,

$\boldsymbol{n}$ – the normal vector, defined as $\boldsymbol{n} = \boldsymbol{o} \times \boldsymbol{a}$.

The measured forces along the X and Y axes are in the range of $-245, 25$N to $245, 25$N, and along Z axis in the range of $-490, 5$N to $490, 5$N. The torques along all axes are the same within the range of $-19, 62$N·m to $19, 62$N·m. Procedure MoveOnLine is declared as

procedure MoveOnLine (pf : Vec5; vwyp : integer; Is_Forces, inpal : Boolean)

with the formal parameters defined below:

pf - a destination point in joint space (Vec5 = array [1..5] of real),

Is_Forces - a flag, **true** if forces are measured and **false** if they are not.

inpal - a flag, **true** if the system checks whether the destination point has been reached and **false** if this check is not performed.
vwyp - the velocity of movement (in mm/s).

Note that procedure MoveOnLine makes it possible to move the manipulator along a straight line to actual position pf with velocity vwyp [mm/s] with optional checking whether the desired forces and torques have been exceeded. The straight line is constructed in joint space. The forces and torques can be preset by the user in array Forces in the main programme. The movement stops when the preset force or torque in the specified direction(s) has been exceeded.

Procedure MoveOnScrew is declared as
procedure MoveOnScrew (vwyp : real; pf: Vec5)
where

vwyp - a slip of the screw drive,
pf - a direction of the movement described in joint space.

Both procedures were tested along straight and helix lines in joint space with force and torque limits in each direction.

Last two procedures are a part of the block in Fig.2.2 – the sensor integration programmes and are a very important part of the software. This block manages data coming from external sensors. One of them is the device which measures positions, velocities, currents, and welding parameters. Details of this can be found in reference [34].

2.3.5 Trajectory planning in Cartesian space for the IRp-6 robot

The other type of trajectories are generated in Cartesian space. One of them is the straight line in space, where position and orientation between two points in space vary linearly. Cartesian straight line motion is difficult to realise in practice, because our robot has only five degrees of freedom. Besides that, the kinematical structure of the robot is quite complex, of a parallelogram form. This makes the inverse kinematic algorithm more involved. With our robot it is almost impossible for the user to determine, given an arbitrary position, whether an orientation belongs to the group of realisable orientations, so it is checked automatically. Two approaches to generate a straight line trajectory have been implemented in our system. One, more accurate, varies all the three components of position linearly and the end-effector moves along a linear path in Cartesian space. Our system also implements the angle-axis representation to specify an orientation with three numbers. We have combined this representation of orientation with 3×1 Cartesian position representation and finally got 6×1 representation of Cartesian position and orientation. This type of interpolation appears to be very accurate. The other interpolation is based on cubic and quintic polynomials for both position and orientation. This interpolation happens to be inaccurate, particularly for fast movements. These problems are related to workspace and singularities. Among them are, for example, unreachable intermediate points, high joint rates near singularity, or start and goal reachable in different solutions. These problems are well-known and we solved them in our robot implementation.

Now we consider trajectory planning in Cartesian space in greater detail. Assume that we want to move the tip of the manipulator along a straight line in Cartesian space. In order to do that we use the frame concept introduced in Section 2.3.1. The coordinate frame which is associated with the tool centre point has three axis which are normal $\boldsymbol{n}$, orientation $\boldsymbol{o}$, and approach $\boldsymbol{a}$, respectively. Taking into account the concept of homogeneous representation we express the frame as in Eq.(2.7), in which the columns of the orientation matrix are written in terms of the vectors $\boldsymbol{n}$, $\boldsymbol{o}$, and $\boldsymbol{a}$.

A programme was developed to move the tool centre point in three-dimensional Cartesian space along a straight line from point A to point B. We assume that these two points are "expressed" in terms of the frame given by Eq.(2.7). It means that each of these points has three position coordinates p_x p_y, and p_z components and at the same time each of these two points constitutes the origin of the coordinate frame which is the tool centre frame.

Now we assume that points A and B in Cartesian space are associated with two vectors $\boldsymbol{p}_A$ and $\boldsymbol{p}_B$ which connect the origin of the base coordinate frame to these points. In order to execute movement between points A and B we propose the following procedure:

- calculate distance d between points A and B,
- calculate the value of the discretised distance according to ds := d/lod, where lod denotes the assumed number of discretisation points along the straight line between A and B,
- calculate an intermediate frame delta for the movement between points A and B according to the following statements:

```
xdelta :=ds * ca + xdelta;
ydelta := ds * cb + ydelta;
zdelta := ds * cg + zdelta;
```

  where

```
ca := ( xB – xA)/ d;
cb := ( yB – yA)/ d;
cg := ( zB – zA)/d;
```

The coordinates of points A and B are x_A, x_B, z_A, and x_B, y_B, z_B, respectively. At the initial point A the elements of the frame are the coordinates of A.

The above definition constitutes a linear interpolation between points A and B which is understood in a usual way. We also suggest to use the linear interpolation for the orientation representation. Here we assume that orientation is represented by the axis of rotation and the equivalent angle [31].

Next denote by xA and xB the axes of rotation associated with the coordinate frames at points A and B, and by angle A and angle B the equivalent angles of the coordinate frames. Recall that the orientation is represented with respect to the base coordinate frame, since we are concerned with trajectory planning in Cartesian space. In this particular representation, the orientation is here equivalent to d_ax and d_angle. The interpolation procedure is carried out in the following form:

```
xd_ax := dx + xd_ax;
yd_ax := dy + yd_ax;
zd_ax := dz + zd_ax;
d_angle := dk + angleA;
```

where

```
dx := (xaxB – xaxA)/lod;
dy := (yaxB – yaxA)/lod;
dz := (zaxB – zaxA)/lod;
dk := (angleB – angleA)/lod.
```

The above statements define the interpolation of the frame in Cartesian space.

At the intermediate points of the interpolation procedure the velocity and acceleration are assumed to be continuous. Calculation of these quantities is subdivided into three groups:

1. in the first discretisation interval:
```
veloc0[1][j] := 0;
velocf[1][j] := (pf[j] – p0[j])/(time * lod);
accel0[1][j] := 2 * velocf[j]/time;
accelf[1][j] := 0
```
2. at the intermediate points:
```
veloc0[i] := velocf[i–1];
velocf[i] := velocf[i–1];
accel0[i] := accelf[i–1];
accelf[i] := accelf[i–1]
```
3. in the last discretisation interval:
```
veloc0[i][j] := (pf[j] – p0[j])/(time * lod);
velocf[i][j] := 0;
accel0[i][j] := 0;
accelf[i][j] := – 2 * velocf[j]/time;
```

where j denotes the axis number (here j= $1, 2, \ldots, 5$) and j is the current discretisation interval, veloc0 and velocf are velocities at the beginning and at the end of the discretisation interval, pf and p0 denote the initial and final positions, and time = 32ms is the constant time interval for the IRp-6 robot.

Note that the above scheme does not guarantee that rotations occur about a single equivalent axis when moving from point to point. Rather, our scheme is a simple one which provides smooth orientation changes, using the same mathematics we developed for planning joint interpolated trajectories. One possible way to avoid the problem with orientation is to fix the axis of rotation in space and apply interpolation only to the equivalent angle, which guarantees existence of the equivalent axis. This procedure was also developed in our programming system. The problems associated with orientation formulation and its transformation from point to point can be solved using one of the methods of Sciavicco and Siciliano [195].

The programme described above uses the bus access interface [144]. Before the programme is executed the robot has to be synchronised, as it was necessary in joint space trajectory planning. The robot can be controlled using a standard keyboard or a joystick. One can also use the sensor ball [90], which is a six-dimensional converter of the forces and torques exerted on it to the position and orientation of the robot. This device is very useful in trajectory planning in Cartesian space.

The user can choose off-line or on-line mode of trajectory generation along a straight line. In the off-line mode the whole trajectory is stored in array traj in terms of the generated positions. The velocities and accelerations are stored in arrays veloc0, velocf, accel0, and accelf. At the same time we check whether all the trajectory points are within the working space of the IRp-6 robot. The corresponding positions are stored as a sequence of the joint generalised positions, calculated using the inverse kinematics algorithm (procedure Inv_Kin). In addition, all the trajectory points are accumulated as the corresponding frames. This data is stored in the host computer. Next the data is transformed to the memory of the controller using procedure MoveIRp6. In the off-line mode it is possible to check whether the subsequent positions are obtainable by the robot. In order to do that we use procedure InPosAll. The results of the execution of the trajectory in off-line mode were reproducible and very accurate.

In the on-line mode the consecutive points are calculated in real time at the host computer and next sent to the robot controller by making use of the same procedures. The InPosAll procedure is used as well.

The software discretises the whole trajectory and then calculates the position and orientation increments. In order to execute the movement in real time all procedures were optimised and we were able to perform all calculations in time less then 32ms, which is the discretisation time of the robot controller. This time can be reduced to 8ms.

A detailed list of all procedures can be found in references [34] and [139]. These procedures can be used to plan trajectories in both joint and Cartesian space. All supporting software (including kinematics, communication between the host computer and the robot controller, software for the keyboard and for the sensor ball) are written in Turbo Pascal.

The part of software described in this and previous sections is strictly connected with a library of move statements (see Fig.2.2). Move operations can be considered in joint and Cartesian space. We can move single joint or all joints, simultaneously or at different times. The operations are specified by move statements with several parameters such as velocity, duration of movement, starting and end position. In all move statements, after the move operation is completed, the final position and orientation are read from the resolvers and compared to the predefined values. Several error subroutines are incorporated into the move statements, which help the user control the move execution, particularly in Cartesian space. The move statement can be monitored during its execution.

The standard programming panel and the six-dimensional sensor ball are connected to the robot. All functions of the programming panel are implemented on the host computer. A simple interpreter was written to allow the user to programme the robot as with the standard programming panel. Instead of the standard joystick, the six-dimensional sensor ball can be used to move the robot in joint space in a flexible way. The sensor ball has several features not present in the standard joystick. For example, we can select either translational or rotational movement. We can programme the sensor ball by introducing 256 8-bit binary codes, which can be used as input flags in the programming or teaching process. This makes the system more flexible and open to the user. We decided to retain the original software of the manufacturer as an option in order not to force all users of the robot to know the new system.

As shown in Fig.2.2, there are two blocks calculating the robot kinematics and dynamics. The inverse and forward kinematics algorithms are required to perform all operations on objects in three-dimensional space in the interactive teaching system as well as in the move statements in Cartesian space. Both algorithms are computationally optimised in order to obtain fast performance of the move statements. The software was tested on an experimental set-up by executing several tasks in Cartesian space. All of them were performed successfully and the same trajectory pattern was repeated with very high accuracy.

Now we consider the elements of the interactive teaching system which is very useful in practice. Here we borrow these structures from the PAL language described by Takase and co-authors [217]. They use a structural approach to robot programming and teaching. These elements are also discussed in references [47, 115, 118, 120, 121, 238].

In Section 2.3.1 we mentioned that in order to avoid the problems with the orientation and frame representation we can use a three or four pointing procedure using the bendy pointer. This procedure consists of the following steps:

1. Mount the bendy pointer at the gripper and move it to a chosen point in Cartesian space (the movement can be executed using the joystick, keyboard, or sensor ball). This particular chosen point is assumed the origin of a new coordinate frame.
2. Point the bendy pointer to two other points. One of them lies on an axis of the new coordinate frame and the other on a plane to which the axis belongs.

3. Based on these three points and using Eq.(2.7) calculate all the axes of the new coordinate frame.

The results of these calculations are summarised in Table 2.2. In this table vectors of orientation $\boldsymbol{o}$, approach $\boldsymbol{a}$, and normal $\boldsymbol{n}$, corresponding vectors $\boldsymbol{v}_1$, $\boldsymbol{v}_2$, and $\boldsymbol{v}_3$, and the axes and planes used in the above procedure are collected.

Table 2.2. Orientation vectors marked using the bendy pointer

	x, xy	x, xz	x, yx	y, yz	z, zx	y, zy
$\boldsymbol{n}$	$\boldsymbol{v}_1$	$\boldsymbol{v}_1$	$\boldsymbol{v}_2$	$\boldsymbol{v}_3$	$\boldsymbol{v}_2$	$\boldsymbol{v}_3$
$\boldsymbol{o}$	$\boldsymbol{v}_2$	$\boldsymbol{v}_3$	$\boldsymbol{v}_1$	$\boldsymbol{v}_1$	$\boldsymbol{v}_3$	$\boldsymbol{v}_2$
$\boldsymbol{a}$	$\boldsymbol{v}_3$	$\boldsymbol{v}_2$	$\boldsymbol{v}_3$	$\boldsymbol{v}_2$	$\boldsymbol{v}_1$	$\boldsymbol{v}_1$

The programme which executes the teaching procedure according to the rules described in Section 2.3.1 makes it possible to define the following frames (see Fig.2.5): Z (associated with the robot itself), E (gripper), and W (bendy pointer). They are necessary for correct calculations of an arbitrary frame chosen by the user. In the interactive teaching system we implemented a set of procedures of the robot programming language POLROB, defined for the ROMIK robot used for educational purposes [178].

2.3.6 An example of palletising

In this section we illustrate results presented in previous sections by a simple task of palletising. This task is presented in Fig.2.5

The task is to pick up and put down small pins at any position on a pallet. A frame called PAL is defined for the pallet with respect to the base coordinate system $x_0y_0z_0$. The pallet is divided into nine equal parts. At the centre of each part a coordinate system P_{ij} is located, which is defined with respect to frame PAL as Pij $= (\boldsymbol{R}, \boldsymbol{p})$, where $\boldsymbol{p} = [(i-0.5)a,\ (j-0.5)b,\ 0]^T$, $\boldsymbol{R}$ is a 3×3 unit matrix, i denotes a row number and j is a column number. Frame PAL is defined by making use of the bendy pointer while P_{ij} is obtained from engineering drawings (assuming that distances a and b are known). The pin has two coordinate systems. First one, named PIN, is defined with respect to PAL and can be found as PIN = PAL$\cdot\mathsf{P}_{ij}$. Second, PCH, is defined with respect to PIN. Frame W describes the coordinate system associated with the bendy pointer. In order to obtain the coordinate system PCH one has to rotate PIN about the z axis by 90 degrees, translate along the z axis by distance d, and finally rotate about the x axis by 180 degrees, which results in the following homogeneous matrix:

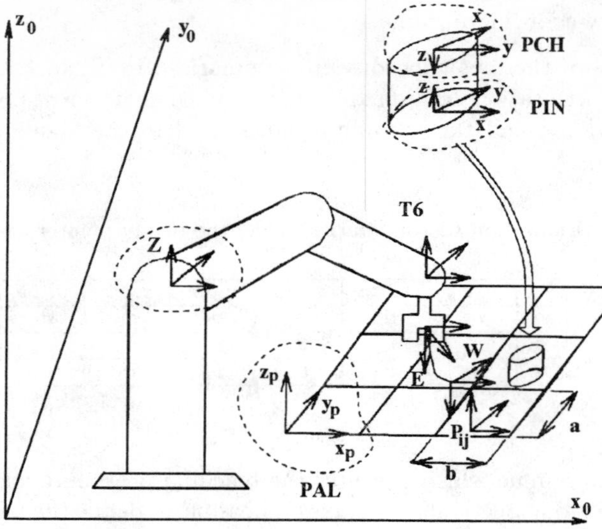

Fig. 2.5. An example of palletising

$$\mathsf{PCH} = \begin{bmatrix} 0 & 1 & 0 & 0 \\ 1 & 0 & 0 & 0 \\ 0 & 0 & -1 & d \\ 0 & 0 & 0 & 1 \end{bmatrix} . \tag{2.25}$$

This coordinate system can be entered into the programme directly from the keyboard. Frame Z can be entered from the keyboard or defined during the teaching process. The last frame which we specify is called PIN_APPROACH and has the same orientation as PCH with vector $\boldsymbol{p} = [0,\ 0,\ a]^T$, where a denotes the distance between the tool centre point and the pin. This frame can be entered from the keyboard. Each of these frames can be changed during the teaching process from the keyboard or by making use of the bendy pointer.

To make it possible for the robot to perform the task, the following position equations have to be satisfied:

$$Z \cdot T6 \cdot E = \mathsf{PAL} \cdot P_{ij} \cdot \mathsf{PIN\_APPROACH} \tag{2.26}$$

and

$$Z \cdot T6 \cdot E = \mathsf{PAL} \cdot P_{ij} \cdot \mathsf{PCH}\ , \tag{2.27}$$

where Z is defined with respect to the base coordinate, T6 with respect to Z to represent the robot wrist, and E is defined with respect to T6 to represent the robot end-effector. These two equations allow the end-effector to approach the pin and then to make contact with it. From these two equations frame T6 can easily be obtained. If any motion of accommodation is made, T6 must be recalculated according to the position equations. To construct the data structure corresponding to the different frames, an external file is defined. The unknown frame T6 is obtained by making use of the operation of frames defined in this section. The proposed method seems very practical and easy to use. The programme uses move statement described in this chapter. If a move statement cannot be executed, the system signals this; the movement can also be stopped at any time by the user.

After collecting all the coordinates of the frames presented above the computer calculates coordinate frame C, which depending on the stage of the operation, is defined with respect to the frame PCH or with respect to frame A, which describes the height of the pin moved above the pallet. Frame C adjusts frame PCH to the desired position and orientation of the gripper. In the palletising programme there is another frame, B, which corrects the position and orientation of the gripper with respect to the pallet. Frames A and B allow smooth movement of the pin carried by the gripper. They are not shown in Fig.2.5.

All operations are stored in the object named cycle, which is an array of records of type OpRec. This type is defined as follows:

```
type
   OpRec = record
      P1, P2 : Frame;
      Pausel, Pause2 : real;
      Vel : real;
   end;
```

Frames P1 and P2 denote a location at which the pin is placed and taken from the pallet. Variables Pausel and Pause2 hold the time interval of the delay (in s) during which the robot stops after taking the pin from and placing it on the parallel. The velocity of the movement of all the links is defined by variable vel. This velocity determines the speed of the sequence of operations.

In the programming system developed at our laboratory we incorporated elements of world modelling which describe the relationship between the objects at the scene in which the robot operates. There are two procedures which facilitate this, Affix and Unfix. The former is declared as

Affix (var F1, F2, TT : Frame; Str : string);

Here F1 and F2 are frames and TT is a transformation. This procedure causes frame F1 to be attached to frame F2. Variable TT is assigned the value of the transformation between F2 and F1.

Depending on string Str in the above procedure, the attachments can be rigid (Str is 'Rigidly') or nonrigid (Str is 'NonRigidly'), also called one-side attachment. In the former case two objects on the scene are glued together or twisted together. Alternatively we can say that frames F1 and F2 are two parts of one body. Therefore in the case when Str = 'Rigidly' it means that every movement of frame F1 causes also the movement of the frame F2 and vice versa. In the latter case we have, for example, one body on a top of another body. In this case moving frame F2 causes frame F1 to move at the same time, but when we change the position of frame F1 the position of frame F2 does not change.

In order to inform the system that the relationship between the bodies is no more of the Affix type we execute the statement named Unfix. Both statements cooperate with the move statement pMove using a flag which constitutes a parametric of the rigid or nonrigid type of attachment. Note that it is equivalent to the close or open operation of the gripper. In the move statement these variables have the value false. When statement Anfix is executed the value of this variable is changed to true. At the same time the system knows about the Affix relationship, which is no longer valid when the statement Unfix is executed.

2.3.7 Virtual programming panel

The presented functions of the particular parts of the host computer level software are incorporated into the VPANEL (virtual panel) programme. This programme was developed to control and programme the IRp-6 robot using an IBM compatible PC computer equipped with the interface of direct access to the resources of the robot controller [144].

The virtual panel supports all the functions of the industrial programming panel of the IRp-6 robot making the work easier for the user as well as extending some programming functions and adding new ones. This programme performs the following functions:

- control robot movement using the keyboard, the sensor ball (RTCB-Robot Teaching and Control Ball [90]), and the joystick of the industrial panel,
- control the pneumatic-driven gripper,
- define tools,
- synchronise the robot,
- handle errors,
- edit and execute the control file, consisting of high-level programming language statements,
- execute command sequences stepwise with monitoring of programme variables,
- collect data from the welding process: current, voltage, and other parameters.

2.4 Hardware description of an experimental set-up consisting of a one link geared robot

In order to avoid difficulties with a closed architecture of the robot controller we built an open system, a one link geared robot. A detailed description of the system can be found in reference [145]. In addition references [128, 135, 137] also give information about the system.

The general structure of the system is presented in Fig.2.6. The system consists of a DC motor M (PZTK 88-35 TOR) with a rate generator (PT in Fig.2.6) installed in the feedback loop, a harmonic drive PH (HDU C2) with the gear ratio of 1:160, a position sensor, a six-dimensional force/torque sensor (JR3), and a load (OBC). The power unit of the system was designed and built at the Robot Control Laboratory. The SBC (single board computer) is a NEC μPD 78310 microprocessor system. It has a 16-bit central unit, RAM memory, 4 channel analog-to-digital (A/D) converter, an RS 232C serial port, two pulse-width modulation (PWM) outputs, a multifunction pulse input/output unit, an interrupt controller, and a watch dog unit. The interrupt controller can perform many functions. It can serve the vector interrupts and interrupt-called context switching. It is also equipped with a fast DMA channel to transmit data between the memory and the input/output devices (this channel is known as the macro service). The maximum clock frequency is 12MHz. The shortest instruction requires three machine cycles.

In order to control the DC motor an input voltage signal for the power controller in the range of $\pm$10V is required. This can be obtained in two different ways: through the D/A converter (PK1 in Fig.2.6) and by converting the output PWM signal of the SBC to the voltage signal. Taking into account the converter price, addressing problems, and latching values at the output, the PWM/U converter was chosen and built at the Robot Control Laboratory.

The converter board is based on a 4-th order active filter and a system which moves the zero of the output signal and amplifies it at the same time (the amplitude of the PWM signal is 5V).

A block diagram of the control system is presented in Fig.2.7 The proposed hardware architecture is designed to be open and allows the programmer to distribute the intelligence between the SBC and a PC compatible computer, which is connected to the SBC through either the RS 232 serial or the parallel board interface. The SBC collects data from different sensors using the A/D converter. The converter, though only 8-bit, is very fast. Therefore it is possible to collect data from a single input four or eight times during one processing cycle of the converter, so that it is possible to average the data at the same time. The SBC collects information about current, position, and velocity. Data coming from the force/torque sensor is collected by the computer installed on the sensor. This data is transmitted through the DMA channel to the host computer. The SBC was originally designed to programme new control algorithms for the system. Unfortunately, a C cross-compiler was not available, so we decided that the SBC would collect data and execute control commands received from the host computer.

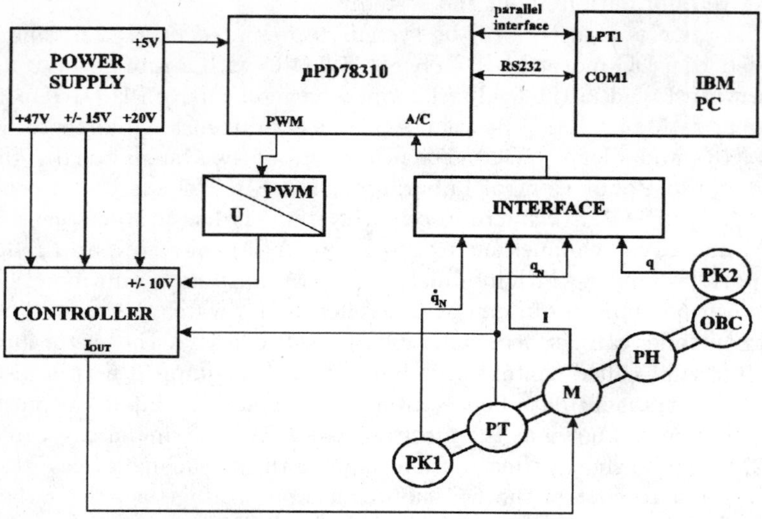

Fig. 2.6. A general overall of the system

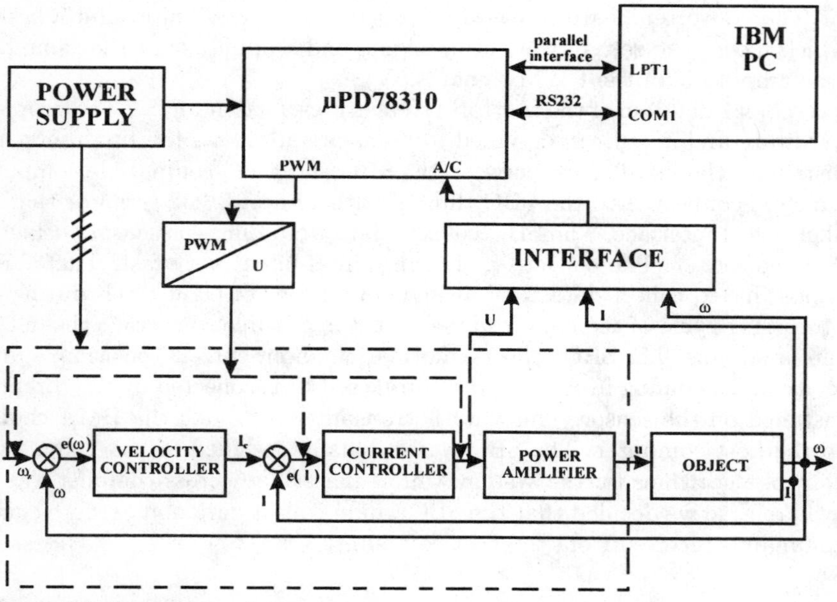

Fig. 2.7. Block diagram of the control system

2.5 Further comments on the robot programming system

The programming system has open architecture and is written in Pascal, which makes it portable. It contains elements of standard programming language statements. The programming system is a set of procedures. The system fully implements the structured position description introduced in the PAL language [217]. Cartesian programming is one of its interesting features. We make use of the idea that sensor integration is handled naturally if the world model can be synchronously or asynchronously modified. In any case the user has full control over synchronisation, whether it is necessary to synchronise the programme flow with the arm motion or vice versa. The programming system is to some extent manipulator independent. The arm dependencies, such as the kinematics and dynamics and the physical capabilities of the arm, are isolated (see Fig.2.2) and can easily be modified. Dependencies such as the working envelope, sweep, reach and kinematic configuration cannot be avoided at the manipulator level programming. Robot programmes are as independent of any particular manipulator as Cartesian programming allows.

At the end of this section we have to comment on a general programming system for robots. From the above description it is clear that the nature of robot programming is quite different from the nature of algorithmic programming in such languages as Pascal, C++, FORTRAN etc. Robot programmes have to do repetitive operations like general programmes, but these operations have to be performed in real changing world. Therefore an operation can sometimes fail, for example, because there is a drop of an oil and the pin slips off from the gripper. Such situation is quite common in industrial practice. The changing external environment can cause many unpredictable problems which are difficult to solve. This does not occur in a numerical programming when we have to solve, for example, a set of equations. Besides that each robot has its own kinematics and dynamics and the inverse and forward problems associated with them cannot be solved universally for all robots. These specific features have to be taken into account and solved separately for each robot. Therefore standardisation at this level is not possible.

On the other hand each robot uses the same data structure in order to describe kinematics, dynamics etc. In fact these structures are universal and can be coded in any high-level programming language. The same applies to move statements in Cartesian space, which conceptually can be programmed in universal way. Here we have to solve a general problem of how to move an object from position A to position B in Cartesian space. This requires the knowledge of the kinematics and dynamics and depends on the particular robot.

Manufacturers develop many programming languages for different type of robots and are, in general, not interested in any standardisation of the software. Moreover, the robot controller has closed structure and the manufacturer does not provide the user with any information about it. Therefore we equipped the IRp-6 robot with the bus access interface which connects the robot controller to the host computer. As a consequence we were able to develop the programming tools at the host computer level. Some parts of this software were analysed in previous sections at quite detailed level in order to show the specific features of robot programming. The programmes developed were used for research purposes such as robot dynamics identification, load identification, and hybrid control. Without these tools this would not be possible because of the restrictions imposed by the manufacturer. It is worth mentioning that we investigated the standardisation problem following references [170, 171], but unfortunately we were not able to realise those ideas. Some valuable comments are included in the M.S. Thesis of Moder [162], who cooperated with the Volkswagen Company in Germany. Also in this case the standardisation failed. We believe that in comparison with any algorithmic language the standardisation in robot programming is a very difficult problem. Some general considerations are included in Zielinski [239].

Finally, it should be pointed out that the part of the software which is responsible for collecting data from different sensors is also very important. It has to be developed separately due to the fact that many different sensors can be used in robot environment. Considerations presented in this chapter have move engineering flavour and show what kind of tools have to be developed in order to conduct the experiments discussed in the second part of the monograph.

CHAPTER 3
ROBOT DYNAMIC MODELS

3.1 Introduction

This chapter describes dynamic models that are used for the identification of robot parameters. Considerations in this chapter apply to two types of robots: robots with gears (as an example see the description of the industrial robot IRp-6 [12]) and direct drive robots (DDA) without gears [9, 185]. These two classes of robots are the most common in industrial applications. We assume in this monograph that geometrical parameters of the robot are known; by geometrical parameters we mean parameters which are defined on the base of the modified Denavit–Hartenberg notation [10, 113]. Each link of a manipulator is characterised by dynamic parameters: mass, the centre of mass (which multiplied by mass represents a moment of the first order) and six parameters of the inertia tensor (which essentially are elements of a moment of the second order). The simplest way to measure these parameters is obviously to take the robot to pieces and then measure all the details thoroughly [1]. In most cases this is not possible, but when it is, it gives valuable benchmarks for other research. Therefore it is necessary to build a dynamic model for the robot itself. In practice these models are quite complicated and highly nonlinear with respect to joint positions, velocities, and accelerations. However, they are linear with respect to the dynamic parameters, which greatly simplifies the problem of identifying the dynamic parameters of the robot. Here we assume that the links of the robot are made to follow a predefined test trajectory. The movement parameters (positions, velocities etc.) can be measured simultaneously and from them the robot dynamic parameters can be calculated. Geared robots are complex mechanisms due to the gears and transmissions. In this chapter the transmission dynamics is neglected. We incorporate only the gear ratios which for our considerations are constant. However, as will be shown in Chapter 5, this is not always the case. We also assume that the generalised friction force acting at joint i is linear with respect to the viscous friction coefficient and the Coulomb friction coefficient. A more general friction model will be considered in Chapter 5.

In order to describe mathematical properties of a manipulator, any classical method known from the analytical mechanics can be used [7, 81, 149, 212, 229]. Many authors use Lagrange equations of motion of the first and second kind [7, 81, 212] to formulate the equations of motion of the manipulator. More popular are Lagrange equations of the second kind and these will be used in this chapter; compare references [1, 8, 12, 31, 58, 102, 117, 158, 163, 225, 235]. Some authors suggest using classical Newton-Euler equations [1, 29, 117, 125, 146, 147, 160]. It is well known that both approaches lead to the same dynamical model. This has been proved in a rigorous manner by Silver [206].

When implementating identification schemes it is important to know which signals can be measured. Usually industrial robots have position sensors, seldom velocity sensors and very rarely acceleration sensors. In practice it is quite difficult to measure these signals as they are corrupted by noise. Besides, in order to implement an identification scheme it is necessary to measure generalised forces acting at the joints. Robots are seldom equipped with torque sensors since they are rather expensive. Usually we measure armature current and we assume that it is proportional to the torque. In the case of direct drive robots, the measured quantities which are necessary to implement identification scheme are generalised joint positions, velocities, accelerations, and torques. In the case of geared robots the situation is more complicated because of the transmission mechanisms and gears: we measure motor quantities such as positions, velocities, accelerations, and torques, and then recalculate them in order to obtain the joint quantities. In order to do that we devide the motor quantities by the gear ratio. Moreover, many geared robots have a parallelogram structure and motor and joint quantities are calculated by use of a nonlinear function. These calculations are nontrivial. Later in this chapter we consider the IRp-6 robot and appropriate nonlinear functions.

Mathematical models which require measurements of positions, velocities, and accelerations are called differential models [59, 99], while models which only require position and velocity measurements are called integral models [122, 183]. In both cases, from the identification point of view, we need to measure joint torques (or motor torques). The integral model is derived from the energy theorem [81]. It will be shown later in this chapter that both differential and integral models are characterised by the same set of dynamic parameters. Generally, all dynamic parameters of a manipulator can be classified into three categories: uniquely identifiable, identifiable in linear combinations, and unidentifiable [108, 109]. The category to which a dynamic parameter belongs is dictated by the kinematic structure of the manipulator and the input trajectory used for the estimation. Here we assume that the robot model is canonical [61, 62, 63, 64, 103, 180, 202, 205], which means that the vector $\boldsymbol{x}_c$ of the parameters consists of the minimum number of parameters which are combinations of the link inertial parameters (mass and the first and second moments of the individual links). Some authors call this vector a vector of base parameters. Categorisation of dynamic parameters makes it possible to determine a minimum set of parameters that affect the equations of motion of an n-degrees-of-freedom manipulator. A straightforward method to categorise the dynamic parameters is to derive the dynamic equations of a manipulator symbolically and then manually classify the parameters into one of the three categories. While it is possible to do this for manipulators having two or three degrees of freedom, extending it to manipulators with six or more degrees of freedom becomes a formidable task. Categorisation can be done symbolically [108, 109] or by applying a set of rules [61, 62, 63, 64] which will be shown in this chapter. Once a minimum set of parameters has been found as a result of the categorisation procedure, a persistently exciting trajectory can be chosen in order to identify this set of parameters. It happens that some of the base parameters do not have a physical meaning because they are linear combinations of the individual link dynamic parameters; it is nevertheless still possible to identify them. The reason why the input trajectory has to satisfy persistently exciting conditions will be explained in Chapter 5. Here we only point out that nonlinear functions which are associated with a set of minimum parameters depend in general on joint positions, velocities, and accelerations. One can imagine that for certain input trajectories some of these functions may become linearly dependent and parameters associated with them cannot be identified. In conclusion, obtaining a canonical model is a necessary but not sufficient condition for successful identification. In this chapter we will show how to obtain a canonical model for a general purpose manipulator.

The chapter is organised as follows. In Section 3.2 the derivation of the differential model is presented, including friction effects. In Section 3.3 the derivation of the integral model is outlined. Comparison of the differential and integral models is the main objective of the next section. In Section 3.5 we present how to obtain a canonical model, including a general procedure for its generation. In the next section we present some comments concerning different possibilities of obtaining dynamic models for the purpose of identification of their parameters which are known in robotics literature. Finally, in the last section we present two examples of the dynamic equations of motion for the two classes of robots discussed in this monograph. One example is concerned with an industrial robot IRp-6, built in Poland under the licence of the Swedish ASEA. The other dynamic model built for identification of dynamic parameters is a 2-link Experimental Direct Drive Arm (EDDA). This direct drive robot was built at the Institut für Robotik und Prozessinformatik, Technische Universität Braunschweig [124, 181, 185]. In both cases differential and integral models are considered and the canonical models are derived. These models will be used for the purpose of identification in Chapter 5.

3.2 Derivation of the differential model

Consider a manipulator which consists of n links presented in Fig. 3.1. Links are numbered in increasing order from the fixed base (link 0) towards the tip (link n). Joint k connects links $(k-1)$ and k. The system contains simple revolute and/or prismatic joints; each link has one degree of freedom. Coordinate frames are assigned according to the modified Denavit-Hartenberg notation [31, 100]. According to this notation each link has a right-handed coordinate frame $x_i y_i z_i$, located at the end of the link that is closer to the manipulator base (as opposed to the standard notation in which the coordinate frame is located at the distal end of each link). The relative position and orientation between two successive links is described by the modified

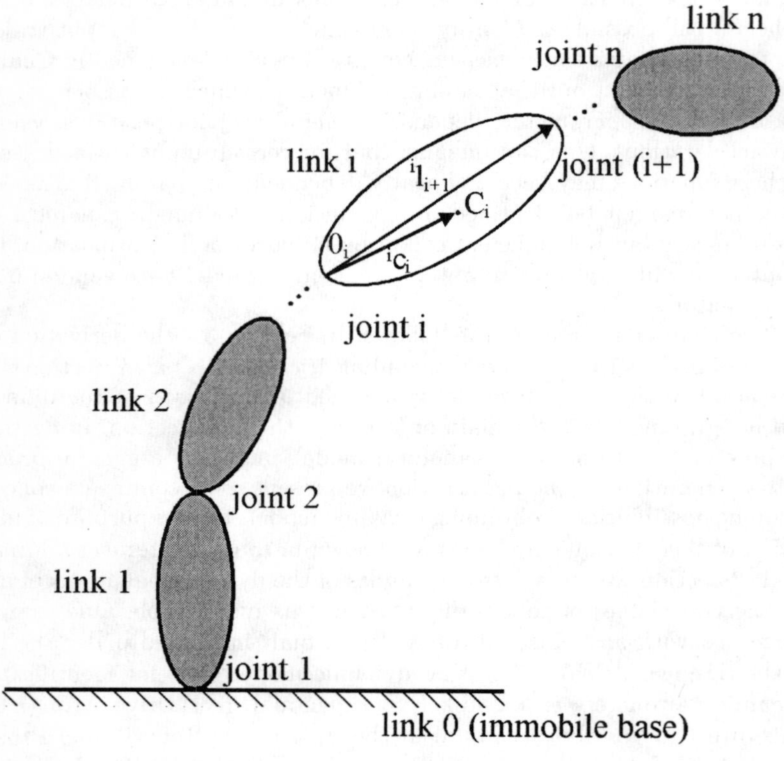

Fig. 3.1. A manipulator with n degrees of freedom

Denavit-Hartenberg parameters. The generalised coordinate which describes the movement of the i-th link is denoted by q_i and can be calculated from the following expression:

$$q_i = (1 - \sigma_i)\theta_i + \sigma_i a_i, \tag{3.1}$$

where $\sigma_i = 0$ if joint is rotational and $\sigma_i = 1$ if joint i is translational. A direction cosine matrix which describes an orientation of the i-th coordinate frame with respect to the $(i-1)$-th coordinate frame is written as

$$
{}^{i-1}_{i}\boldsymbol{R} = \begin{bmatrix} \cos\theta_i & -\sin\theta_i & 0 \\ \cos\alpha_i \sin\theta_i & \cos\alpha_i \cos\theta_i & -\sin\alpha_i \\ \sin\alpha_i \sin\theta_i & \sin\alpha_i \cos\theta_i & \cos\alpha_i \end{bmatrix} . \tag{3.2}
$$

The vector which connects the origin of the $(i-1)$-th coordinate frame to the origin of the i-th coordinate frame is

$$
{}^{i-1}\boldsymbol{l}_i = \begin{bmatrix} d_i, & -a_i \sin\alpha_i, & a_i \cos\alpha_i \end{bmatrix}^T , \tag{3.3}
$$

where $(\cdot)^T$ denotes the transposition operation. Parameters θ_i, α_i, a_i, and d_i constitute the modified Denavit-Hartenberg notation [31, 100]. Angle α_i is sometimes called an offset angle, and a_i is an offset distance. Referring to Fig. (3.1) C_1, C_2, $\cdots$, C_n denote the centres of mass of individual links and O_1, O_2, $\cdots$, O_n are the origins of the coordinate frames associated with the links of the manipulator. The vector which connects the origin coordinate frame i to the centre of mass of link i is denoted by ${}^i\boldsymbol{c}_i$. This vector is usually expressed in the i-th coordinate frame. Now we can write Lagrange equations of the second kind for a kinematic chain presented in Fig. (3.1) :

$$
\frac{d}{dt}\frac{\partial \mathcal{L}}{\partial \dot{q}_i} - \frac{\partial \mathcal{L}}{\partial q_i} = \tau_i - Q_i(\dot{q}_i) , \quad i = 1, 2, \cdots, n. \tag{3.4}
$$

In the above equation τ_i stands for the generalised force (which can be a force or a moment in general) which acts along the z_i axis of the i-th coordinate frame. For a direct drive robot this force is the force of the i-th motor of the actuator. $\mathcal{L}$ is a Lagrange function (which is the difference between the total kinetic and potential energy of the manipulator), and $Q_i(\dot{q}_i)$ represents the friction force acting at the i-th joint. This force has the following simplified form

$$
Q_i(\dot{q}_i) = F_{iv}\dot{q}_i + F_{is}\mathrm{sgn}(\dot{q}_i) , \tag{3.5}
$$

where F_{iv} is the viscous friction coefficient of the i-th link while F_{is} is the coefficient of the Coulomb friction (independent of the magnitude of the velocity). A more general friction model will be considered in Chapter 5.

The total kinetic energy of the manipulator from Fig. (3.1) can be written in the following form [59, 102]:

$$
E_{kc} = \frac{1}{2}\sum_{i=1}^{n} \left({}^i\boldsymbol{\omega}_i^T \, {}^i\boldsymbol{I}_i \, {}^i\boldsymbol{\omega}_i + m_i \, {}^i\boldsymbol{v}_i^T \, {}^i\boldsymbol{v}_i + 2 \, {}^i\boldsymbol{v}_i^T (\, {}^i\boldsymbol{\omega}_i \times m_i^i\boldsymbol{c}_i) \right) . \tag{3.6}
$$

In the above ${}^i\boldsymbol{\omega}_i$ is the vector of the angular velocity (of dimensions 3×1) of the i-th link, and ${}^i\boldsymbol{v}_i$ is the linear velocity of the origin of the i-th coordinate frame assigned to the i-th link (with the same dimensions as ${}^i\boldsymbol{\omega}_i$).

The appearance of the third term in Eq. (3.6) is a consequence of the fact that the local coordinate frame assigned to the i-th link is not located at its centre of mass. The summation in Eq. (3.6) includes all links of the manipulator. The expression $m_i^i\boldsymbol{c}_i$ stands for the first moment of link i. The three components of this vector are $m_i^i c_{ix}$, $m_i^i c_{iy}$, $m_i^i c_{iz}$. ${}^i\boldsymbol{I}_i$ denotes the inertia tensor of the i-th link and has the form

$$ {}^i\boldsymbol{I}_i = \begin{bmatrix} I_{ixx} & I_{ixy} & I_{ixz} \\ I_{ixy} & I_{iyy} & I_{iyz} \\ I_{ixz} & I_{iyz} & I_{izz} \end{bmatrix} . \qquad (3.7) $$

All the quantities that appear in Eq. (3.6) are expressed in the local i-th coordinate frame, which is denoted by the left i superscript. Note that the dynamic parameters of link i appear linearly in Eq. (3.6).

The angular and linear velocities are calculated proceeding from the base of the manipulator towards its tip according to the following recursions:

$$ {}^i\boldsymbol{\omega}_i = {}^{\;\;i}_{i-1}\boldsymbol{R}\,{}^{i-1}\boldsymbol{\omega}_{i-1} + \bar{\sigma}_i\,\boldsymbol{z}_0\,\dot{q}_i \,, \qquad (3.8) $$

$$ {}^i\boldsymbol{v}_i = {}^{\;\;i}_{i-1}\boldsymbol{R}\,{}^{i-1}\boldsymbol{v}_{i-1} + {}^{\;\;i}_{i-1}\boldsymbol{R}\left({}^{i-1}\boldsymbol{\omega}_{i-1} \times {}^{i-1}\boldsymbol{l}_i\right) + \sigma_i\,\boldsymbol{z}_0\,\dot{q}_i \qquad (3.9) $$

in which we assume the initial conditions ${}^0\boldsymbol{\omega}_0 = {}^0\boldsymbol{v}_0 = \boldsymbol{0}$, $\bar{\sigma}_i = 1 - \sigma_i$, and $\boldsymbol{z}_0 = [0\ 0\ 1]^T$ is the unit vector.

The expression for the total potential energy is [59, 102]

$$ E_{pc} = -\sum_{i=1}^{n} m_i \boldsymbol{g}^T \left({}^0\boldsymbol{l}_i + {}^0_i\boldsymbol{R}\,{}^i\boldsymbol{c}_i \right) , \qquad (3.10) $$

where m_i denotes the mass of the i-th link and $\boldsymbol{g}$ is the gravity vector $\boldsymbol{g} = [0,\ 0,\ g_0]^T$, with $g_0 = -9.81\text{m/s}^2$. Notice that ${}^0\boldsymbol{l}_i$ is a vector which connects the origin of coordinate frame 0 to the origin of coordinate frame i. Equation (3.10) describes the potential energy of all the links of the manipulator, expressed with respect to the immobile base coordinate frame. It can be seen clearly from equation (3.6) that the total kinetic energy depends on the vector of generalised positions and velocities and the potential energy depends only on the vector of joint displacements (because ${}^0_i\boldsymbol{R}$ matrix is a function of the joint coordinates). The Lagrange function (shortly Lagrangian) of the manipulator can be written as $\mathcal{L} = E_{kc} - E_{pc}$.

The inertial parameters and the friction coefficients of the manipulator can be expressed by the $12n \times 1$ vector $\boldsymbol{x}'$ of robot dynamic parameters:

$$ \begin{aligned} \boldsymbol{x}' = {} & [I_{1xx}, I_{1xy}, I_{1xz}, I_{1yy}, I_{1yz}, I_{1zz}, m_1, m_1c_{1x}, m_1c_{1y}, m_1c_{1z}, \\ & F_{1v}, F_{1s}, \cdots, I_{nxx}, I_{nxy}, I_{nxz}, I_{nyy}, I_{nyz}, I_{nzz}, \\ & m_n, m_nc_{nx}, m_nc_{ny}, m_nc_{nz}, F_{nv}, F_{ns}]^T . \end{aligned} \qquad (3.11) $$

Note that as all the dynamic parameters are expressed in local coordinate frames, they are constant when the manipulator moves along a predefined trajectory.

In addition, by vector $\boldsymbol{x}$ with dimensions $10n \times 1$ we denote the vector of dynamic parameters which does not include viscous and Coulomb friction coefficients.

Since the total energy of the robot is linear with respect to these parameters, formal differentiation of the Lagrangian leads to the following vector equation:

$$\boldsymbol{\tau} = \boldsymbol{D}(\boldsymbol{q}, \dot{\boldsymbol{q}}, \ddot{\boldsymbol{q}})\, \boldsymbol{x} + \boldsymbol{\tau}_{fs} + \boldsymbol{\tau}_{fv}, \tag{3.12}$$

where $\boldsymbol{q} = [q_1, q_2, \cdots, q_n]^T$, is the vector of generalised coordinates q_i (angular or linear displacements), $\dot{\boldsymbol{q}} = [\dot{q}_1, \dot{q}_2, \cdots, \dot{q}_n]^T$, $\ddot{\boldsymbol{q}} = [\ddot{q}_1, \ddot{q}_2, \cdots, \ddot{q}_n]^T$, $\boldsymbol{\tau} = [\tau_1, \tau_2, \cdots, \tau_n]^T$ is the vector of generalised forces τ_i (force or torque, depending on the type of the joint), and $\boldsymbol{D}$ is a $(n \times 10n)$ matrix, whose elements depend on $\boldsymbol{q}$, $\dot{\boldsymbol{q}}$, and $\ddot{\boldsymbol{q}}$. Thus described robot dynamics model is called the differential model. Equation (3.12) is linear with respect to the vector $\boldsymbol{x}$. In Eq.(3.12) we take into account two components which include the friction forces $\boldsymbol{\tau}_{fs}$ and $\boldsymbol{\tau}_{fv}$. These two components give the dissipation of the energy due to the viscous and Coulomb friction. Since the manipulator consists of the rotational joints only (which is not a restriction) we use terminology viscous friction torque and Coulomb friction torque. Representation (3.12) is particularly useful if we want to compare differential and integral models.

The compact form of Eq.(3.12) is suitable for identification of dynamic parameters. Sometimes it is more convenient to put Eq.(3.12) in a different form. In order to do this we rewrite the expression for the total kinetic energy (3.6) as follows:

$$E_{kc} = \frac{1}{2}\dot{\boldsymbol{q}}^T \boldsymbol{A}\dot{\boldsymbol{q}} = \frac{1}{2}\sum_{i=1}^{n}\sum_{j=1}^{n} A_{ij}\dot{q}_i\dot{q}_j\ , \tag{3.13}$$

in which $\boldsymbol{A}$ stands for the manipulator inertia matrix with components A_{ij}. Eq.(3.13) clearly follows when we substitute Eqs.(3.8) and (3.9) into Eq.(3.6). Substituting Eq.(3.13) into the Lagrangian and formal differentiation of it, leads to the following vector equation:

$$\boldsymbol{A}(\boldsymbol{q})\ddot{\boldsymbol{q}} + \boldsymbol{C}(\boldsymbol{q}, \dot{\boldsymbol{q}})\dot{\boldsymbol{q}} + \boldsymbol{G}(\boldsymbol{q}) = \boldsymbol{\tau}_e\ , \tag{3.14}$$

where

$$\boldsymbol{G}(\boldsymbol{q}) = \frac{dE_{pc}}{d\boldsymbol{q}}\ . \tag{3.15}$$

In Eq.(3.14) $\boldsymbol{A}(\boldsymbol{q})$ is the $n \times n$ symmetric positive definite manipulator inertia matrix, $\boldsymbol{C}(\boldsymbol{q}, \dot{\boldsymbol{q}})\dot{\boldsymbol{q}}$ is the $n \times 1$ vector of centrifugal and Coriolis torques (shortly the Coriolis term),$\boldsymbol{G}(\boldsymbol{q})$ is the $n \times 1$ vector of gravitational torques, and $\boldsymbol{\tau}_e$ is the vector of non-potential forces (compare with the derivation of the integral model which will be discussed in the next section). Equation (3.14) can be rewritten in the component form as follows:

$$\sum_{j=1}^{n} A_{ij}\ddot{q}_j + \sum_{j=1}^{n}\sum_{k=1}^{n} h_{ijk}\dot{q}_j\dot{q}_k + G_i = \tau_{ie} \quad i = 1, 2, \ldots, n\ , \tag{3.16}$$

where

$$h_{ijk} = \frac{\partial A_{ij}}{\partial q_k} - \frac{1}{2}\frac{\partial A_{jk}}{\partial q_i} , \tag{3.17}$$

and G_i is the i-th component of the vector of gravitational torques. It is important to note that interactive inertia torques $A_{ij}\ddot{q}_j$ $(j \neq i)$ result from the off–diagonal elements of the manipulator inertia matrix and that the Coriolis and centrifugal torques $h_{ijk}\dot{q}_j\dot{q}_k$ arise because the inertia tensor is configuration–dependent. From Eqs.(3.14) and (3.16) it is clear that elements C_{ij} of $n \times n$ matrix $\boldsymbol{C}$ satisfy the equation

$$\sum_{j=1}^{n} C_{ij}\dot{q}_j = \sum_{j=1}^{n}\sum_{k=1}^{n} h_{ijk}\dot{q}_j\dot{q}_k . \tag{3.18}$$

Friction torques can be incorporated into Eq.(3.14) in a similar way as in Eq.(3.12). Terms involving a product of the type $\dot{q}_i^2$ are called centrifugal, while those involving a product of the type $\dot{q}_k\dot{q}_j$, where $k \neq j$, are called Coriolis terms.

Next we write the equations of motion in differential form, but we separate Corriolis and centrifugal coefficients. These equations are written in the vector form as follows [103]:

$$\boldsymbol{A}(\boldsymbol{q})\ddot{\boldsymbol{q}} + \boldsymbol{B}(\boldsymbol{q})\dot{\boldsymbol{q}}\dot{\boldsymbol{q}} + \bar{\boldsymbol{C}}(\boldsymbol{q})\dot{\boldsymbol{q}}^2 + \boldsymbol{G}(\boldsymbol{q}) + \boldsymbol{\tau}_{fs} + \boldsymbol{\tau}_{fv} = \boldsymbol{\tau} , \tag{3.19}$$

where $\boldsymbol{A}$ is, as before, the $n \times n$ symmetric and nonsingular inertia matrix, $\boldsymbol{B}$ is the $n \times (n\,(n-1)\,/2)$ Coriolis coefficient matrix, $\bar{\boldsymbol{C}}$ is the $n \times n$ centrifugal coefficient matrix,

$$\dot{\boldsymbol{q}}\dot{\boldsymbol{q}} = \left[\dot{q}_1\dot{q}_2, \dot{q}_1\dot{q}_3 \ldots, \dot{q}_1\dot{q}_n, \dot{q}_2\dot{q}_3, \ldots, \dot{q}_2\dot{q}_n, \ldots, \dot{q}_{n-1}\dot{q}_n\right]^T ,$$

and

$$\dot{\boldsymbol{q}}^2 = \left[\dot{q}_1^2, \dot{q}_2^2, \ldots, \dot{q}_n^2\right]^T .$$

In general, the elements of $\boldsymbol{B}(\boldsymbol{q})$ and $\bar{\boldsymbol{C}}(\boldsymbol{q})$ can be obtained from the elements of $\boldsymbol{A}$ using the following relations:

$$B_{ijk} = \frac{\partial A_{ij}}{\partial q_k} + \frac{\partial A_{ik}}{\partial q_j} - \frac{\partial A_{jk}}{\partial q_i} , \tag{3.20}$$

$$\bar{C}_{ij} = \frac{\partial A_{ij}}{\partial q_j} - \frac{1}{2}\frac{\partial A_{jj}}{\partial q_i} . \tag{3.21}$$

B_{ijk} is the element of $\boldsymbol{B}$ in the i-th line that is multiplied by $\dot{q}_j\dot{q}_k$. Comparing Eqs.(3.12) and (3.14) it can easily be noted that

$$\boldsymbol{D}(\boldsymbol{q}, \dot{\boldsymbol{q}}, \ddot{\boldsymbol{q}})\boldsymbol{x} = \boldsymbol{A}(\boldsymbol{q})\ddot{\boldsymbol{q}} + \boldsymbol{B}(\boldsymbol{q})\dot{\boldsymbol{q}}\dot{\boldsymbol{q}} + \bar{\boldsymbol{C}}(\boldsymbol{q})\dot{\boldsymbol{q}}^2 + \boldsymbol{G}(\boldsymbol{q}). \tag{3.22}$$

Now denote by $\boldsymbol{D}^i$ the $n \times 1$ vector composed of the i-th column of the $\boldsymbol{D}$ matrix. Then the contribution of the dynamic parameter X_i which appears in Eq.(3.11) to the generalised force can be written as follows:

$$\boldsymbol{D}^i X_i = \boldsymbol{A}^i \ddot{q} + \boldsymbol{B}^i \dot{q}\dot{q} + \bar{\boldsymbol{C}}^i \dot{q}^2 + \boldsymbol{G}^i \,, \tag{3.23}$$

where $\boldsymbol{A}^i$, $\boldsymbol{B}^i$, $\bar{\boldsymbol{C}}^i$, $\boldsymbol{G}^i$ represent the terms of $\boldsymbol{A}$, $\boldsymbol{B}$, $\bar{\boldsymbol{C}}$, $\boldsymbol{G}$ containing the inertial parameter X_i. Therefore we have

$$\boldsymbol{A} = \sum_{i=1}^{10n} \boldsymbol{A}^i, \quad \boldsymbol{B} = \sum_{i=1}^{10n} \boldsymbol{B}^i, \quad \bar{\boldsymbol{C}} = \sum_{i=1}^{10n} \bar{\boldsymbol{C}}^i, \quad \boldsymbol{G} = \sum_{i=1}^{10n} \boldsymbol{G}^i \,.$$

The elements of matrix $\boldsymbol{A}$ can be calculated from the general expression for the total kinetic energy of the manipulator, Eq.(3.6). In order to do so one has to substitute the angular and linear velocities of the i-th link given by Eqs.(3.8) and (3.9) into Eq.(3.6). It is rather straightforward though laborious to calculate these coefficients. They can be calculated automatically by the algorithm presented in reference [102]. According to this algorithm, matrices $\boldsymbol{A}^i$ associated with X_i can be calculated for all dynamic parameters. Next, the elements of $\boldsymbol{B}(\boldsymbol{q})$ and $\bar{\boldsymbol{C}}(\boldsymbol{q})$ are calculated according to Eqs.(3.20) and (3.21). Often this calculation can be done by hand.

From the above it is clear that elements B_{ijk} and $\bar{C}_{ij}$ of matrices $\boldsymbol{B}(\boldsymbol{q})$ and $\bar{\boldsymbol{C}}(\boldsymbol{q})$ can be calculated from the inertia matrix $\boldsymbol{A}(\boldsymbol{q})$ of the manipulator . The elements of vector $\boldsymbol{G}(\boldsymbol{q})$ are calculated directly from the potential energy. It will be shown later that the same applies to the integral model. The whole knowledge about the system is hidden in the manipulator inertia matrix and its potential energy. As suggested by the form of Eq.(3.18), many authors combine Coriolis and centrifugal coefficients in one matrix, which is denoted by $\boldsymbol{C}(\boldsymbol{q}, \dot{\boldsymbol{q}})$. Note that in Eq.(3.19) matrices $\boldsymbol{B}(\boldsymbol{q})$ and $\bar{\boldsymbol{C}}(\boldsymbol{q})$ depend only on a vector of generalised coordinates while matrix $\boldsymbol{C}(\boldsymbol{q}, \dot{\boldsymbol{q}})$ in Eq.(3.14) is a function of a vector of generalised positions and velocities.

So far in our considerations we have not included dynamics of the actuators and transmissions for robots with gears. A general framework for these calculations can be found in references [91, 165, 167, 204]. In those papers the concept of spatial algebra is used. That concept was originally introduced by Rodriguez [190] in order to write equations of motion of complicated mechanisms in a very compact form. In this monograph we do not describe this idea, as we mainly focus on experimental identification of robot dynamics and we keep the theory only at a sufficient level to facilitate the understanding of complicated problems associated with practical implementations of robot dynamics identification. The dynamics of the actuators certainly cannot be neglected. The simplest way to include it is to multiply the motor inertia along the axis of rotation by the square of the gear ratio and add to the link inertia along the axis of rotation of the link itself. Sometimes the gear ratio is a nonlinear function of link coordinates due to the fact that some manipulators have a parallelogram structure. The above assumption is very realistic and does not lead to wrong results as far as identification is concerned. Even if the function describing the gear ratio is nonlinear we can assume that for a wide range of movements it is constant. The above assumptions have been used by many authors [29, 59, 60, 61] and positively verified in practice. Neglecting the cross inertia moments of the actuators seems to be a good approximation. In this monograph we rather focus on joint flexibility assuming both non-ideal and idealised actuator models. Non-ideal actuator models account for such effects as joint elasticity, friction, and backlash, in addition to contributions from motor inertias. Idealised actuator models ignore effects such as joint elasticity and backlash and assume ideal transmission. This will be not discussed here and for general theoretical considerations of robot dynamic models the reader is referred to references [91, 165, 167, 204]. We also concentrate in this monograph on joint and gear mechanism friction, which cannot be neglected in practice. This will be discussed in Chapter 5. On the other hand each complicated mechanical structure requires specific consideration which will be illustrated in further chapters.

3.3 Derivation of the integral model

Now we proceed to derive an integral model of a manipulator in a general form [21, 54, 62]. First note that based on Eqs.(3.6) and (3.10) we can write the following differential expressions:

$$E_{kc} = \sum_{i=1}^{m} \frac{\partial E_{kc}}{\partial X_i} X_i = \sum_{i=1}^{m} DE_{ki}\, X_i \ , \tag{3.24}$$

$$E_{pc} = \sum_{i=1}^{m} \frac{\partial E_{pc}}{\partial X_i} X_i = \sum_{i=1}^{m} DE_{pi}\, X_i \ , \tag{3.25}$$

where X_i is the i-th inertial parameter, $m = 10n$, DE_{ki} is a function of the vector of generalised positions and velocities $\boldsymbol{q}$, $\dot{\boldsymbol{q}}$, and geometric parameters, and DE_{pi} is a function of $\boldsymbol{q}$ and geometric parameters. Eqs.(3.24) and (3.25) are a consequence of the fact that total kinetic and potential energy depend linearly on dynamic parameters. Making use of the energy theorem [81], which states that the work of non-potential generalised forces applied to the system is equal to the change of the total energy of the system, it can be written that

$$\int_{t_1}^{t_2} \boldsymbol{\tau}_e{}^T \dot{\boldsymbol{q}} dt = (E_{kc}(t_2) + E_{pc}(t_2)) - (E_{kc}(t_1) + E_{pc}(t_1)) = H(t_2) - H(t_1) \ , \tag{3.26}$$

where $\boldsymbol{\tau}_e$ is the vector of non-potential forces (or equivalently non-dissipative forces at the system), $E_{kc}(t_i)$ and $E_{pc}(t_i)$ are the kinetic and potential energy, respectively, of the manipulator at time instant t_i, and $H(t_i) = E_{kc}(t_i) + E_{pc}(t_i)$ is the total energy of the system at time instant t_i.

In order to keep our consideration general we assume that vector $\boldsymbol{\tau}$ which appears in Eq.(3.12) is the sum of two vectors: the vector of generalised forces and torques acting at the actuators (it is the vector of non-potential forces) and the vector of friction torques. Therefore we can write

$$\boldsymbol{\tau} = \boldsymbol{\tau}_e + \boldsymbol{\tau}_f \tag{3.27}$$

where $\boldsymbol{\tau}_f$ is the vector of friction torques. Taking into account Eq.(3.5) notice that $\boldsymbol{\tau}_f$ is the sum of two vectors with Coulomb and viscous friction torques components, namely

$$\boldsymbol{\tau}_f = \boldsymbol{\tau}_{fs} + \boldsymbol{\tau}_{fv} \ . \tag{3.28}$$

Let the Coulomb and viscous friction coefficients, describing appropriate friction torques at the i-th joint, be denoted by F_{is} and F_{iv}. Then the following equations can be written

$$\tau_{ifs} = -F_{is}\,\mathrm{sgn}(\dot{q}_i) \ , \tag{3.29}$$

$$\tau_{ifv} = -F_{iv}\dot{q}_i \ , \tag{3.30}$$

where sgn$(\cdot)$ denotes the sign function.

Taking into account Eqs.(3.24) and (3.25) and using the energy theorem described by Eq.(3.26) we can write

$$H(t_2) - H(t_1) = \boldsymbol{dl}^T \boldsymbol{x} \ , \tag{3.31}$$

where

$$\boldsymbol{dl}^T = [DL_1(t_2) - DL_1(t_1), \ \cdots \ , DL_m(t_2) - DL_m(t_1)] \tag{3.32}$$

and

$$DL_i(t_k) = DE_{ki}(t_k) + DE_{pi}(t_k). \tag{3.33}$$

Note that in the above equation, components $DE_{ki}(t_k)$ and $DE_{pi}(t_k)$ appear in Eqs.(3.24) and (3.25) which are written at time instant t_k.

Let define the following vectors consisting of friction coefficients:

$$\boldsymbol{f_s} = [F_{1s}, \ \cdots \ , F_{ns}]^T \ , \tag{3.34}$$

and

$$\boldsymbol{f_v} = [F_{1v}, \ \cdots \ , F_{nv}]^T \ . \tag{3.35}$$

In order to build a linear representation of the friction torques in the integral model define the following vector functions:

$$\boldsymbol{df_s}^T = \left[\int_{t_1}^{t_2} |\dot{q}_1| dt, \ \cdots \ , \int_{t_1}^{t_2} |\dot{q}_n| dt \right] \ , \tag{3.36}$$

and

$$\boldsymbol{df_v}^T = \left[\int_{t_1}^{t_2} \dot{q}_1^2 dt, \ \cdots \ , \int_{t_1}^{t_2} \dot{q}_n^2 dt \right] \ . \tag{3.37}$$

Substitution of Eqs.(3.34), (3.35) and (3.27) into Eq.(3.26) results in

$$y = \int_{t_1}^{t_2} \boldsymbol{\tau}^T \dot{\boldsymbol{q}} dt = \boldsymbol{dl}^T \boldsymbol{x} + \boldsymbol{df}_s^T \boldsymbol{f}_s + \boldsymbol{df}_v^T \boldsymbol{f}_v \; . \tag{3.38}$$

Now define a vector $\boldsymbol{x}'$ with dimensions $12n \times 1$,

$$\boldsymbol{x}'^T = [\boldsymbol{x}^T, \; \boldsymbol{f}_s^T, \; \boldsymbol{f}_v^T] \; , \tag{3.39}$$

in which $\boldsymbol{x}'$ is the vector of dynamic parameters defined by Eq.(3.11) written in a slightly different manner. Substitution of Eq.(3.39) into Eq.(3.38) leads to

$$y = \boldsymbol{d}^T \boldsymbol{x}' \; , \tag{3.40}$$

where

$$\boldsymbol{d}^T = [\boldsymbol{dl}^T, \boldsymbol{df}_s^T, \boldsymbol{df}_v^T] \; . \tag{3.41}$$

We do not distinguish between a vector $\boldsymbol{x}'$ used in Eqs.(3.11) and (3.40) since it contains the same set of parameters. The robot dynamic parameters in Eq.(3.40), are represented in a linear form and the vector of functions $\boldsymbol{d}$ depends on the vector of generalised positions and velocities. On the other hand y is a scalar which depends on the product of generalised forces and generalised velocities. Thus we have obtained a model, described by a scalar equation (3.40), which depends linearly on a vector of dynamic parameters. Many authors [21, 54, 124, 183] refer to this model as integral, as opposed to the model described by the vector equation (3.12), which as it has already been mentioned, is called differential. Equation (3.40) incorporates the Coulomb and viscous friction coefficients. Here we assume a very simple friction model with the advantage of having linear friction coefficients. A more complicated friction model will be discussed in Chapter 5.

Here we make two observations. The integral model defined by Eq.(3.40) can be obtained formally from the differential model by multiplying the generalised forces $\boldsymbol{\tau}$ by the vector of generalised velocities $\dot{\boldsymbol{q}}$ and integrating the product from t_1 to t_2. It is clear that the vector of dynamic parameters $\boldsymbol{x}'$ appearing in both representations is exactly the same. This will be discussed in the next section.

The other observation is associated with the number of equations. The differential model is represented by a set of n equations and the integral model by only one equation. This has consequences in the practical implementation of any identification scheme. The differential model is richer in information than the integral model; in the last one all information is hidden in only one equation. Generally, while one can derive the integral model from the differential one as described above, calculation of the differential model from the integral one is not possible due to the fact that it is very difficult to follow how energy is distributed between the links of the manipulator. Many components of the energy described by the right-hand side of Eq.(3.40) cancel out. Note that we calculate the sum of the kinetic and potential energy and these energies can vanish for some configurations of the manipulator. This has been verified by experimental results during identification of the robot dynamic parameters.

Finally we present expressions for elements DE_{ki}^j and DE_{pi}^j (with j changing from 1 to 10), which appear in Eqs.(3.24) and (3.25). Making use of the following notation:

$$\left[\frac{\partial E_{kc}}{\partial I_{xx}}, \frac{\partial E_{kc}}{\partial I_{xy}}, \frac{\partial E_{kc}}{\partial I_{xz}}, \frac{\partial E_{kc}}{\partial I_{yy}}, \frac{\partial E_{kc}}{\partial I_{yz}}, \right.$$
$$\left. \frac{\partial E_{kc}}{\partial I_{zz}}, \frac{\partial E_{kc}}{\partial m}, \frac{\partial E_{kc}}{\partial mc_x}, \frac{\partial E_{kc}}{\partial mc_y}, \frac{\partial E_{kc}}{\partial mc_z}\right]_i , \tag{3.42}$$

$$\left[DE_{ki}^1,\ DE_{ki}^2,\ \cdots\ DE_{ki}^{10}\right]_i , \tag{3.43}$$

and analysing the expression for the total kinetic energy (3.6) and Eqs.(3.8) and (3.9) we get

$$\begin{aligned} DE_{ki}^1 &= \frac{1}{2}\omega_{ix}^2,\ DE_{ki}^2 = \omega_{ix}\omega_{iy},\ DE_{ki}^3 = \omega_{ix}\omega_{iz}, \\ DE_{ki}^4 &= \frac{1}{2}\omega_{iy}^2,\ DE_{ki}^5 = \omega_{iy}\omega_{iz},\ DE_{ki}^6 = \frac{1}{2}\omega_{iz}^2, \\ DE_{ki}^7 &= \frac{1}{2}\ {}^i\boldsymbol{v}_i^T\ {}^i\boldsymbol{v}_i,\ DE_{ki}^8 = \omega_{iz}v_{iy} - \omega_{iy}v_{iz}, \\ DE_{ki}^9 &= \omega_{ix}v_{iz} - \omega_{iz}v_{iy},\ DE_{ki}^{10} = \omega_{iy}v_{ix} - \omega_{ix}v_{iy} . \end{aligned} \tag{3.44}$$

In Eq.(3.44) ω_{ix}, ω_{iy}, ω_{iz} are the components of vector ${}^i\boldsymbol{\omega}_i$ and v_{ix}, v_{iy}, v_{iz} are the components of vector ${}^i\boldsymbol{v}_i$. Based on the expression for the potential energy (3.10)we can write

$$\begin{aligned} DE_{pi}^1 = DE_{pi}^2 = DE_{pi}^3 = DE_{pi}^4 = DE_{pi}^5 = DE_{pi}^6 = 0 , \\ DE_{pi}^7 = -\boldsymbol{g}^T\ {}^0\boldsymbol{l}_i , \end{aligned} \tag{3.45}$$

and

$$\left[DE_{pi}^8,\ DE_{pi}^9,\ DE_{pi}^{10}\right] = -\boldsymbol{g}^T\ {}^0_i\boldsymbol{R} ,$$

where ${}^0_i\boldsymbol{R} = {}^0_1\boldsymbol{R}\,{}^1_2\boldsymbol{R}\cdots {}^{i-1}_{\ \ i}\boldsymbol{R}$. Based on Eqs.(3.44), (3.45), (3.36), and (3.37) we are able to calculate the right-hand side of the integral model given by Eq.(3.40). A complete derivation of the integral model can be found in [60].

Here we make one comment. For the purpose of identification, a sufficient number of equations has to be calculated based on Eq.(3.38) between different time intervals. One possibility is to calculate the k-th equation in the time interval $(t_1, t_2)_k$. Another one is to calculate Eq.(3.38) in the time interval which starts at $t_0 = 0$ and ends at t_f, where t_f denotes the final time of calculation. In the former case we are dealing with a short time integral, while in the latter we have a long time integral. We will use these definitions often in Chapter 5 when describing experimental results. These definitions were first introduced in reference [124] and later in references [136, 140].

3.4 Comparison of the differential and integral models

In this section we compare the differential and integral models. Recall that the Lagrangian of the manipulator is the difference between total kinetic and potential energy, $\mathcal{L} = E_{kc} - E_{pc}$. The energy theorem uses the function $H = E_{kc} + E_{pc}$ which is the sum of the kinetic and potential energy. First, we calculate a time derivative of the function H:

$$\frac{d}{dt}H = \frac{d}{dt}E_{kc} + \frac{d}{dt}E_{pc} = \frac{1}{2}\frac{d}{dt}\dot{\boldsymbol{q}}^T\boldsymbol{A}(\boldsymbol{q})\dot{\boldsymbol{q}} + \left[\frac{\partial E_{pc}}{\partial \boldsymbol{q}}\right]^T \dot{\boldsymbol{q}} =$$
$$= \dot{\boldsymbol{q}}^T\boldsymbol{A}(\boldsymbol{q})\ddot{\boldsymbol{q}} + \frac{1}{2}\dot{\boldsymbol{q}}^T\frac{d\boldsymbol{A}(\boldsymbol{q})}{dt}\dot{\boldsymbol{q}} + \dot{\boldsymbol{q}}^T\frac{\partial E_{pc}}{\partial \boldsymbol{q}} . \tag{3.46}$$

In the above equation $\frac{\partial E_{pc}}{\partial \boldsymbol{q}}$ is a column vector which consists of partial derivatives of the total potential energy with respect to the generalised positions, q_i. Now we calculate the product $\dot{\boldsymbol{q}}^T\boldsymbol{\tau}_e$. Recall that

$$\boldsymbol{\tau}_e = \frac{d}{dt}\frac{\partial}{\partial \dot{\boldsymbol{q}}}E_{kc} - \frac{\partial}{\partial \boldsymbol{q}}\left[E_{kc} - E_{pc}\right] = \frac{d}{dt}\left[\boldsymbol{A}(\boldsymbol{q})\dot{\boldsymbol{q}}\right] - \frac{\partial}{\partial \boldsymbol{q}}E_{kc} + \frac{\partial E_{pc}}{\partial \boldsymbol{q}} . \tag{3.47}$$

Calculating the time derivative of the first component of the above equation leads to

$$\frac{d}{dt}\left[\boldsymbol{A}(\boldsymbol{q})\dot{\boldsymbol{q}}\right] = \frac{d}{dt}\left[\boldsymbol{A}(\boldsymbol{q})\right]\dot{\boldsymbol{q}} + \boldsymbol{A}(\boldsymbol{q})\ddot{\boldsymbol{q}} , \tag{3.48}$$

where

$$\frac{d}{dt}\left[\boldsymbol{A}(\boldsymbol{q})\right]\dot{\boldsymbol{q}} = \left[\dot{q}_1\frac{\partial \boldsymbol{A}}{\partial q_1} + \dot{q}_2\frac{\partial \boldsymbol{A}}{\partial q_2} + \cdots + \dot{q}_n\frac{\partial \boldsymbol{A}}{\partial q_n}\right]\dot{\boldsymbol{q}} . \tag{3.49}$$

Note also that

$$\frac{\partial E_{kc}}{\partial \boldsymbol{q}} = \frac{1}{2}\left[\dot{\boldsymbol{q}}^T\frac{\partial \boldsymbol{A}}{\partial q_1}\dot{\boldsymbol{q}}, \dot{\boldsymbol{q}}^T\frac{\partial \boldsymbol{A}}{\partial q_2}\dot{\boldsymbol{q}}, \cdots, \dot{\boldsymbol{q}}^T\frac{\partial \boldsymbol{A}}{\partial q_n}\dot{\boldsymbol{q}}\right]^T . \tag{3.50}$$

Substitution of Eqs.(3.48)–(3.50) into (3.47) leads to

$$\boldsymbol{\tau}_e = \left[\dot{q}_1\frac{\partial \boldsymbol{A}}{\partial q_1} + \dot{q}_2\frac{\partial \boldsymbol{A}}{\partial q_2} + \cdots + \dot{q}_n\frac{\partial \boldsymbol{A}}{\partial q_n}\right]\dot{q} + \boldsymbol{A}(\boldsymbol{q})\ddot{\boldsymbol{q}} +$$
$$-\frac{1}{2}\left[\dot{\boldsymbol{q}}^T\frac{\partial \boldsymbol{A}}{\partial q_1}\dot{\boldsymbol{q}} + \dot{\boldsymbol{q}}^T\frac{\partial \boldsymbol{A}}{\partial q_2}\dot{\boldsymbol{q}} + \cdots + \dot{\boldsymbol{q}}^T\frac{\partial \boldsymbol{A}}{\partial q_n}\dot{\boldsymbol{q}}\right]^T + \frac{\partial E_{pc}}{\partial \boldsymbol{q}} . \tag{3.51}$$

In order to calculate the product $\dot{\boldsymbol{q}}^T\boldsymbol{\tau}_e$ we have to left multiply the above equation by $\dot{\boldsymbol{q}}^T$:

$$\dot{\boldsymbol{q}}^T\boldsymbol{\tau}_e = \dot{\boldsymbol{q}}^T\left[\dot{q}_1\frac{\partial \boldsymbol{A}}{\partial q_1} + \dot{q}_2\frac{\partial \boldsymbol{A}}{\partial q_2} + \cdots + \dot{q}_n\frac{\partial \boldsymbol{A}}{\partial q_n}\right]\dot{\boldsymbol{q}} + \dot{\boldsymbol{q}}^T\boldsymbol{A}(\boldsymbol{q})\ddot{\boldsymbol{q}} +$$
$$-\frac{1}{2}\dot{\boldsymbol{q}}^T\left[\dot{\boldsymbol{q}}^T\frac{\partial \boldsymbol{A}}{\partial q_1}\dot{\boldsymbol{q}}, \dot{\boldsymbol{q}}^T\frac{\partial \boldsymbol{A}}{\partial q_2}\dot{\boldsymbol{q}}, \cdots, \dot{\boldsymbol{q}}^T\frac{\partial \boldsymbol{A}}{\partial q_n}\dot{\boldsymbol{q}}\right]^T + \dot{\boldsymbol{q}}^T\frac{\partial E_{pc}}{\partial \boldsymbol{q}} . \tag{3.52}$$

The third term on the right–hand side above can be written as

$$\frac{1}{2}\left[\dot{\boldsymbol{q}}^T\frac{\partial \boldsymbol{A}}{\partial q_1}\dot{q}_1\dot{\boldsymbol{q}}+\dot{\boldsymbol{q}}^T\frac{\partial \boldsymbol{A}}{\partial q_2}\dot{q}_2\dot{\boldsymbol{q}}+\cdots+\dot{\boldsymbol{q}}^T\frac{\partial \boldsymbol{A}}{\partial q_n}\dot{q}_n\dot{\boldsymbol{q}}\right]=\frac{1}{2}\dot{\boldsymbol{q}}^T\frac{d\boldsymbol{A}}{dt}\dot{\boldsymbol{q}}\,.$$

Therefore Eq.(3.52) can be rewritten as follows:

$$\dot{\boldsymbol{q}}^T\boldsymbol{\tau}_e=\dot{\boldsymbol{q}}^T\boldsymbol{A}(\boldsymbol{q})\ddot{\boldsymbol{q}}+\frac{1}{2}\dot{\boldsymbol{q}}^T\frac{\partial \boldsymbol{A}}{\partial \boldsymbol{q}}\dot{\boldsymbol{q}}+\dot{\boldsymbol{q}}^T\frac{\partial E_{pc}}{\partial \boldsymbol{q}}\,. \tag{3.53}$$

Comparing Eqs.(3.46) and (3.53) we obtain

$$\frac{d}{dt}H=\dot{\boldsymbol{q}}^T\boldsymbol{\tau}_e\,, \tag{3.54}$$

which results from the energy theorem given by Eq.(3.26), where we have assumed that the integral is indefinite. From the above derivation it is clear that calculation of the product $\dot{\boldsymbol{q}}^T\boldsymbol{\tau}_e$ results in the time derivative of the sum of the kinetic and potential energy. In practice instead of calculating this product we rather calculate the right–hand side of Eq.(3.38) and use Eqs.(3.44) and (3.45), which result directly from the kinetic and potential energy. These calculations are straightforward and much simpler. We have shown that both differential and integral models are equivalent and that the latter can be derived from the former and vice–versa. The derivation described above shows how to calculate the product $\dot{\boldsymbol{q}}^T\boldsymbol{\tau}_e$. Inclusion of the friction coefficients is straightforward. From the above derivation it is clear that the dynamic parameters that appear in the differential model and those which appear in the integral model are exactly the same. Therefore one can use the integral model for identification and the identified parameters can then be used in the differential model, for example, for control. Elsefari and Khalil [54] show that the integral model can also be used for control. However, the integral model does not possess the skew–symmetric property which is exhibited by the differential model, where the matrix $\dot{\boldsymbol{A}}-(\boldsymbol{C}^T+\boldsymbol{C})$ is skew–symmetric (here $\boldsymbol{C}$ represents the Coriolis and centrifugal coefficients combined in one matrix).

3.5 Canonical models

In Sections 3.2 and 3.3 we introduced differential and integral models given by Eqs.(3.12) and (3.38), respectively. It was shown also that both models are described in terms of the same set of dynamic parameters. These parameters are the inertial parameters of individual links - their mass, centre of mass, and second moment of inertia. All of them are expressed in local coordinate frames assigned to individual links. Looking at the kinematics of the robot, some of the inertial parameters may have no effect on the joint generalised forces and some others may be regrouped together as linear combinations of inertial parameters of the individual links. Looking at Eq.(3.23) one can observe the following two general rules [103].

1. An inertial parameter X_i has no effect on the dynamic model if the corresponding vector $\boldsymbol{D}^i$ which is associated with this parameter is equal to zero.
2. An inertial parameter X_i can be regrouped to some other parameters $(X_{i1}, X_{i2}, \ldots, X_{ir})$ if $\boldsymbol{D}^i$ is linearly dependent on $\boldsymbol{D}^{i1}, \boldsymbol{D}^{i2}, \ldots, \boldsymbol{D}^{ir}$, i.e. if
$$\boldsymbol{D}^i = \alpha_{i1}\boldsymbol{D}^{i1} + \cdots + \alpha_{ik}\boldsymbol{D}^{ik} + \cdots + \alpha_{ir}\boldsymbol{D}^{ir} , \tag{3.55}$$
where α_{ik} is constant $(k = 1, 2, \ldots, r)$.

Khalil and Kleinfinger [103] proposed to eliminate X_i and $\boldsymbol{D}^i$ as a result of the regrouping process. The effect of X_i on the dynamic model is maintained by putting X_{ikR} in place of X_{ik}, where

$$X_{ikR} = X_{ik} + \alpha_{ik} X_i . \tag{3.56}$$

The necessary and sufficient conditions to satisfy the relation (3.55) are given by

$$\boldsymbol{A}'^i = \alpha_{i1}\boldsymbol{A}'^{i1} + \cdots + \alpha_{ir}\boldsymbol{A}'^{ir} \tag{3.57}$$

and

$$\boldsymbol{G}'^i = \alpha_{i1}\boldsymbol{G}'^{i1} + \cdots + \alpha_{ir}\boldsymbol{G}'^{ir} , \tag{3.58}$$

where

$$\boldsymbol{A}'^i = \boldsymbol{A}^i / X^i \quad \text{and} \quad \boldsymbol{G}'^i = \boldsymbol{G}^i / X_i . \tag{3.59}$$

The proof comes from the fact that if Eq.(3.57) is satisfied, taking into account Eqs.(3.20) and (3.21), we conclude that similar relations for $\boldsymbol{B}^i$ and $\bar{\boldsymbol{C}}^i$ are also satisfied. Therefore, only the expressions for $\boldsymbol{A}^i$ and $\boldsymbol{G}^i$ are needed to carry out the regrouping of the inertial parameters. Khalil and Kleinfinger developed software which generates automatically the expressions for matrices $\boldsymbol{A}'^i$ and $\boldsymbol{G}'^i$.

A dynamic model described by a minimum set of parameters is referred to as a canonical model. Such a model can be used for the purpose of identification of its parameters. As mentioned before, the dynamic parameters of the canonical model are linear combinations of dynamic parameters of individual links. The requirement of the dynamic model to be canonical is a necessary condition for a successful identification. Note that functions $\boldsymbol{D}(\boldsymbol{q}, \dot{\boldsymbol{q}}, \ddot{\boldsymbol{q}})$ or $\boldsymbol{d}(\boldsymbol{q}, \dot{\boldsymbol{q}})$ associated with the dynamic parameters may vanish at certain points, depending on the predefined test trajectory defined in terms of $\boldsymbol{q}, \dot{\boldsymbol{q}}$, and $\ddot{\boldsymbol{q}}$, and we cannot identify them based on measurements of these functions (compare Eqs.(3.12) and (3.40). This problem will be discussed in Chapter 5.

Gautier and Khalil [61] defined practical rules which make it possible to regroup the inertial parameters. The rules can be derived based on the considerations presented above. Here we cite those rules.

1. If joint 1 is translational then parameters I_{1xx}, I_{1xy}, I_{1xz}, I_{1yy}, I_{1yz}, I_{1zz}, $m_1 c_{1x}$, $m_1 c_{1y}$ and $m_1 c_{1z}$ have no effect on the dynamic model.

2. If join 1 is rotational then parameters I_{1xx}, I_{1xy}, I_{1xz}, I_{1yy}, I_{1yz}, m_1c_{1z}, m_1 have no effect on the dynamic model. Moreover, if the axis of the first joint is aligned with the direction of the acceleration of gravity then parameters m_1c_{1x} and m_1c_{1y} have no effect as well.
3. In order for a parameter of link j to be regrouped to one or more parameters of link $j-2$, this parameter has first to be combined with one or more parameters of link $j-1$, which can then be combined with the parameters of link $j-2$.
4. If joint j is rotational then parameters I_{jyy}, m_jc_{jz}, m_j can be eliminated using the following relations:

$$
\begin{aligned}
&X_{jR} = I_{jxx} - I_{jyy} \, , \\
&X_{1(j-1)R} = I_{(j-1)xx} + I_{jyy} + 2a_jm_jc_{jz} + a_j^2m_j \, , \\
&X_{2(j-1)R} = I_{(j-1)yy} + \cos^2\alpha_jI_{jyy} + 2a_j\cos^2\alpha_jm_jc_{jz} + \\
&+(d_j^2 + a_j^2\cos^2\alpha_j)m_j \, , \\
&X_{3(j-1)R} = I_{(j-1)zz} + \sin^2\alpha_jI_{jyy} + 2a_j\sin^2\alpha_jm_jc_{jz} + \\
&+(d_j^2 + a_j^2\sin^2\alpha_j)m_j \, , \\
&X_{4(j-1)R} = I_{(j-1)xy} + d_j\sin\alpha_jm_jc_{jz} + d_ja_j\sin\alpha_jm_j \, , \\
&X_{5(j-1)R} = I_{(j-1)xz} - d_j\cos\alpha_jm_jc_{jz} - d_ja_j\cos\alpha_jm_j \, , \qquad (3.60) \\
&X_{6(j-1)R} = I_{(j-1)yz} + \cos\alpha_j\sin\alpha_jI_{jyy} + 2a_j\cos\alpha_j\sin\alpha_jm_jc_{jz} + \\
&a_j^2\cos\alpha_j\sin\alpha_jm_j \, , \\
&X_{7(j-1)R} = m_{j-1}c_{(j-1)x} + d_jm_j \, , \\
&X_{8(j-1)R} = m_{j-1}c_{(j-1)y} - \sin\alpha_jm_jc_{jz} - a_j\sin\alpha_jm_j \, , \\
&X_{9(j-1)R} = m_{j-1}c_{(j-1)z} + \cos\alpha_jm_jc_{jz} + a_j\cos\alpha_jm_j \, , \\
&X_{10(j-1)R} = m_{j-1} + m_j \, .
\end{aligned}
$$

The symbols on the right–hand side above denote individual dynamic parameters of link $(j-1)$ and j, while on the left–hand side we have notation according to Eq.(3.56). We use modified Denavit-Hartenberg parameters between two successive links $(j-1)$ and j.
5. If joint j is translational then parameters I_{jxx}, I_{jxy}, I_{jxz}, I_{jyy}, I_{jyz}, I_{jzz} can be eliminated by using the following equation, which sums up two inertia matrices (of type given by Eq.(3.7)) of two successive links $(j-1)$ and j:

$$
\boldsymbol{I}_{(j-1)R} = {}^{j-1}\boldsymbol{I}_{j-1} + {}^{j-1}_{\;\;j}\boldsymbol{R}\,{}^{j}\boldsymbol{I}_j\,{}^{\;\;j}_{j-1}\boldsymbol{R}. \qquad (3.61)
$$

All the quantities here are 3×3 matrices. Note that six scalar equations can be written easily from this equation, due to the symmetry property of the inertia matrix.
6. If for all the parameters of link $j-1$ the appropriate $DE_{ki} \neq 0$ or $DE_{pi} \neq$ const, then Eqs.(3.60) and (3.61) detect all the regrouping parameters for links $j, \ldots, n$. Moreover, no elimination is possible if

$DE_{ki} = 0$ and $DE_{pi} =$ const. This rule is a consequence of the detailed analysis of Eqs.(3.24) and (3.25), which is straightforward though rather laborious. Details can be found in reference [61].
Moreover, Gautier and Khalil state, based on the above rules 1, 2, 4, and 5, that the number of minimum inertial parameters is less than or equal to

$$7n_r + 4n_t - 3 - \bar{\sigma_1} \,, \tag{3.62}$$

where n_r denotes the number of rotational joints and n_t denotes the number of translational joints which are present at the linkage. The above practical rules have been successfully applied by hand to the dynamic models considered in this monograph.
Note that analytical derivation of the minimum number of link parameters is carried out by considering the total kinetic and potential energy for the differential and integral model, respectively. This is obvious because the kinetic and potential energy both depend on the dynamic parameters of the individual links of a manipulator. The friction coefficients are not included in these considerations since they appear linearly in both models with known functions associated with them.

Similar considerations were carried by Mayeda and co-authors [159], where a set of base parameters in a closed form solution is described and the exact number of those parameters for kinematically simple manipulators is given. The results are not discussed since they are restricted to manipulators with rotational joints. Khosla [108] proposes a symbolic procedure for calculating a set of independent inertial parameters. Another symbolic procedure can be found in reference [104]; the results are included in those obtained by Gautier and Khalil [61].

Now we show that the dynamic parameters which appear in the integral model can be used in the dynamic equations of motion. The two types of equations are functions of the dynamic parameters, which are not exactly measurable and need to be identified. Although the integral model (or energy difference equation) is much easier to formulate than the dynamic equations of motion (differential model), it is the latter that is used for manipulator control. Thus it must be shown formally that these two equations are equivalent, in the sense that they contain the same dynamic parameters. In this way we are assured that the results of the estimation procedure based on the integral model can be used for control; nevertheless the differential model is the most popular model used for that purpose. The presented derivation is based on references [203, 204, 205]; it is extended here to the point when the dissipated energy contains the Coulomb friction effect.

Now recall the definition of the Lagrangian function $\mathcal{L}$ and the definition of the total energy H of the system which is the sum of the kinetic and potential energy. They are related to each other by the following equation:

$$\mathcal{L} = E_{kc} - E_{pc} = H - 2E_{pc}. \tag{3.63}$$

Taking into account Eqs.(3.10) and (3.31) gives

$$E_{pc}(\boldsymbol{q},\boldsymbol{x}) = H(\boldsymbol{q},\boldsymbol{0},\boldsymbol{x}) = \boldsymbol{dl}^T\boldsymbol{x}. \tag{3.64}$$

In Eq.(3.64) it is shown explicitly that the total energy of the system H depends on the vector of generalised positions and velocities as well as on a vector of dynamic parameters, $\boldsymbol{x}$. Next substitute in H the generalised velocity, $\dot{\boldsymbol{q}} = \boldsymbol{0}$ (since the total potential energy does not depend on $\dot{\boldsymbol{q}}$). Therefore the Lagrangian function can be written as

$$\mathcal{L}(\boldsymbol{q},\dot{\boldsymbol{q}},\boldsymbol{x}) = H(\boldsymbol{q},\dot{\boldsymbol{q}},\boldsymbol{x}) - 2H(\boldsymbol{q},\boldsymbol{0},\boldsymbol{x}) = \left[\boldsymbol{dl}^T(\boldsymbol{q},\dot{\boldsymbol{q}}) - 2\boldsymbol{dl}^T(\boldsymbol{q},\boldsymbol{0})\right]\boldsymbol{x} \triangleq \boldsymbol{d}^T\boldsymbol{x}. \tag{3.65}$$

Now write the Lagrange equations of motion given by Eq.(3.4) (which represent the differential model in a slightly different form):

$$\frac{d}{dt}\frac{\partial\mathcal{L}}{\partial\dot{\boldsymbol{q}}} - \frac{\partial\mathcal{L}}{\partial\boldsymbol{q}} + \boldsymbol{\tau}_{fs} + \boldsymbol{\tau}_{fv} = \boldsymbol{\tau}\,. \tag{3.66}$$

Introduce the following notation:

$$\boldsymbol{C}_v = diag\,[\dot{q}_1,\dot{q}_2,\ldots,\dot{q}_n] \quad\text{and}\quad \boldsymbol{C}_s = diag\,[sgn(\dot{q}_1),sgn(\dot{q}_2),\ldots,sgn(\dot{q}_n)]\,. \tag{3.67}$$

Then taking into account the above definitions, Eqs.(3.34), (3.35), and Eq.(3.66) can be written as

$$\frac{d}{dt}\frac{\partial\mathcal{L}}{\partial\dot{\boldsymbol{q}}} - \frac{\partial\mathcal{L}}{\partial\boldsymbol{q}} + \boldsymbol{C}_v\boldsymbol{f}_v + \boldsymbol{C}_s\boldsymbol{f}_s = \boldsymbol{\tau}. \tag{3.68}$$

Using the definition of vector $\boldsymbol{x}'$ given by Eq.(3.39) and using Eq.(3.65) the last equation can be written in the following form:

$$\boldsymbol{\tau}(\boldsymbol{q},\dot{\boldsymbol{q}},\boldsymbol{x}') = \left(\frac{d}{dt}\frac{\partial\boldsymbol{d}}{\partial\dot{\boldsymbol{q}}} - \frac{\partial\boldsymbol{d}}{\partial\boldsymbol{q}}\right)^T\boldsymbol{x} + \boldsymbol{C}_v\boldsymbol{f}_v + \boldsymbol{C}_s\boldsymbol{f}_s \tag{3.69}$$

or

$$\boldsymbol{\tau}(\boldsymbol{q},\dot{\boldsymbol{q}},\boldsymbol{x}') = \boldsymbol{\Omega}\boldsymbol{x}'\,, \tag{3.70}$$

where

$$\boldsymbol{\Omega} = \left[\left(\frac{d}{dt}\frac{\partial\boldsymbol{d}}{\partial\dot{\boldsymbol{q}}} - \frac{\partial\boldsymbol{d}}{\partial\boldsymbol{q}}\right)^T,\ \boldsymbol{C}_v\,,\ \boldsymbol{C}_s\right] \quad\text{and}\quad \boldsymbol{x}' = \left[\boldsymbol{x}^T,\boldsymbol{f}_v^T,\boldsymbol{f}_s^T\right]^T.$$

From Eq.(3.70) it is clear that the dynamic equations are linear in terms of the dynamic parameter vector $\boldsymbol{x}'$.

Often, due to kinematical constraints, only p of the manipulator joints motion are independent. This is, for example, the case with manipulators with closed kinematic chains. The above and subsequent considerations still hold true. One has to choose the independent joints, possibly those, where the actuators are located and find the related equations which describe the transformation between new independent coordinates and the remaining coordinates. The Jacobian matrix between the two sets of coordinates should be nonsingular.

Now take Eq.(3.40) for the integral model and corresponding Eqs.(3.36) and (3.37), and change the limits, namely assume the starting point t_0 and the final point t. Then one can write

$$[\boldsymbol{d}(t) - \boldsymbol{d}(t_0)]^T \boldsymbol{x} + \sum_{i=1}^{n} F_{iv} \int_{t_0}^{t} \dot{q}_i^2 dt + \sum_{i=1}^{n} F_{is} \int_{t_0}^{t} |\dot{q}_i| dt = \int_{t_0}^{t} \boldsymbol{\tau}^T \dot{\boldsymbol{q}} dt. \quad (3.71)$$

To simplify this equation we use the following notation:

$$y(t_0, t) = \int_{t_0}^{t} \boldsymbol{\tau}^T \dot{\boldsymbol{q}} dt, \quad dl_{iv} = \int_{t_0}^{t} \dot{q}_i^2 dt, \quad dl_{is} = \int_{t_0}^{t} |\dot{q}_i| dt$$

and consequently

$$\boldsymbol{df}_v(t_0, t) = [dl_{1v}(t_0, t), dl_{2v}(t_0, t), \ldots, dl_{nv}(t_0, t)]^T$$

and

$$\boldsymbol{df}_s(t_0, t) = [dl_{1s}(t_0, t), dl_{2s}(t_0, t), \ldots, dl_{ns}(t_0, t)]^T .$$

Note that the last two definitions are closely related to Eqs.(3.36) and (3.37). Taking into account the above definitions we rewrite Eq.(3.71) as follows:

$$[\boldsymbol{d}(t) - \boldsymbol{d}(t_0)]^T \boldsymbol{x} + \boldsymbol{df}_v^T(t_0, t) \boldsymbol{f}_v + \boldsymbol{df}_s^T(t_0, t) \boldsymbol{f}_s = y(t_0, t). \quad (3.72)$$

Eq.(3.72) can be written in a more compact form, namely

$$\boldsymbol{\Phi}(t_0, t)^T \boldsymbol{x}' = y(t_0, t) , \quad (3.73)$$

where $\boldsymbol{\Phi}(t_0, t) = \begin{bmatrix} \boldsymbol{d}(t) - \boldsymbol{d}(t_0) \\ \boldsymbol{df}_v \\ \boldsymbol{df}_s \end{bmatrix}$ and $\boldsymbol{x}' = \begin{bmatrix} \boldsymbol{x} \\ \boldsymbol{f}_v \\ \boldsymbol{f}_s \end{bmatrix}$.

Equation (3.73) is the energy difference equation written for instances of time t_0 and t, which is dissipated and that which is supplied by the actuators as opposed to Eq.(3.26), which is written for instances t_1 and t_2. Now we have to prove that both Eqs.(3.70) and (3.73) make use of the same components of $\boldsymbol{x}'$. We start by reformulating the above equations in terms of the inertial parameter vector $\boldsymbol{x}$ and the frictional coefficients vectors $\boldsymbol{f}_v$ and $\boldsymbol{f}_s$. Vector $\boldsymbol{x}'$ consists of individual dynamic parameters of all the links and friction coefficients, so its dimensions are $12n \times 1$. The null space of these equations is defined as the set of $\boldsymbol{x}'$ such that for all t

$$[\boldsymbol{d}(t) - \boldsymbol{d}(t_0)]^T \boldsymbol{x} + \boldsymbol{df}_v^T(t_0, t) \boldsymbol{f}_v + \boldsymbol{df}_s^T(t_0, t) \boldsymbol{f}_s \equiv 0 \quad (3.74)$$

and

$$\left(\frac{d}{dt} \frac{\partial \boldsymbol{d}}{\partial \dot{\boldsymbol{q}}} - \frac{\partial \boldsymbol{d}}{\partial \boldsymbol{q}} \right)^T \boldsymbol{x} + \boldsymbol{C}_v \boldsymbol{f}_v + \boldsymbol{C}_s \boldsymbol{f}_s \equiv \boldsymbol{0} . \quad (3.75)$$

Note that in Eq.(3.74) all the elements of the vector $\boldsymbol{d}(t) - \boldsymbol{d}(t_0)$ are linearly independent of the elements of vectors $\boldsymbol{df}_v(t_0, t)$ and $\boldsymbol{df}_s(t_0, t)$. Therefore Eq.(3.74) can be decoupled as

$$[\boldsymbol{d}(t) - \boldsymbol{d}(t_0)]^T \boldsymbol{x} \equiv 0, \quad \boldsymbol{df}_v^T(t_0, t)\boldsymbol{f}_v \equiv 0 \text{ and } \boldsymbol{df}_s^T(t_0, t)\boldsymbol{f}_s \equiv 0 . \tag{3.76}$$

In Eq.(3.75), all the elements of the matrix $\frac{d}{dt}\frac{\partial \boldsymbol{d}}{\partial \dot{\boldsymbol{q}}} - \frac{\partial \boldsymbol{d}}{\partial \boldsymbol{q}}$ are also linearly independent of the elements of matrices $\boldsymbol{C}_v$ and $\boldsymbol{C}_s$. Thus Eq.(3.75) can be decoupled as

$$\left(\frac{d}{dt}\frac{\partial \boldsymbol{d}}{\partial \dot{\boldsymbol{q}}} - \frac{\partial \boldsymbol{d}}{\partial \boldsymbol{q}}\right)^T \boldsymbol{x} \equiv \boldsymbol{0}, \quad \boldsymbol{C}_v \boldsymbol{f}_v \equiv \boldsymbol{0} \text{ and } \boldsymbol{C}_s \boldsymbol{f}_s \equiv \boldsymbol{0} . \tag{3.77}$$

Therefore the dynamic parameter vector $\boldsymbol{x}'$ can always be written as the sum of the two vectors

$$\boldsymbol{x}' = \begin{bmatrix} \boldsymbol{x} \\ \boldsymbol{0} \\ \boldsymbol{0} \end{bmatrix} + \begin{bmatrix} \boldsymbol{0} \\ \boldsymbol{f}_v \\ \boldsymbol{f}_s \end{bmatrix} \triangleq \boldsymbol{x}_d + \boldsymbol{x}_f . \tag{3.78}$$

Now we denote by S_x and S_f the subspaces spanned by vectors $\boldsymbol{x}_d$ and $\boldsymbol{x}_f$, respectively. Then the vector space $R^{12n} = S_x \oplus S_f$. Here for simplicity we consider the friction coefficients as a single vector $\boldsymbol{x}_f$ which combines the velocity and Coulomb friction coefficients.

We know from the preceding considerations that not all of the dynamic parameters affect the dynamic equations. This means that less than $12n$ basis vectors are required to span a particular solution of the dynamic equations. Now define the null space of the dynamic equations of motion, $\mathcal{N}(\boldsymbol{\tau})$ as the set of dynamic parameter vectors $\boldsymbol{x}'$ which satisfy $\boldsymbol{\tau}(\boldsymbol{q}, \dot{\boldsymbol{q}}, \boldsymbol{x}') = \boldsymbol{\Omega}\boldsymbol{x}' \equiv \boldsymbol{0}\ \forall \boldsymbol{q}, \dot{\boldsymbol{q}}$. Consequently the subspace $\mathcal{N}(\boldsymbol{\tau})^\perp$ is the orthogonal element of the null space. Define $E_{x'} \triangleq \mathcal{N}(\boldsymbol{\tau})^\perp$ and call it the essential parameter space (EPS) after Shih-Ying Sheu and Walker [203, 204, 205]. The projection of $\boldsymbol{x}'$ into this space, $\boldsymbol{x}_\tau$, is called the essential parameter vector (EPV) and its components are called the essential parameters. The EPV is the particular solution of the dynamic equations which has the minimum length, since it does not contain the null space components. Therefore, the basis of the EPS is the minimum set of parameters to span a particular solution. The dimension of the EPV is equal to the number of independent dynamic parameters in the differential model (or in the dynamic equations of motion).

In a similar way define the null space of the energy difference equation, $\mathcal{N}(y)$, as the set of dynamic parameter vector $\boldsymbol{x}'$ which satisfy $y(\boldsymbol{q}, \dot{\boldsymbol{q}}, \boldsymbol{x}') \equiv 0$, $\forall \boldsymbol{q}, \dot{\boldsymbol{q}}$. It is clear that the components of $\boldsymbol{x}' \in \mathcal{N}(y)$ are not identifiable. As before, only the components belonging to $\mathcal{N}(y)^\perp$ are identifiable. Let us define $I_{x'} \triangleq \mathcal{N}(y)^\perp$ and call it the identifiable parameter space (IPS) [203, 204, 205]. The projection of $\boldsymbol{x}'$ into this space, $\boldsymbol{x}'_y$, is called the identifiable parameter vector (IPV), and its components are called the identifiable parameters.

Now we prove that both the identifiable parameter space and the essential parameter space are equivalent. The IPS and EPS are the orthogonal components of the null spaces $\mathcal{N}(y)$ and $\mathcal{N}(\boldsymbol{\tau})$, respectively. Therefore, the IPS and EPS are the same if both $\mathcal{N}(y)$ and $\mathcal{N}(\boldsymbol{\tau})$ are the same. In addition, since $R^{12n} = S_x \oplus S_f$, we can prove that $\mathcal{N}(y) = \mathcal{N}(\boldsymbol{\tau})$ by showing independently that $\mathcal{N}(y_e) = \mathcal{N}(\boldsymbol{\tau}_e)$ and $\mathcal{N}(y_f) = \mathcal{N}(\boldsymbol{\tau}_f)$, where

$$y_e \triangleq [\boldsymbol{d}(t) - \boldsymbol{d}(t_0)]^T \boldsymbol{x} \ ,$$

$$y_f \triangleq \left[\boldsymbol{df}_v^T, \boldsymbol{df}_s^T\right] \boldsymbol{f} \ , \quad \text{where} \quad \boldsymbol{f} = \begin{bmatrix} \boldsymbol{f}_v \\ \boldsymbol{f}_s \end{bmatrix} ,$$

$$\boldsymbol{\tau}_e \triangleq \left(\frac{d}{dt}\frac{\partial \boldsymbol{d}}{\partial \dot{\boldsymbol{q}}} - \frac{\partial \boldsymbol{d}}{\partial \boldsymbol{q}} \right)^T \boldsymbol{x} \ ,$$

and

$$\boldsymbol{\tau}_f \triangleq [\boldsymbol{C}_v, \boldsymbol{C}_s] \, \boldsymbol{f} .$$

The above equations give the inertial energy, the total dissipated energy due to viscous and Coulomb friction, the inertial force, and the total frictional forces.

First, we prove that $\mathcal{N}(y_e) = \mathcal{N}(\boldsymbol{\tau}_e)$. The proof is borrowed from references [203, 204, 205]. Taking into account the definition of the Lagrangian given by Eq.(3.65) we can write

$$\mathcal{L}(\boldsymbol{q}, \dot{\boldsymbol{q}}, \boldsymbol{x}) - \mathcal{L}(\boldsymbol{q}_0 {\cdot} \dot{\boldsymbol{q}}_0, \boldsymbol{x}) = y_e(\boldsymbol{q}, \dot{\boldsymbol{q}}, \boldsymbol{x}) - 2y_e(\boldsymbol{q}, \boldsymbol{0}, \boldsymbol{x}). \tag{3.79}$$

Now take any $\boldsymbol{x} \in \mathcal{N}(y_e)$. From the last equation it is clear that $\mathcal{L}(\boldsymbol{q}, \dot{\boldsymbol{q}}, \boldsymbol{x})$ is a constant function of the vectors of generalised positions and velocities, $\boldsymbol{q}$ and $\dot{\boldsymbol{q}}$. Therefore all its derivatives are identically equal to zero. Hence, from Eq.(3.68) we have that $\boldsymbol{x} \in \mathcal{N}(\boldsymbol{\tau}_e)$, which implies that $\mathcal{N}(y_e) \subset \mathcal{N}(\boldsymbol{\tau}_e)$. In order to prove that $\mathcal{N}(\boldsymbol{\tau}_e) \subset \mathcal{N}(y_e)$ we calculate the time derivative of the total energy of the manipulator, namely

$$\frac{dH(\boldsymbol{q}, \dot{\boldsymbol{q}}, \boldsymbol{x})}{dt} = \boldsymbol{\tau}_e^T \dot{\boldsymbol{q}} = \varepsilon(\boldsymbol{q}, \dot{\boldsymbol{q}}, \ddot{\boldsymbol{q}}, \boldsymbol{x}). \tag{3.80}$$

Obviously scalar ε is a function of the vector of joint positions, velocities, and accelerations and of vector $\boldsymbol{x}$. Take any $\boldsymbol{x} \in \mathcal{N}(\boldsymbol{\tau}_e)$; then $\boldsymbol{\tau}_e(\boldsymbol{x})$ is identically equal to zero, $\boldsymbol{\tau}_e(\boldsymbol{x}) \equiv \boldsymbol{0}$. Therefore from Eq.(3.80) $\varepsilon(\boldsymbol{x}) \equiv 0$ and $\boldsymbol{x} \in \mathcal{N}(\varepsilon)$. This implies that $\mathcal{N}(\boldsymbol{\tau}_e) \subset \mathcal{N}(\varepsilon)$. However, if $\boldsymbol{x} \in \mathcal{N}(\varepsilon)$, then it follows from Eq.(3.80) that $dH/dt \equiv 0$, and as a consequence $H(\boldsymbol{q}, \dot{\boldsymbol{q}}, \boldsymbol{x})$ is constant. From Eq.(3.26) in which we substitute $t_1 = t_0$ and $t_2 = t$, we obtain $\boldsymbol{x} \in \mathcal{N}(y_e)$, which in turn implies $\mathcal{N}(\varepsilon) \subset \mathcal{N}(y_e)$, and hence $\mathcal{N}(\boldsymbol{\tau}_e) \subset \mathcal{N}(y_e)$. This proves that $\mathcal{N}(y_e) = \mathcal{N}(\boldsymbol{\tau}_e)$.

Now we extend the results presented in [204] to the case considered here and we prove that $\mathcal{N}(y_f) = \mathcal{N}(\boldsymbol{\tau}_f)$. First assume that $\boldsymbol{f} \in \mathcal{N}(y_f)$; then $y_f = \left[\boldsymbol{df}_v^T, \boldsymbol{df}_s^T\right] \boldsymbol{f} \equiv 0$ which implies that

$$\sum_{i=1}^{n}\left[\int_{t_0}^{t}\dot{q}_i^2 f_{iv}dt + \int_{t_0}^{t}|\dot{q}_i|f_{is}dt\right] \equiv 0 \text{ or } \int_{t_0}^{t}\left(\sum_{i=1}^{n}\dot{q}_i^2 f_{iv} + \sum_{i=1}^{n}|\dot{q}_i|f_{is}\right)dt \equiv 0 .$$

Since time t is arbitrary,

$$\sum_{i=1}^{n}\dot{q}_i^2 f_{iv} + \sum_{i=1}^{n}|\dot{q}_i|f_{is} = \dot{\boldsymbol{q}}^T\left[\boldsymbol{C}_v, \boldsymbol{C}_s\right]\boldsymbol{f} \equiv 0 \tag{3.81}$$

Because $\dot{q}_i$ $(i = 1, 2, \ldots, n)$ are independent joint variables, from Eq.(3.81) we can conclude that $\boldsymbol{\tau}_f = [\boldsymbol{C}_v, \boldsymbol{C}_s]\,\boldsymbol{f} \equiv \boldsymbol{0}$. Therefore $\boldsymbol{f} \in \mathcal{N}(\boldsymbol{\tau}_f)$, which implies that $\mathcal{N}(y_f) \subset \mathcal{N}(\boldsymbol{\tau}_f)$. Conversely, if $\boldsymbol{f} \in \mathcal{N}(\boldsymbol{\tau}_f)$ then $y_f \equiv 0$ since $\int_{t_0}^{t}\dot{\boldsymbol{q}}^T\boldsymbol{\tau}_f dt = y_f$. This implies $\boldsymbol{f} \in \mathcal{N}(\boldsymbol{y_f})$ and therefore $\mathcal{N}(\boldsymbol{\tau}_f) \subset \mathcal{N}(y_f)$. This proves that $\mathcal{N}(y_f) = \mathcal{N}(\boldsymbol{\tau}_f)$. This last theorem implies that $\mathcal{N}(y_f)^{\perp} = \mathcal{N}(\boldsymbol{\tau}_f)^{\perp}$. The identifiable frictional coefficient space $I_f \triangleq \mathcal{N}(y_f)^{\perp}$ is then equivalent to the essential frictional coefficient space $E_f \triangleq \mathcal{N}(\boldsymbol{\tau}_f)^{\perp}$. Thus we have proved that the IPS $(I_{x'} = I_x \oplus I_f)$ is equivalent to the EPS $(E_{x'} = E_x \oplus E_f)$. This completes the proof that the set of independent parameters in both differential and integral models is exactly the same.

So far we have proposed a symbolic procedure to find the minimum set of inertial parameters. This set can be obtained numerically for both differential and integral models by use of such methods as singular value decomposition (SVD) [72]. An, Atkenson and Hollerbach [1] apply the SVD to the differential model in order to estimate the inertial parameters. Applying the SVD method to Eq.(3.12) and neglecting friction torques we can write $\boldsymbol{D}(\boldsymbol{q}, \dot{\boldsymbol{q}}, \ddot{\boldsymbol{q}}) = \boldsymbol{U}\sum\boldsymbol{V}^T$, where $\sum = diag\{\sigma_i\}$ and $\boldsymbol{U}$ and $\boldsymbol{V}^T$ are orthogonal matrices. For each column of $\boldsymbol{V}$ there corresponds a singular value σ_i which if non–zero indicates that the linear combination of parameters $\boldsymbol{V}_i^T\boldsymbol{x}$ is identifiable. The unidentifiable parameters will have zero singular values associated with them. Since $\boldsymbol{D}(\boldsymbol{q}, \dot{\boldsymbol{q}}, \ddot{\boldsymbol{q}})$ is a function of only the geometry of the arm and the predefined trajectory (here we assume that the trajectory is sufficiently rich), it can be generated exactly by simulation rather than by actually moving the real arm and recording data with the concomitant and inevitable noise. For completely unidentifiable parameters the corresponding columns of $\boldsymbol{D}$ can be deleted without affecting the generalised forces. For parameters that are identifiable in linear combinations, all columns except one in linear combination can also be deleted. As a result we get a differential model with a reduced set of parameters, from which these parameters can easily be estimated. The SVD method can be used for the integral model as well [62, 63, 68, 203, 205]. Any other numerical method, for example, QR can also be applied.

It is worth noting that not every parameter from the set of base parameters necessarily has a clear physical interpretation. Individual parameters of the links do have such interpretation but an arbitrary combination of link inertial parameters does not carry any physical meaning. This, however, does not have any impact on model based robot control schemes.

3.6 Further comments on robot dynamics modelling for identification of their parameters

In this section we comment on other methods which are used to derive the equations of motion for robot manipulators. In previous sections we have made use of the Lagrangian formalism. Robot dynamic models can also be derived based on the Newton-Euler formalism [55, 165, 167, 190]. This derivation is recursive in nature [190] and consists of two recursions. One recursion is from the base of the manipulator towards its tip and transforms spatial velocities and accelerations. It is assumed that the base is immobile (but it is not a restriction). The spatial forces which consist of interaction forces and torques are transformed in the opposite direction, i.e. from the tip of the manipulator towards its base. As an initial condition for this recursion we assume a spatial force acting at the tip of the manipulator. From a practical point of view, this spatial force can be measured by means of a force and torque sensor mounted at the wrist of the manipulator. This technique will be used later for load identification (see Chapter 6). In order to calculate the joint forces or torques, the spatial forces are projected at the joints. As it was mentioned before, the spatial algebra concept is very useful in the formulation of robot dynamics by making use of the Newton-Euler formalism. An, Atkeson, and Hollerbach [1] use also the Newton-Euler formalism for the purpose of dynamic parameter identification.

Silver [206] proved that both the Lagrangian and the Newton-Euler formalism lead to the same dynamic equations when written in the closed form. This result is obvious from the physical point of view but is quite difficult to prove it rigorously.

Note that, the recursive form of the equations of motion is very efficient from the implementation point of view. All algorithms which are recursive in nature have a straightforward computer implementation. This has been observed by many authors [21, 31, 60, 108, 188, 224, 235], who implemented recursive dynamics as an $O(n)$ order algorithm. For the purpose of identification the recursive robot dynamics has to be written in the closed form. As a result we get a vector of spatial interaction forces acting at all the links on the left-hand side and an upper triangular matrix (compare Eqs.(3.92) and (3.93) discussed later in this section) which depends on the generalised positions, velocities, and accelerations multiplied by a vector of all the link dynamic parameters, namely vector $\boldsymbol{x}'$. This equation is linear in the unknown parameters, but the left hand side is composed of a full force torque vector at each joint. From this representation it is clear that we need a full measurement of forces and torques acting at all joints. This requirement cannot be satisfied in practice since we need these measurements for each joint and this would be too expensive. Besides that, only the joint generalised force can be measured which reduces the dimension of the problem. The other components of the interaction forces are compensated for by the mechanical structure. From this, is it clear that we are not able to identify the dynamic parameters of all links due to the lack of the measurements of the joint spatial forces. This observation is important and shows that it is not only kinematical structure that is important for successful identification of robot dynamics. The kinematical structure is somehow an inherent feature of a particular robot and determines strictly parameters which can be identified. As a consequence it also enables us to find a minimum set of dynamic parameters. On top of that, exciting predefined trajectories and abilities of an experimental set-up impose practical restrictions in the identification process.

Yang and Tzeng [234] use the above information in simplification and linearization of manipulator dynamics by the design of inertia distribution. Their method is based on the elimination of coefficients of nonlinear terms in the system's kinetic and potential energy equations. Accordingly, a set of design criteria regarding the link's inertia distribution can be established for each robot type. A robot designed to satisfy these criteria will result in much simplified dynamics. Yang and Tzeng found that for some configurations of three and four–link robots, it is possible to design completely linearised dynamic equations. Consequently we get a robot dynamic model with less dynamic parameters, which simplifies the practical identification greatly. However the robots proposed by Yang and Tzeng are difficult to built in practice, which shows that real robots always have coupling effects. Along these lines Asada and Youcef-Toumi [9] propose an alternative design which simplifies robot dynamics. They use the differential model which is derived by making use of the Lagrangian formalism. They formulate the kinetic energy in terms of angular and linear velocities of the centre of mass of each link. Based on the total kinetic and potential energy they designed a manipulator which is mechanically decoupled. In their design the manipulator has a mass matrix which is configuration independent and in the dynamic model there are no centrifugal and Coriolis coefficients. As discussed in the previous work, these manipulators cannot be built in practice. This shows evidently that usually we have to deal with the full dynamic model which is difficult to identify in practice.

Lee and co–authors [147, 148] in constructing the dynamic model use generalised d'Alembert equations because generalised forces are formally associated with the d'Alembert forces. They formulate the total kinetic and potential energy based on the linear velocity of the centre of mass of each link with respect to the base coordinate frame. Angular velocities are expressed in local coordinate frames associated with links of the manipulator. Next they differentiate separately the kinetic energy, which results from a linear movement of the manipulator, and the kinetic energy of the angular movement. As a consequence they obtain dynamic equations of motion which are known as structured equations of motion. These equations contain four terms. The first term is the mass matrix of the manipulator multiplied by the vector of generalised accelerations. The next two components are the results of the differentiation of the kinetic energy of the linear and angular movement, respectively. The last term results from the appropriate differentiation of the total potential energy. The equations of motion derived by Lee and co–authors [147, 148] are not suitable for the purpose of identification because the same dynamic parameters appear in each term; therefore we have to do some non–trivial transformations of equations of motion in order to obtain the differential form as in Eq.(3.12).

So far, in formulating the dynamic equations we have used direction cosine matrices and the distances between two successive coordinate frames as given by Eqs.(3.2) and (3.3) according to the modified Denavit–Hartenberg notation. From a historical point of view in formulating the kinematics and dynamics many authors used a concept of homogeneous transformation which combines definitions given by Eqs.(3.2) and (3.3) in one 4×4 matrix denoted by ${}^{i-1}_{\ i}\boldsymbol{T}$. This concept can be found in Paul [176]. Paul uses homogeneous transformation matrices because one can perform any algebraic operation on them quite easily in solving problems in robotics. Kinetic and potential energy can be expressed in terms of the homogeneous transformation matrices. The Lagrangian has the following form

$$\begin{aligned} L &= \frac{1}{2}\sum_{i=1}^{n}\sum_{j=1}^{i}\sum_{k=1}^{i} Tr\left(\left(\frac{\partial {}^0_i\boldsymbol{T}}{\partial q_i}\right) {}^i\boldsymbol{J}_i \left(\frac{\partial {}^0_i\boldsymbol{T}}{\partial q_k}\right)^T\right) \dot{q}_j \dot{q}_k + \\ &+ \sum_{i=1}^{n} m_i \bar{\boldsymbol{g}}^T\, {}^0_i\boldsymbol{T}\, {}^i\bar{\boldsymbol{c}}_i. \end{aligned} \tag{3.82}$$

In the last equation Tr denotes trace of the matrix, ${}^0_i\boldsymbol{T}$ is the homogeneous transformation matrix relating the coordinate frame of the i-th link to the base coordinate frame, $\bar{\boldsymbol{g}} = [g_x, g_y, g_z, 1]^T$ is the gravity vector with respect to the base coordinate frame, ${}^i\bar{\boldsymbol{c}}_i = [c_{ix}, c_{iy}, c_{iz}, 1]^T$ is the position vector of the centre of the i-th coordinate frame, and ${}^i\boldsymbol{J}_i$ is the pseudo–inertia matrix of the i-th link which has the following form

$${}^i\boldsymbol{J}_i = \begin{bmatrix} \frac{-I_{ixx}+I_{iyy}+I_{izz}}{2} & I_{ixy} & I_{ixz} & m_i c_{ix} \\ I_{ixy} & \frac{I_{ixx}-I_{iyy}+I_{izz}}{2} & I_{izy} & m_i c_{iy} \\ I_{ixz} & I_{izy} & \frac{I_{ixx}+I_{iyy}-I_{izz}}{2} & m_i c_{iz} \\ m_i c_{ix} & m_i c_{iy} & m_i c_{iz} & m_i \end{bmatrix}. \tag{3.83}$$

Differentiation of the Lagrangian function leads to the following equations of motion

$$\begin{aligned} \tau_i &= \sum_{j=i}^{n}\sum_{k=1}^{j} Tr\left(\frac{\partial\, {}^0_j\boldsymbol{T}}{\partial q_i}\, {}^j\boldsymbol{J}_j \left(\frac{\partial\, {}^0_j\boldsymbol{T}}{\partial q_k}\right)^T\right) \ddot{q}_k + \\ &+ \sum_{j=1}^{n}\sum_{k=1}^{j}\sum_{l=1}^{j} Tr\left(\frac{\partial\, {}^0_j\boldsymbol{T}}{\partial q_i}\, {}^j\boldsymbol{J}_j \left(\frac{\partial^2\, {}^0_j\boldsymbol{T}}{\partial q_k \partial q_l}\right)^T\right) \dot{q}_k \dot{q}_l - \end{aligned} \tag{3.84}$$

$$- \sum_{j=1}^{n} m_j \bar{\boldsymbol{g}}^T \frac{\partial\, {}^0_j\boldsymbol{T}}{\partial q_i} {}^j\boldsymbol{c}_j + \tau_{ifs} + \tau_{ifv} \quad \text{for} \quad i = 1, 2, \ldots, n.$$

Ha and co–authors of the paper [82] use equations of motion described by (3.84) and prove that the model is linear with respect to dynamic parameters. Using the concept of the homogeneous transformation Ha and co–authors [82] find a minimum set of dynamic parameters.

One of the practical problems with experimental identification arises when we cannot measure generalised acceleration signals. This problem can be solved by formulating the integral model described in Section 3.3. The acceleration signal $\ddot{q}_i$ can be obtained by passing the data of $\dot{q}_i$ through a band–limited differentiator. When we desire to avoid the differentiation of $\dot{q}_i$, the following approach can be undertaken. Integrating both sides of Eq.(3.12) from the initial time $t_0 = 0$ to time t results in, the so-called filtered (or implicit [69]) differential model. In order to do it formally, consider Eq.(3.4). Integrating two sides of this equation from 0 to t results in

$$\frac{\partial \mathcal{L}}{\partial \dot{\boldsymbol{q}}}(t) - \frac{\partial \mathcal{L}}{\partial \dot{\boldsymbol{q}}}(0) = \int_0^t \left(\frac{\partial \mathcal{L}}{\partial \boldsymbol{q}} + \boldsymbol{\tau} - \boldsymbol{Q}(\dot{\boldsymbol{q}}) \right) dt \, , \tag{3.85}$$

where the last equation is written in a vector form. Now taking into account the definition of the total kinetic energy of the manipulator given by Eq.(3.13) and the time derivative of the total potential energy (compare Eq.(3.15)), the last equation can be rewritten in the following equivalent form:

$$\boldsymbol{A}(\boldsymbol{q})\dot{\boldsymbol{q}} - \boldsymbol{A}(\dot{\boldsymbol{q}}_0)\dot{\boldsymbol{q}}_0 + \int_0^t \left[\boldsymbol{G}(\boldsymbol{q}) - \frac{1}{2}\frac{\partial}{\partial \boldsymbol{q}} \left[\dot{\boldsymbol{q}}^T \boldsymbol{A}(\boldsymbol{q})\dot{\boldsymbol{q}}\right] + \boldsymbol{Q}(\dot{\boldsymbol{q}}) \right] dt = \int_0^t \boldsymbol{\tau} dt \, . \tag{3.86}$$

Obviously in the last equation, the manipulator inertia matrix $\boldsymbol{A}(\boldsymbol{q})$ and vector $\boldsymbol{G}(\boldsymbol{q})$ are functions of a vector $\boldsymbol{x}$ of dynamic parameters and a component $\boldsymbol{Q}(\dot{\boldsymbol{q}})$ depends linearly on a vector of viscous and Coulomb friction coefficients. We write the last equation in a more compact form as

$$\boldsymbol{Y}(\boldsymbol{q}_0, \dot{\boldsymbol{q}}_0, \boldsymbol{q}, \dot{\boldsymbol{q}}, \boldsymbol{F}(\boldsymbol{q}, \dot{\boldsymbol{q}}))\boldsymbol{x}' = \int_0^t \boldsymbol{\tau} dt \, , \tag{3.87}$$

which is more suitable for the purpose of identification. $\boldsymbol{F}(\boldsymbol{q}, \dot{\boldsymbol{q}})$ denotes an integral term which results from integration of the third term in Eq.(3.86). Function $\boldsymbol{Y}$ is referred to as a regression matrix (or regressive factor). Model given by Eq.(3.87) is called as an implicit dynamic model written in the integral form [69]. Obviously this model has the same inertial set of parameters as the differential model. This approach was implemented by authors of a previously cited reference. This method was used by authors of reference [1] but its method is not preferred because an integrator is an infinite–gain filter at zero frequency. This means that large errors can result from small low–frequency errors such as offsets. To overcome this shortcoming, a low–pass filter $G(s)$ as in reference [87], with unit gain at zero frequency, is applied to both sides of Eq.(3.12) as follows

$$(\boldsymbol{\tau})_l = (\boldsymbol{D})_l x + (\boldsymbol{\tau}_{fs})_l + (\boldsymbol{\tau}_{fv})_l \ , \tag{3.88}$$

where

$$(\cdot)_l = L^{-1}\left[G(s)L\left[\cdot\right]\right] = L^{-1}\left[\frac{l}{l+s}L\left[\cdot\right]\right] , \tag{3.89}$$

where $L[\,]$ and $L^{-1}[\,]$ represent the Laplace transformation and the inverse Laplace transformation, respectively. In $G(s)$, l is a positive constant. Then, $(\boldsymbol{D})_l$ can be obtained without the measurement of accelerations. Proof of this statement can be found in reference [154]. The model given by Eq.(3.88) is also know as the differential filtered model. Integration of the differential model was also proposed by Schaefers and co–authors [191] without any filter matrix $G(s)$. From considerations presented in Section 3.2 it is clear that the differential model is very popular for the purpose of identification. Some authors [26, 194] reformulate this model for the purpose of sequential identification. In order to do that we take into account Eq.(3.66) which by making use of Eq.(3.65) can be written as

$$\tau_i = \sum_{j=1}^{n} \boldsymbol{Y}_{ij}^T \boldsymbol{x}'_j \ , \tag{3.90}$$

where the $\boldsymbol{Y}_{ij}$ are given as

$$\boldsymbol{Y}_{ij}^T = \left[\frac{d}{dt}\left(\frac{\partial \boldsymbol{d}_j^T}{\partial \dot{q}_i}\right) - \frac{\partial \boldsymbol{d}_j^T}{\partial q_i},\ \dot{q}_i,\ \mathrm{sgn}(\dot{q}_i)\right] \ . \ \forall i, j \tag{3.91}$$

Here we have included the friction model representation given by Eq.(3.5). $\boldsymbol{x}'_j$ represents the set of dynamic parameters of j-th link including viscous and Coulomb friction coefficients, namely $\boldsymbol{x}'_j = [I_{jxx}, I_{jxy}, I_{jxz}, I_{jyy}, I_{jyz}, I_{jzz}, m_j, m_j c_{jx}, m_j c_{jy}, m_j c_{jz}, F_{jv}, F_{js}]^T$. Note that the partial derivatives of the $\boldsymbol{d}_j^T$ with respect to the $q_i's$ are zero for all $j < i$ since kinetic energy of the i-th link is only a function of $q_1, q_2, \ldots, q_i$ and $\dot{q}_1, \dot{q}_2, \ldots, \dot{q}_i$ for any $i \leq j$ and potential energy is a function of $q_1, q_2, \ldots, q_i$. Therefore the robot dynamics equations have the following form

$$\begin{bmatrix} \tau_1 \\ \tau_2 \\ \vdots \\ \tau_n \end{bmatrix} = \begin{bmatrix} \boldsymbol{Y}_{11}^T & \boldsymbol{Y}_{12}^T & \ldots & \boldsymbol{Y}_{1n}^T \\ \boldsymbol{0} & \boldsymbol{Y}_{22}^T & \ldots & \boldsymbol{Y}_{2n}^T \\ \vdots & \vdots & \vdots & \vdots \\ \boldsymbol{0} & \ldots & \ldots & \boldsymbol{Y}_{nn}^T \end{bmatrix} \begin{bmatrix} \boldsymbol{x}'_1 \\ \boldsymbol{x}'_2 \\ \vdots \\ \boldsymbol{x}'_n \end{bmatrix} \tag{3.92}$$

or more compactly

$$\boldsymbol{\tau} = \boldsymbol{Y}(\boldsymbol{q}, \dot{\boldsymbol{q}}, \ddot{\boldsymbol{q}})\boldsymbol{x}' \ , \tag{3.93}$$

where $\boldsymbol{Y}(\boldsymbol{q}, \dot{\boldsymbol{q}}, \ddot{\boldsymbol{q}})$ is the $n \times 12n$ information matrix or regression matrix which is a function of the link joint positions, velocities, and accelerations. As before the total inertial parameter vector $\boldsymbol{x}'$ is the concatenation of all the $\boldsymbol{x}_i's$ and its dimension is $1 \times 12n$.

Possible linear combinations among the elements of $\boldsymbol{Y}_{ij}$ and $\boldsymbol{Y}_{ik}$ give rise to a new parametrization

$$\tau_i = \sum_{j=1}^{n} \boldsymbol{Y}_{ij}^T \boldsymbol{x}_j' = \sum_{j=1}^{n} \boldsymbol{\Phi}_{ij}^T \boldsymbol{x}_{jc} , \tag{3.94}$$

in terms of the reduced order vector $\boldsymbol{x}_c$ (dimension of $\boldsymbol{x}_i' \geq$ dimension of $\boldsymbol{x}_{ic}$) and the new upper triangular matrix $\boldsymbol{\Phi}$

$$\begin{bmatrix} \tau_1 \\ \tau_2 \\ \vdots \\ \tau_n \end{bmatrix} = \begin{bmatrix} \boldsymbol{\Phi}_{11}^T & \boldsymbol{\Phi}_{12}^T & \cdots & \boldsymbol{\Phi}_{1n}^T \\ \boldsymbol{0} & \boldsymbol{\Phi}_{22}^T & \cdots & \boldsymbol{\Phi}_{2n}^T \\ \vdots & \vdots & \vdots & \vdots \\ \boldsymbol{0} & \cdots & \cdots & \boldsymbol{\Phi}_{nn}^T \end{bmatrix} \begin{bmatrix} \boldsymbol{x}_{1c} \\ \boldsymbol{x}_{2c} \\ \vdots \\ \boldsymbol{x}_{nc} \end{bmatrix} , \tag{3.95}$$

where a new vector $\left[\boldsymbol{x}_{1c}^T, \boldsymbol{x}_{2c}^T, \ldots, \boldsymbol{x}_{nc}^T\right]^T = \boldsymbol{x}_c$ can be written as a linear combination of the vector $\boldsymbol{x}'$. This parametrization avoids the structural identifiability problems and reduces the dimension of the unknown parameters. The equation set (3.95) can be compactly written as

$$\boldsymbol{\tau} = \boldsymbol{\Phi}(\boldsymbol{q}, \dot{\boldsymbol{q}}, \ddot{\boldsymbol{q}})\boldsymbol{x}_c , \tag{3.96}$$

where $\boldsymbol{\Phi}$ is the new regression matrix of dimension $n \times m_c$ (recall that m_c denotes a number of parameters of the canonical model).

There are more papers on robot dynamics formulation but from the identification point of view the above described models are the most frequently used in experimental identification.

3.7 Examples of robot dynamics models for experimental identification

In this section we present two examples of robot dynamic models in order to illustrate the theoretical considerations presented in Sections 3.2 and 3.3. The first example describes the dynamic model of the EDDA robot while the second concerns the IRp-6 dynamic model. They are described in the following sections.

3.7.1 The EDDA dynamic model

First we consider a simpler example of a robot which represents a class of direct drive robots. The name of this robot is EDDA (Experimental Direct Drive Robot) [124, 181]. This robot was built in the Institut für Robotik und Prozessinformatik, Technische Iniversität Braunschweig, Germany. The main purpose of the design of EDDA was to build an experimental set-up to test dynamic interactions between links and to evaluate different control algorithms based on an identified model of this robot. Industrial robots are not useful for this kind of research as the associated control systems are usually based on pure inverse and forward kinematics schemes. Dynamic effects are usually neglected because industrial robots are geared mechanisms and gears partly decouple nonlinear dynamics.

The EDDA robot has two parallel joints and is equipped with two megatorque motors. Two links of this robot are designed in such a way that dynamic interactions are dominant and cannot be neglected in any control strategy. Motor 1, which is the large one, is mounted on a steel construction. This motor directly drives link 1. On the top of link 1 there is motor 2 which moves link 2. The axes of motor 1 and 2 are parallel. Motor 1 is mounted on the construction and can be turned from a vertical to a horizontal position. The two configurations are double pendulum and scara, respectively. In other words, we can turn the gravity on and off or change it between $0 \geq g_0 \geq -9.8$ m/s^2. The link lengths have been optimised in order to achieve high coupling effects. Two variable reluctance motors are used to actuate the links. The maximum velocity for both motors is 1.3 2π/s. The motors have high resolution position encoders with $6 \cdot 10^5$ inc/2π. The drive units get a desired torque signal and produce the corresponding motor currents. Measuring the torques was found a ripple error with 150 ripples per revolution was found but it is assumed that measured currents are proportional to torques. This problem will be discussed later with the experimental results in Chapter 5.

This two–degrees–of–freedom robot with associated coordinate frames according to the modified Denavit-Hartenberg notation is depicted in Fig. 3.2. These parameters are summarised in Table 3.1 Now we calculate total kinetic and potential energy based on Eqs.(3.6) and (3.10), respectively. First based on Eq.(3.2) we calculate

Table 3.1. Modified Denavit-Hartenberg parameters of EDDA

i	θ_i	d_i	a_i	α_i	σ_i
1	q_1	0	0	−90	0
2	q_2	0	l_1	0	0

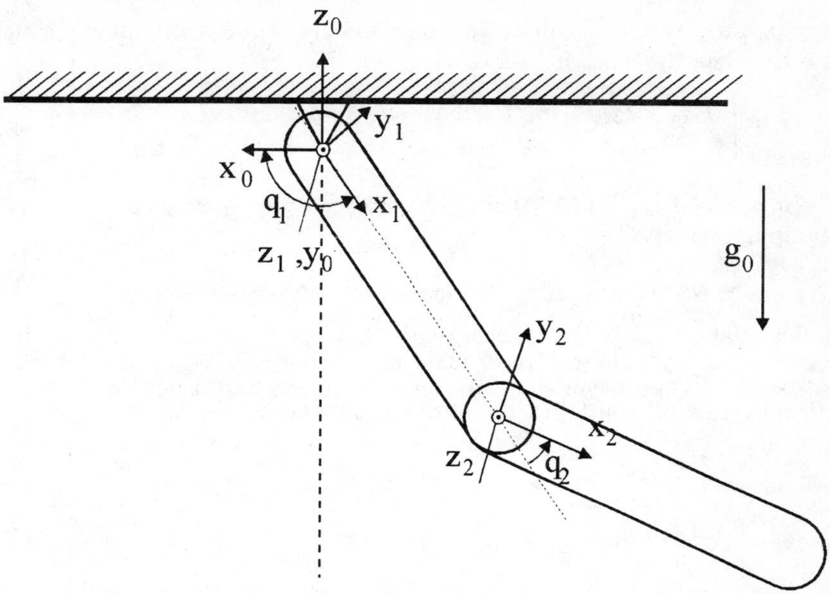

Fig. 3.2. EDDA with associated coordinate frames

$$
{}^{0}_{1}\boldsymbol{R} = \begin{bmatrix} \cos q_1 & -\sin q_1 & 0 \\ 0 & 0 & 1 \\ -\sin q_1 & -\cos q_1 & 0 \end{bmatrix} \quad \text{and} \quad {}^{2}_{1}\boldsymbol{R} = \begin{bmatrix} \cos q_2 & -\sin q_2 & 0 \\ \sin q_2 & \cos q_2 & 0 \\ 0 & 0 & 1 \end{bmatrix}. \tag{3.97}
$$

A sequence of angular velocities is calculated from Eq.(3.8):

$$
{}^{1}\boldsymbol{\omega}_1 = \begin{bmatrix} 0 \\ 0 \\ \dot{q}_1 \end{bmatrix} \quad \text{and} \quad {}^{2}\boldsymbol{\omega}_2 = \begin{bmatrix} 0 \\ 0 \\ \dot{q}_1 + \dot{q}_2 \end{bmatrix}. \tag{3.98}
$$

Linear velocities are calculated from Eq.(3.9) as follows:

$$
{}^{1}\boldsymbol{v}_1 = \begin{bmatrix} 0 \\ 0 \\ 0 \end{bmatrix} \quad \text{and} \quad {}^{2}\boldsymbol{v}_2 = \begin{bmatrix} l_1\dot{q}_1 \sin q_2 \\ l_1\dot{q}_1 \cos q_2 \\ 0 \end{bmatrix}. \tag{3.99}
$$

Substitution of Eqs.(3.98) and (3.99) into Eq.(3.6) gives the following expression for the total energy of the manipulator:

$$
E_{kc} = \frac{1}{2}I_{1zz}\dot{q}_1^2 + \frac{1}{2}I_{2zz}\left(\dot{q}_1 + \dot{q}_2\right)^2 + \frac{1}{2}m_2 l_1^2 \dot{q}_1^2 + m_2 c_{2x} l_1 \dot{q}_1 \left(\dot{q}_1 + \dot{q}_2\right)\cos q_2 . \tag{3.100}
$$

Here we have assumed that the only coordinate of the centre of mass of the second link, c_{2c}, is non–zero. Based on the total kinetic energy of the manipulator we can calculate the mass matrix of the manipulator, which in this case has the following form:

$$A = \begin{bmatrix} I_{1zz} + I_{2zz} + m_2 l_1^2 + 2m_2 c_{2x} l_1 \cos q_2 & I_{2zz} + m_2 c_{2x} l_1 \cos q_2 \\ I_{2zz} + m_2 c_{2x} l_1 \cos q_2 & I_{2zz} \end{bmatrix}. \tag{3.101}$$

Now based on Eq.(3.10) we calculate the total potential energy of the manipulator:

$$E_{pc} = m_1 c_{1x} g_0 \sin q_1 + m_2 l_1 g_0 \sin q_1 + m_2 c_{2x} g_0 \sin(q_1 + q_2). \tag{3.102}$$

In formulating the potential energy of the first link we have assumed that its c_{1x} coordinate of the centre of mass is non–zero. Based on the kinetic and potential energy we can calculate the Lagrangian and using Eq.(3.4) we can calculate the differential model of the manipulator:

$$\begin{aligned} \begin{bmatrix} \tau_1 \\ \tau_2 \end{bmatrix} &= \begin{bmatrix} \ddot{q}_1 \\ 0 \end{bmatrix} \left(I_{1zz} + I_{2zz} + l_1^2 m_2\right) + \\ &+ \begin{bmatrix} g_0 \cos q_1 \\ 0 \end{bmatrix} (l_1 m_2 + m_1 c_{1x}) + \begin{bmatrix} \ddot{q}_2 \\ \ddot{q}_1 + \ddot{q}_2 \end{bmatrix} I_{2zz} + \\ &+ \begin{bmatrix} l_1 \left[(2\ddot{q}_1 + \ddot{q}_2)\cos q_2 - \left(\dot{q}_2^2 + 2\dot{q}_1\dot{q}_2\right)\sin q_2\right] + g_0 \cos(q_1 + q_2) \\ l_1 \left(\dot{q}_1^2 \sin q_2 + \ddot{q}_1 \cos q_2\right) + g_0 \cos(q_1 + q_2) \end{bmatrix} m_2 c_{2x} + \\ &+ \begin{bmatrix} \operatorname{sgn}(\dot{q}_1) \\ 0 \end{bmatrix} F_{1s} + \begin{bmatrix} 0 \\ \operatorname{sgn}(\dot{q}_2) \end{bmatrix} F_{2s} + \begin{bmatrix} \dot{q}_1 \\ 0 \end{bmatrix} F_{1v} + \\ &+ \begin{bmatrix} 0 \\ \dot{q}_2 \end{bmatrix} F_{2v}. \end{aligned} \tag{3.103}$$

In the last equation $g_0 = -9.81\text{m/s}^2$. Due to the construction of the EDDA link inertia elements I_{1zz} and I_{2zz} include motor inertias. We do not recognise the link inertias, and I_{1zz}, I_{2zz} represent both link and motor inertias. This situation applies only to direct drive robots. Notice that in Eq.(3.103) there are eight dynamic parameters which are linear combinations of the link dynamic parameters, namely: I_{1zz}, I_{2zz}, m_2, $m_1 c_{1x}$. Other parameters I_{2zz}, $m_2 c_{2x}$, F_{1s}, F_{2s}, F_{1v} and F_{2v} do not appear as linear combinations in the dynamic equations of motion. In the dynamic model given by Eq.(3.103) we included Coulomb and viscous friction coefficients similar to Eq.(3.12). The dynamic model given by Eq.(3.103) is a canonical model with the minimum number of dynamic parameters. In the model given by Eq.(3.12) each individual dynamic parameter has its representation in the final equations of motion. In the canonical model generally it is not true. With eight dynamic parameters there are eight D^i functions which are written as column vectors. Matrix D in the equations of motion given by (3.103) has dimensions 2×4. Notice that D^i columns depend on joint positions, velocities and accelerations, which is one of the features of the differential model. Therefore the differential model of the EDDA robot has the following dynamic parameters

$$X_1 = I_{1zz} + I_{2zz} + l_1^2 m_2, \quad X_2 = l_1 m_2 + m_1 c_{1x}, \quad X_3 = I_{2zz},$$
$$X_4 = m_2 c_{2x}, \quad X_5 = F_{1s}, \quad X_6 = F_{2s}, \quad X_7 = F_{1v}, \quad X_8 = F_{2v}. \tag{3.104}$$

Now taking into account the expressions for the total kinetic and potential energy given by Eqs.(3.100) and (3.102) and using equations (3.31) and (3.32) one can calculate the following integral model of the EDDA robot

$$\begin{aligned} y &= \frac{1}{2}\dot{q}_1^2 X_1 + g_0 \sin(q_1) X_2 + \frac{1}{2}\left(2\dot{q}_1\dot{q}_2 + \dot{q}_2^2\right) X_3 + \\ &+ \left[\sin\left(q_1 + q_2\right) g_0 + l_1 \dot{q}_1^2 \cos q_2 + l_1 \dot{q}_1 \dot{q}_2 \cos q_2\right] X_4 + \\ &+ X_5 \int |\dot{q}_1| dt + X_6 \int |\dot{q}_2| dt + X_7 \int \dot{q}_1^2 dt + X_8 \int \dot{q}_2^2 dt. \end{aligned} \tag{3.105}$$

As a result we get one equation which represents a canonical integral model of the EDDA robot. As before we have eight dynamic parameters which are exactly the same as the dynamic parameters in the differential model. With each dynamic parameter there is an associated function d^i and in general this function depends on the generalised positions and velocities. The integral model given by Eq.(3.105) is of the form of Eq.(3.40).

As mentioned in Section 3.5 both models are equivalent. Namely we can calculate a product $\int \boldsymbol{\tau}^T \dot{\boldsymbol{q}} dt$ based on Eq.(3.103) in order to get the integral model. Calculating the first component of Eq.(3.103) we get

$$\int (I_{1zz} + I_{2zz} + l_1^2 m_2)\ddot{q}_1\dot{q}_1 dt = \frac{1}{2}(I_{1zz} + I_{2zz} + l_1^2 m_2)\dot{q}_1^2 ,$$

which is exactly the same as the first component of Eq.(3.105). As a result of the integration of the second component of Eq.(3.103) we have

$$\int g_0(l_1 m_2 + m_1 c_{1x}) \cos q_1 \dot{q}_1 dt = g_0(l_1 m_2 + m_1 c_{1x}) \sin q_1 ,$$

which is exactly the same as the second element of Eq.(3.105). Next continuing the same with the third element of Eq.(3.103) we get

$$\begin{aligned} & I_{2zz} \int \left[\ddot{q}_2\dot{q}_1 + (\ddot{q}_1 + \ddot{q}_2)\dot{q}_2\right] dt = I_{2zz} \int \frac{d}{dt}(\dot{q}_1\dot{q}_2) dt + I_{2zz} \int \ddot{q}_2\dot{q}_2 dt = \\ & = I_{2zz}\frac{1}{2}(2\dot{q}_1\dot{q}_2 + \dot{q}_2^2) , \end{aligned}$$

which is the same as the third component of Eq.(3.105). Now we calculate the product of the fourth term of Eq.(3.103). First we take the components with gravity, namely

$$\begin{aligned} & m_2 c_{2x} g_0 \int \cos(q_1 + q_2)\dot{q}_1 dt + m_2 c_{2x} g_0 \int \cos(q_1 + q_2)\dot{q}_2 dt = \\ & = m_2 c_{2x} g_0 \int \cos(q_1 + q_2)\frac{d}{dt}(q_1 + q_2) dt = g_0 m_2 c_{2x} \sin(q_1 + q_2) , \end{aligned}$$

which is a part of the fourth term, associated with gravity, at Eq.(3.103). Other components result in

$$
\begin{aligned}
& m_2 c_{2x} \int l_1 \left[(2\ddot{q}_1 + \ddot{q}_2) \cos q_2 - (\dot{q}_2^2 + 2\dot{q}_1 \dot{q}_2) \sin q_2 \right] \dot{q}_1 dt + \\
& + m_2 c_{2x} \int \left[l_1 \dot{q}_1^2 \sin q_2 + \ddot{q}_1 \cos q_2 \right] \dot{q}_2 dt = \\
& = m_2 c_{2x} \left[\int 2 l_1 \ddot{q}_1 \cos(q_2) \dot{q}_1 dt + \int l_1 \frac{d}{dt} (\dot{q}_1 \dot{q}_2) \cos q_2 dt \right. \\
& \left. - \int l_1 (\dot{q}_2^2 \dot{q}_1 + \dot{q}_1^2 \dot{q}_2) \sin q_2 dt \right] = \left[l_1 \dot{q}_1^2 \cos q_2 + l_1 \dot{q}_1 \dot{q}_2 \cos q_2 \right] m_2 c_{2x}.
\end{aligned}
$$

In order to obtain the last equality we have integrated by parts integrals in the last equation. It is rather straightforward to show that, integrating the last four elements from Eq.(3.103) results in the last four components in Eq.(3.105).

Taking into account the last calculations we have got an equivalent integral model derived from the different model. Considering Section 3.4 this is obvious. Here we have shown that the integral model can be derived directly from the differential model. As we mentioned before, calculations in reverse order are not possible because we cannot predict how the total kinetic and potential energy is distributed through the links of the manipulator. It can be seen clearly from the above calculations that many terms cancel out and assuming we have an integral model we cannot derive from it the differential model. Therefore one can state from the above derivations that the integral model has inherently less information that the differential model. This fact was observed earlier.

Finally, we check the last property of the differential model, namely we verify that the matrix $\dot{\boldsymbol{A}} - 2\boldsymbol{C}$ is skew–symmetric. In order to do that we calculate the time derivative of the mass matrix of the manipulator given by Eq.(3.101) and get

$$
\dot{\boldsymbol{A}} = \begin{bmatrix} -2 l_1 m_2 c_{2x} \sin(q_2) \dot{q}_2 & -m_2 c_{2x} l_1 \sin(q_2) \dot{q}_2 \\ -m_2 c_{2x} l_1 \sin(q_2) \dot{q}_2 & 0 \end{bmatrix}. \tag{3.106}
$$

Now taking into account the equations of motion (3.103) we can write the following $\boldsymbol{C}(\boldsymbol{q}, \dot{\boldsymbol{q}})$ matrix (remember this matrix summarises Coriolis and centrifugal effects)

$$
\boldsymbol{C}(\boldsymbol{q}, \dot{\boldsymbol{q}}) = \begin{bmatrix} -m_2 c_{2x} l_1 \dot{q}_2 \sin q_2 & -m_2 c_{2x} l_1 (\dot{q}_1 + \dot{q}_2) \sin q_2 \\ m_2 c_{2x} l_1 \dot{q}_1 \sin q_2 & 0 \end{bmatrix}. \tag{3.107}
$$

Straightforward calculations based on Eq.(3.106) and (3.107) shows that $\dot{\boldsymbol{A}} - 2\boldsymbol{C}$ is skew-symmetric which is the desired result.

3.7.2 The IRp-6 dynamic model

The second example considered in this monograph, represents, the differential and integral model of the IRp-6 industrial robot which belongs to a class of geared robots. This robot, which is described in references [73] and [74], has five degrees of freedom. Its kinematical structure is quite complicated due to its parallelogram structure. All five motors are DC motors and actuate links through harmonic gears (first, fourth and fifth links) and hypoid gear (second and third links).

Since the drive system for the second and third links are the most complicated, we have illustrated these two drive systems. The second drive system is presented in Fig.3.3a [73]. In Fig.3.3a location of the second motor is indi-

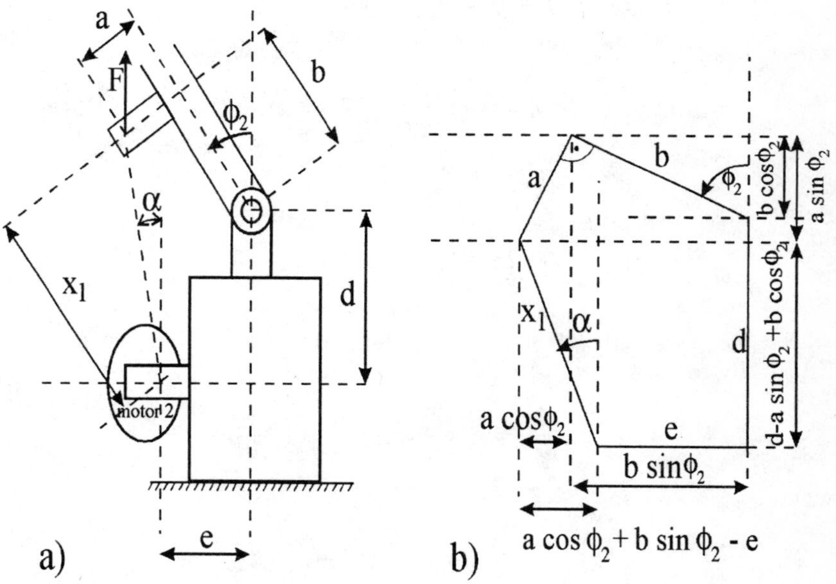

Fig. 3.3. The drive system for the second link and its simplified scheme

cated. Its stator forms an angle α with gravity direction. A joint variable for the second link is indicated as ϕ_2. Based on Fig.3.3a we can write

$$x_1 = f_2(\phi_2) = n_2^{-1}\Psi_2. \tag{3.108}$$

In Eq.(3.108) n_2 denotes a gear ratio of the screw drive and Ψ_2 is a joint angle associated with a rotor of the DC motor. Note that in Fig.3.3a F is a force due to the displacement of the screw drive. This force allows the second link of the manipulator to move. Taking into account the notation introduced in Fig.3.3b we can write

$$f_2(\phi_2) = \left[(d - a\sin\phi_2 + b\cos\phi_2)^2 + (-e + a\cos\phi_2 + b\sin\phi_2)^2\right]^{\frac{1}{2}} \ , \tag{3.109}$$

which represents a nonlinear function of the joint variable ϕ_2 and constant elements of the geometrical distances depicted from Fig.3.3b. Note also that an angle α (which is a slope angle of the stator) can be calculated from the following expression:

$$\alpha = \arctan\frac{-e + a\cos\phi_2 + b\sin\phi_2}{d - a\sin\phi_2 + b\cos\phi_2} . \tag{3.110}$$

Now we rearrange the function $F_2(\phi_2)$ in a different form which is more suitable for subsequent considerations. Namely we expand the quadratic terms and use simplified trigonometric formulae in order to get

$$f_2(\phi_2) = \sqrt{B^2 - A^2\sin(\beta - \phi_{20} + \phi_2)} \ , \tag{3.111}$$

where

$$\beta = \arctan\frac{e}{d}, \quad \phi_{20} = \arctan\frac{b}{a}, \quad B^2 = a^2 + b^2 + d^2 + e^2, \quad \text{and} \quad A^2 = 2\sqrt{a^2 + b^2}\sqrt{d^2 + e^2}. \tag{3.112}$$

Substitution of Eq.(3.111) into (3.108) results in

$$\Psi_2 = n_2 f_2(\phi_2) = \frac{2\pi}{SK}\sqrt{B^2 - A^2\sin(\beta - \phi_{20} + \phi_2)} \ , \tag{3.113}$$

where $n_2 = \frac{2\pi}{SK}$ denotes a slip of the screw drive. Now we differentiate Eq.(3.113) with respect to time

$$\frac{d\Psi_2}{dt} = n_2 \frac{df_2}{d\phi_2}\frac{d\phi_2}{dt} = \frac{dF_2}{d\phi_2}\frac{d\phi_2}{dt} \ , \tag{3.114}$$

where

$$\frac{dF_2}{d\phi_2} = -\frac{\pi}{SK}\frac{A^2\cos(\beta - \phi_{20} + \phi_2)}{\sqrt{B^2 - A^2\sin(\beta - \phi_{20} + \phi_2)}} . \tag{3.115}$$

Here $\frac{d\Psi_2}{dt}$ and $\frac{d\phi_2}{dt}$ denote angular motor and joint generalised velocities, respectively. In the last equation $\frac{dF_2}{d\phi_2}$ is a gear ratio between both velocities and it depends on the generalised position ϕ_2. Integration of Eq.(3.114) with respect to time results in

$$\Psi_2 = \frac{2\pi}{SK}\sqrt{B^2 - A^2\sin(\beta - \phi_{20} + \phi_2)} + C \tag{3.116}$$

where the integration constant C can be derived from the initial condition stating that $\phi_2 = 0$ for $\Psi_2 = 0$. Note that function Ψ_2 given by Eq.(3.116) differs from the function described by Eq.(3.113) by constant C which results idirectly from the relationship between motor and link velocities. Therefore we get

$$C = -\frac{2\pi}{SK}\sqrt{B^2 - A^2 \sin(\beta - \phi_{20})}\,. \tag{3.117}$$

From Eq.(3.116) one can calculate explicitly ϕ_2 as follows

$$\phi_2 = -\beta + \phi_{20} + \arctan\frac{E}{\sqrt{1-E^2}}\,, \quad \text{where} \quad E = \frac{-(\frac{(\Psi_2 - C)SK}{2\pi})^2 + B^2}{A^2}\,. \tag{3.118}$$

From Eq.(3.118) it can be seen that ϕ_2 can be recovered from Ψ_2 through nontrivial calculations.

The next step is to describe a drive system for the third link. This situation is depicted in Fig.3.4a. Note the location of the third motor and the

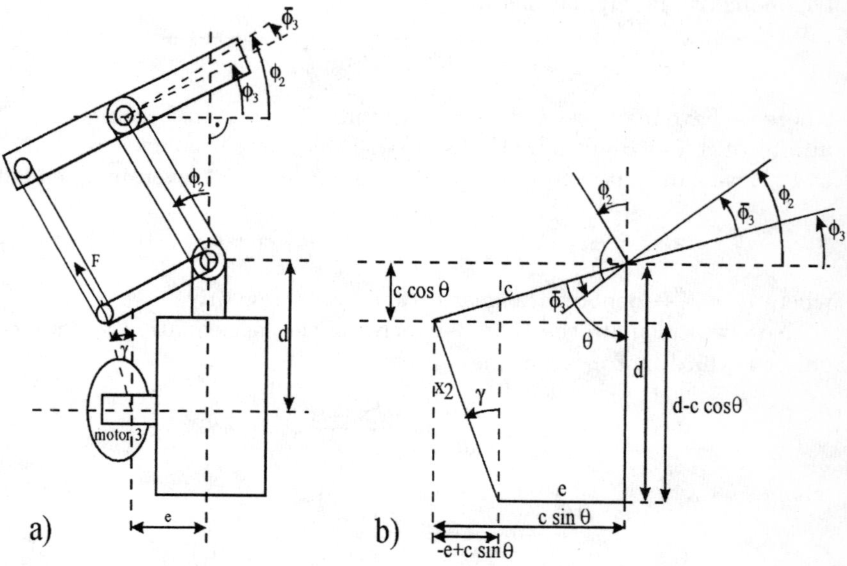

Fig. 3.4. The drive system for the third link and its simplified scheme

parallelogram structure of the manipulator. The third motor drives the third and second links due to the selection of the joint coordinate $\bar{\phi_3}$. Based on Fig.3.4b we can write the following equation:

$$x_2 = f_3(\phi_2, \bar{\phi_3}) = n_3^{-1}\Psi_3, \tag{3.119}$$

which relates the motor coordinate Ψ_3 to the displacement of the screw drive mechanism which moves one vertex of the parallelogram structure of the manipulator. As before n_3 denotes a gear ratio of the screw drive. The applied force F moves the vertex and at the same time changes the position of the third link only.

Taking into account the notation indicated in Fig.3.4b we can write (note that $\theta = 90 + \bar{\phi}_3 - \phi_2$)

$$f_3(\phi_2, \bar{\phi}_3) = n_3^{-1}\Psi_3 = \left[(d + c\sin(\bar{\phi}_3 - \phi_2))^2 + (-e + c\cos(\bar{\phi}_3 - \phi_2))^2\right]^{\frac{1}{2}} . \tag{3.120}$$

which represents a nonlinear function of joint coordinates ϕ_2 and $\bar{\phi}_3$. An angle β which represents a slope angle of the stator measured with respect to the gravity can be written in the following form:

$$\gamma = \arctan \frac{-e + c\cos(\bar{\phi}_3 - \phi_2)}{d + c\sin(\bar{\phi}_3 - \phi_2)} . \tag{3.121}$$

As before we rearrange the function $f_3(\phi_2, \bar{\phi}_3)$ and after some algebraic and trigonometric manipulations we get:

$$f_3(\phi_2, \bar{\phi}_3) = \sqrt{B^2 + A^2 \sin(\bar{\phi}_3 - \phi_2 - \beta)} , \tag{3.122}$$

where we have made used of $c^2 = a^2 + b^2$ and $\beta = \arctan \frac{e}{d}$ (compare the definitions of A and B given by Eq.(3.113)). Taking into account Eqs.(3.120) and (3.122) we can write (note that links 2 and 3 are not driven independently)

$$\Psi_3 = n_3 f_3(\phi_2, \bar{\phi}_3) = \frac{2\pi}{SK}\sqrt{B^2 + A^2 \sin(\bar{\phi}_3 - \phi_2 - \beta)} , \tag{3.123}$$

where $n_3 = \frac{2\pi}{SK}$ denotes the gear ratio of the screw drive.

Now we calculate the time derivative of the last equation assuming that only the third motor is turning

$$\frac{d\psi_3}{dt} = \frac{\partial F_3}{\partial \bar{\phi}_3}\frac{d\bar{\phi}_3}{dt} , \tag{3.124}$$

where

$$\frac{\partial F_3}{\partial \bar{\phi}_3} = \frac{\pi}{SK}\frac{A^2 \cos(\bar{\phi}_3 - \beta_1)}{\sqrt{B^2 + A^2 \sin(\bar{\phi}_3 - \beta_1)}} \tag{3.125}$$

with $\beta_1 = \phi_2 + \beta$. Equation (3.125) describes a nonlinear function which is formally the gear ratio relating third motor and link velocity. Gear ratios given by Eqs.(3.115) and (3.125) are important in calculations of inertia motors. In practice these functions do not change significantly and this will be discussed later in experimental identification of the dynamic parameters of the IRp-6 robot. One can integrate Eq.(3.124) which results in

$$\psi_3 = \frac{2\pi}{SK}\sqrt{B^2 + A^2 \sin(\bar{\phi}_3 - \beta_1)} + C , \tag{3.126}$$

where C is the integration constant which is calculated from the initial condition $\bar{\phi}_3 = 0$ for $\Psi_3 = 0$:

$$C = -\frac{2\pi}{SK}\sqrt{B^2 - A^2 \sin\beta_1} . \tag{3.127}$$

From Eq.(3.126) we can easily calculate $\bar{\phi}_3$ as follows:

$$\bar{\phi}_3 = \beta_1 + \arctan \frac{-E}{\sqrt{1-E^2}} . \tag{3.128}$$

It will be shown at the end of this section that $\bar{\phi}_3$ is not a good choice of joint coordinate. From the above considerations it is clear that the joint coordinates ϕ_2 and $\bar{\phi}_3$ cannot be changed independently. It is due to the complicated kinematical structure of the IRp-6 robot.

In order to complete the kinematical considerations we summarise the results for other joints

$$\begin{aligned} \Psi_1 &= -n_1\phi_1 , \\ \Psi_4 &= -n_4(\phi_2 + \phi_3 + \phi_4) , \\ \Psi_5 &= -n_4(\phi_2 + \phi_3 + \phi_4) - n_4 n_5 \phi_5 , \end{aligned} \tag{3.129}$$

where n_1, n_4, n_5 are gear ratios for the first, fourth and fifth link, respectively.

Similarly, Ψ_1, Ψ_4, and Ψ_5 denote motor displacements for the first, fourth, and fifth motor, respectively. The angular displacements of ϕ_1, ϕ_4 and ϕ_5 correspond to the first, fourth, and fifth link (compare Table 3.2). All these calculations are important for kinematical considerations and for the purpose of control at the motor's level.

Dynamical considerations in this monograph are restricted to the first three links and motors of the IRp-6 robot. We have assumed that the wrist is an integral part of the third link and that the dynamic coupling for the last two links can be neglected. This assumption comes from practical considerations. Links 4 and 5 are light and move with low velocities. Besides that we are dealing with a geared robot in which we can assume that the dynamic coupling can be partly neglected. In general this is not true. The most important part for dynamic model considerations is a friction model which will be discussed later in Chapter 5. From now on we assume for dynamic modelling that the IRp-6 robot has three degrees of freedom. This is shown in Fig.3.5. Note that the generalised coordinates ϕ_2 and ϕ_3 in Fig.3.5 have a different sign in comparison to those presented in Fig.3.4a. Joint torques (which are calculated in terms of the generalised angles ϕ_1, ϕ_2, ϕ_3 and their first and second derivatives, see Fig.(3.5)) are calculated according to the following expressions:

$$\begin{aligned} \tau_1 = \big[&J_1 + J_6 + (J_2 + m_3 s_{22}^2)\sin^2\phi_2 + 2s_{22}x_3\sin\phi_2\cos\phi_3 + \\ &+ J_3\cos^2\phi_3\big]\ddot{\phi}_1 + \big[(J_2 + m_3 s_{22}^2)\sin(2\phi_2) + 2s_{22}x_3\cos\phi_2\cos\phi_3\big]\dot{\phi}_1\dot{\phi}_2 + \\ &+ [-2s_{22}x_3\sin\phi_2\sin\phi_3 - J_3\sin(2\phi_3)]\,\dot{\phi}_1\dot{\phi}_3 + b_1\dot{\phi}_1 + \\ &+ c_1\mathrm{sgn}(\dot{\phi}_1) , \end{aligned}$$

$$\tau_2 = \left[J_2 + m_3 s_{22}^2 + J_4\left(\frac{dF_2}{d\phi_2}\right)^2\right]\ddot{\phi}_2 -$$

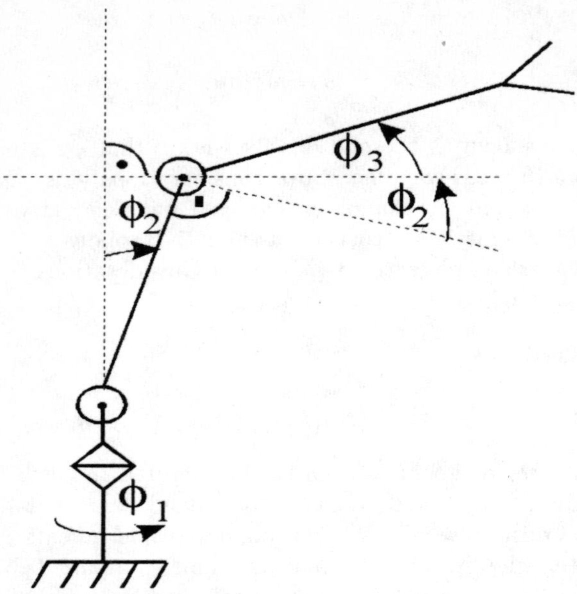

Fig. 3.5. Simplified kinematical structure of the IRp-6 robot.

$$
\begin{aligned}
&-s_{22}x_3\sin(\phi_2+\phi_3)\ddot{\phi}_3 - s_{22}x_3\cos(\phi_2 + \\
&+\phi_3)\dot{\phi}_3^2 - 0.5\left[(J_2+m_3s_{22}^2)\sin 2\phi_2 + 2s_{22}x_3\cos\phi_2\cos\phi_3\right]\dot{\phi}_1^2 + \\
&J_4\tfrac{dF_2}{d\phi_2}\tfrac{d^2F_2}{d\phi_2^2}\dot{\phi}_2^2 + b_2\dot{\phi}_2 + c_2\mathrm{sgn}(\dot{\phi}_2) - \\
&-g_0(m_2L_2+m_3c_{22})\sin\phi_2\ ,
\end{aligned}
\tag{3.130}
$$

$$
\begin{aligned}
\tau_3 = &-s_{22}x_3\sin(\phi_2+\phi_3)\ddot{\phi}_2 + \\
&+\left[J_3 + J_5\left(\tfrac{dF_3}{d\phi_3}\right)^2\right]\ddot{\phi}_3 - s_{22}x_3\cos(\phi_2+\phi_3)\dot{\phi}_2^2 + \\
&+\tfrac{1}{2}\left[2s_{22}x_3\sin\phi_2\sin\phi_3 + J_3\sin 2\phi_3\right]\dot{\phi}_1^2 + \\
&+J_5\tfrac{dF_3}{d\phi_3}\tfrac{d^2F_3}{d\phi_3^2}\dot{\phi}_3^2 + b_3\dot{\phi}_3 + c_3\mathrm{sgn}(\dot{\phi}_3) - g_0m_3L_3\cos\phi_3\ .
\end{aligned}
$$

The gear ratios can be calculated as follows:

$$n_1 = 160\ , \tag{3.131}$$

$$\frac{dF_2}{d\phi_2} = \frac{\pi}{SK}\frac{A^2\cos(\beta-\phi_{20}-\phi_2)}{\sqrt{B^2 - A^2\sin(\beta-\phi_{20}-\phi_2)}}\ , \tag{3.132}$$

$$\frac{dF_3}{d\phi_3} = \frac{\pi}{SK}\frac{A^2\cos(\phi_3+\beta)}{\sqrt{B^2 - A^2\sin(\phi_3+\beta)}}\ . \tag{3.133}$$

The constant parameters have the following values: $\phi_{20} = 0,26\text{rad}$, $c = 140\text{mm}$, $d = 183\text{mm}$, $e = 155\text{mm}$ and $SK = 5\text{mm}$. In order to obtain Eq.(3.132) we have to change angle sign in Eg.(3.113) and differentiate it with respect to ϕ_2. Equation (3.133) will be derived at the end of this section. Now we explain all quantities which appear in Eqs.(3.130)–(3.133).

- J_1 – inertia of the first link which includes inertia of the non–moving motors mounted on the link (this link is called column) with respect to its axis of rotation,
- J_2 – inertia of the second link (which is called arm) with respect to its axis of rotation,
- J_3 – inertia of the third link (which is called upper arm) with respect to its axis of rotation,
- J_4 – inertia of the rotor of the second motor with respect to its axis of rotation,
- J_5 – inertia of the rotor of the third motor with respect to its axis of rotation,
- J_6 – inertia which contributes to the inertia of the column due to rotating rotor and rate generator, which is transformed via a harmonic drive with a gear ratio n_1,
- m_2, m_3 – mass of the second and third link, respectively,
- x_3 – first order moment of the third link (which is essentially link and counterbalance) with respect to its axis of rotation,
- s_{22} – length of the second link,
- L_2, L_3 – distance between the centre of mass and the axis of rotation for the second and third link, respectively,
- g_0 – gravity measured with respect to the base coordinate frame,
- b_1, b_2, b_3 – velocity dependent friction coefficients associated with the first, second and third joint, respectively,
- c_1, c_2, c_3 – Coulomb friction coefficients for the first, second, and third joint, respectively.

The details of the dynamic model of the IRp-6 robot as quoted above, are taken from reference [73]. We keep the original notation due to the fact that this model was first derived by researchers from the Institute of Control Engineering, Technical University of Warsaw. Its complete derivation can be found in reference [73] and we also have made use of this model in practice. It is essentially the differential dynamic model discussed in Section 3.2.

Now extend the results presented in [73] by considering a more detailed differential model. This will be done in comparison with the integral model and its equivalence to the differential model described above. In order to do that, consider a full kinematical scheme of the IRp-6 robot, which is presented in Fig.3.6a. Parameters are described by making use of the modified Denavit-Hartenberg parameters for all the links (namely five degrees of freedom). Here we need only three degrees of freedom for dynamic considerations but in Chapter 6 we will use the full kinematical structure for the purpose of load identification. For both models we find a minimum set of parameters which are important when we use any model based control scheme. In Fig.3.6a we did not indicate the kinematics of the drive systems. This can be found in Figs.3.3 and 3.4. Parameters of the modified Denavit-Hartenberg notation of the IRp-6 robot are summarised in Table 3.2. Note that in Table

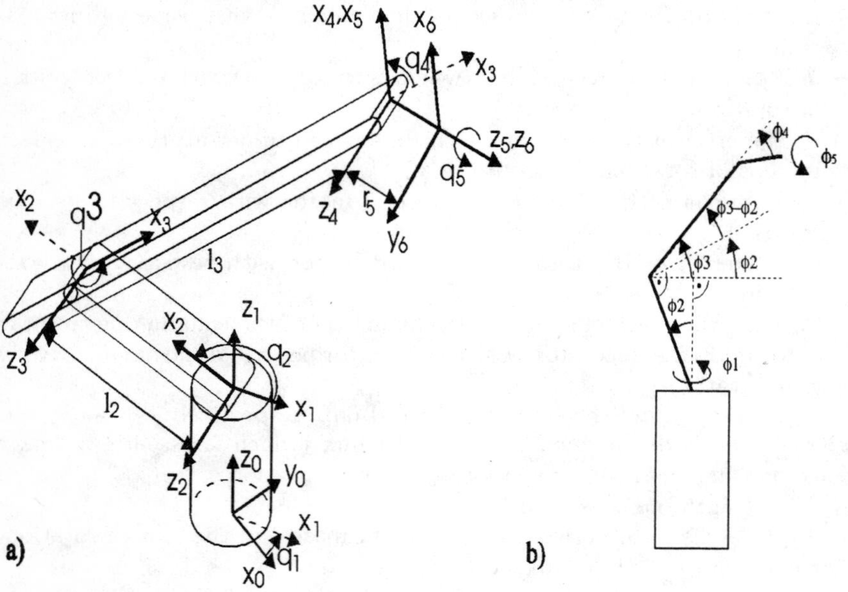

Fig. 3.6. Full kinematical model of the IRp-6 robot

Table 3.2. Modified Denavit-Hartenberg parameters of the IRp-6 robot

i	α_i	d_i	θ_i	r_i
1	0	0	ϕ_1	0
2	$\pi/2$	0	$\pi/2 + \phi_2$	0
3	0	l_2	$3/2\pi - \phi_2 + \phi_3$	0
4	0	l_3	$\pi/2 - \phi_4$	0
5	$\pi/2$	0	$q_5 = \phi_5$	0
6	0	0	0	r_5

3.2 we included parameters of the gripper which is attached to the tip of the manipulator. It is simply a displacement, r_5, from the tip of the manipulator to the tool centre point of the gripper. Here we denote by q_i the generalised coordinates. In Fig.3.6b we indicated another set of generalised coordinates ϕ_1, ϕ_2, ϕ_3, ϕ_4, and ϕ_5 which can be used alternatively to derive equations of motion (compare reference [73]). In Table 3.2 we summarised the relationship between both sets of generalised coordinates. First, we formulate an expression for the total kinetic energy of the three links of the IRp-6 robot.

The angular velocities are calculated based on Eq.(3.8) and are as follows:

$$
{}^1\boldsymbol{\omega}_1 = \begin{bmatrix} 0 \\ 0 \\ \dot{q}_1 \end{bmatrix}, \quad {}^2\boldsymbol{\omega}_2 = \begin{bmatrix} \dot{q}_1 \sin q_2 \\ \dot{q}_1 \cos q_2 \\ \dot{q}_2 \end{bmatrix}, \quad \text{and} \quad {}^3\boldsymbol{\omega}_3 = \begin{bmatrix} \dot{q}_1 \sin(q_2 + q_3) \\ \dot{q}_1 \cos(q_2 + q_3) \\ \dot{q}_2 + \dot{q}_3 \end{bmatrix}. \tag{3.134}
$$

The linear velocities are calculated based on Eq.(3.9) and have the following form

$$
{}^1\boldsymbol{v}_1 = \begin{bmatrix} 0 \\ 0 \\ 0 \end{bmatrix}, \quad {}^2\boldsymbol{v}_2 = \begin{bmatrix} 0 \\ 0 \\ 0 \end{bmatrix}, \quad \text{and} \quad {}^3\boldsymbol{v}_3 = \begin{bmatrix} l_2 \dot{q}_2 \sin q_3 \\ l_2 \dot{q}_2 \cos q_3 \\ -l_2 \dot{q}_1 \cos q_2 \end{bmatrix}. \tag{3.135}
$$

Now we calculate the kinetic energy of the first link. It has the following form

$$
E_{k1} = \frac{1}{2} \left(I_{1zz} + n_1^2 I_{1a} \right) \dot{q}_1^2 \tag{3.136}
$$

where I_{1a} denotes inertia of the rotor of the first link. In calculations of the kinetic energy of the second link we assume that the cross inertia products in the inertia matrix ${}^2\boldsymbol{I}_2$ are zero. Therefore we can write

$$
E_{k2} = \frac{1}{2} \left[I_{2zz} + I_{2a} \left(\frac{dF_2}{dq_2} \right)^2 \right] \dot{q}_2^2 + \frac{1}{2} \dot{q}_1^2 I_{2xx} \sin^2 q_2 + \frac{1}{2} \dot{q}_1^2 I_{2yy} \cos^2 q_2 . \tag{3.137}
$$

An exact expression for $\frac{dF_2}{dq_2}$ will be calculated later in this section and I_{2a} denotes inertia of the rotor of the second link. Now we find the kinetic energy of the third link. The assumption about the elements of the inertia matrix is the same as in the previous case. In addition we assume that components of the first order moment of the third link are $m_3 \, {}^3\boldsymbol{c}_3 = [m_3 c_{3x}, 0, 0]^T$ (compare also Fig.3.6a to verify this assumption). Therefore the third component in Eq.(3.6) is non–zero. Finally, the kinetic energy of the third link has the following form

$$
\begin{aligned}
E_{k3} = & \frac{1}{2} \left[I_{3zz} + I_{3a} \left(\frac{\partial F_3}{\partial q_3} \right)^2 \right] (\dot{q}_2 + \dot{q}_3)^2 + \frac{1}{2} I_{3xx} \dot{q}_1^2 \sin^2 (q_2 + q_3) + \\
& + \frac{1}{2} I_{3yy} \dot{q}_1^2 \cos^2 (q_2 + q_3) + \frac{1}{2} \left(l_2^2 \dot{q}_2^2 + l_2^2 \dot{q}_1^2 \cos^2 q_2 \right) m_3 \\
& + m_3 c_{3x} \left[l_2 \dot{q}_2 \left(\dot{q}_2 + \dot{q}_3 \right) \cos q_3 + l_2 \dot{q}_1^2 \cos q_2 \cos \left(q_2 + q_3 \right) \right]
\end{aligned} \tag{3.138}
$$

in which I_{3a} denotes inertia of the rotor of the third link and the nonlinear function $\frac{\partial F_3}{\partial q_3}$ will be calculated later in this section.

The total kinetic energy of the manipulator is a sum of kinetic energies given by Eq.(3.136), (3.137), and (3.138), therefore we get

$$E_{kc} = \frac{1}{2}\left(I_{1zz} + n_1^2 I_{1a}\right)\dot{q}_1^2 + \frac{1}{2}\left[I_{2zz} + I_{2a}\left(\frac{dF_2}{dq_2}\right)^2\right]\dot{q}_2^2 +$$
$$+\frac{1}{2}\dot{q}_1^2 I_{2xx}\sin^2 q_2 + \frac{1}{2}\dot{q}_1^2 I_{2yy}\cos^2 q_2 +$$
$$+\frac{1}{2}\left[I_{3zz} + I_{3a}\left(\frac{\partial F_3}{\partial q_3}\right)^2\right]\left(\dot{q}_2 + \dot{q}_3\right)^2 + \frac{1}{2}I_{3xx}\dot{q}_1^2\sin^2\left(q_2 + q_3\right) +$$
$$+\frac{1}{2}I_{3yy}\dot{q}_1^2\cos^2\left(q_2 + q_3\right) + \frac{1}{2}\left(l_2^2\dot{q}_2^2 + l_2^2\dot{q}_2^2\cos^2 q_2\right)m_3 +$$
$$+m_3 c_{3x}\left[l_2\dot{q}_2\left(\dot{q}_2 + \dot{q}_3\right)\cos q_3 + l_2\dot{q}_1^2\cos q_2\cos\left(q_2 + q_3\right)\right]. \qquad (3.139)$$

Based on the expression for the total kinetic energy we can calculate the elements of the manipulator mass matrix, $\boldsymbol{A}$, which in this case has dimensions 3×3. We expand the quadratic terms in Eq.(3.139) which results in the following elements of the mass matrix:

$$A_{11} = I_{1zz} + n_1^2 I_{1a} + I_{2xx}\sin^2 q_2 + I_{2yy}\cos^2 q_2 + I_{3xx}\sin^2\left(q_2 + q_3\right) +$$
$$+I_{3yy}\cos^2\left(q_2 + q_3\right) + m_3 l_2^2\cos^2 q_2 + 2l_2 m_3 c_{3x}\cos q_2\cos\left(q_2 + q_3\right),$$
$$A_{22} = I_{2zz} + I_{2a}\left(\frac{dF_2}{dq_2}\right)^2 + I_{3zz} + l_2^2 m_3 +$$
$$2m_3 c_{3x} l_2\cos q_3 + I_{3a}\left(\frac{\partial F_3}{\partial q_3}\right)^2, \qquad (3.140)$$
$$A_{33} = I_{3zz} + I_{3a}\left(\frac{\partial F_3}{\partial q_3}\right)^2,$$
$$A_{32} = A_{23} = I_{3zz} + I_{3a}\left(\frac{\partial F_3}{\partial q_3}\right)^2 + m_3 c_{3x} l_2\cos q_3,$$

and $A_{12} = A_{13} = A_{21} = A_{31} = 0$.

It is be worth comparing the symbols used in the above equations to those used in the differential model given by Eq.(3.130). This comparison is given in Table 3.3. The most important is the relationship between generalised coordinates ϕ_1, ϕ_2, ϕ_3, and q_1, q_2, and q_3. This relationship can be easily deducted from Figs. 3.5 and 3.6b. Notice that a linear transformation exists between two sets of coordinates. Recall that the kinetic energy is invariant with respect to a linear transformation of generalised coordinates [81]. In the above calculations we have assumed that $\frac{\partial F_3}{\partial q_2}\frac{\partial^2 F_3}{\partial q_3 \partial q_2} = \frac{\partial F_3}{\partial q_3}\frac{\partial^2 F_3}{\partial q_3^2}$ whenever it was necessary, which will be discussed later in this section, in order to simplify the equations of motion in their final form. Here based on Table 3.3 we write the transformation between joint coordinates. It means that we can calculate the kinetic energy in terms of new coordinates and then differentiate it with respect to the new coordinates, in order to formulate equations of motion.

Table 3.3. Comparison between the different quantities for the differential model of the IRp-6 robot

Differential model given by Eq. (3.130)	Differential model in terms of the modified D–H parameters
J_1	I_{1zz}
J_2	I_{2zz}
J_3	I_{3zz}
J_4	I_{2a}
J_5	I_{3a}
J_6	$I_{1a}n_1^2$
s_{22}	l_2
L_2	m_2c_{2x}/m_2
L_3	m_3c_{3x}/m_3
x_3	m_3c_{3x}
b_1	F_{1v}
b_2	F_{2v}
b_3	F_{3v}
c_1	F_{1s}
c_2	F_{2s}
c_3	F_{3s}
ϕ_1	q_1
ϕ_2	$\frac{\pi}{2} - q_2$
ϕ_3	$\frac{3}{2}\pi + q_2 + q_3$

$$\begin{array}{rll} q_1 = \phi_1 & \dot{q}_1 = \dot{\phi}_1 & \ddot{q}_1 = \ddot{\phi}_1 \\ q_2 = \frac{\pi}{2} - \phi_2 & \dot{q}_2 = \dot{\phi}_2 & \ddot{q}_2 = -\ddot{\phi}_2 \\ q_3 = \frac{3}{2}\pi + \bar{\phi}_3 = \frac{3}{2}\pi + \phi_2 + \phi_3 & \dot{q}_3 = \dot{\phi}_2 + \dot{\phi}_3 & \ddot{q}_3 = \ddot{\phi}_2 + \ddot{\phi}_3 . \end{array} \tag{3.141}$$

Here we also wrote the relationship between two sets of velocities and accelerations. Substitution of the above transformation rules to the expression for the kinetic energy given by Eq.(3.139) leads to the kinetic energy presented in reference [73]:

$$\begin{aligned} E_{kc}(\phi_1, \phi_2, \phi_3) &= \frac{1}{2}\left(I_{1zz} + I_{1a}n_1^2\right)\dot{\phi}_1^2 + \frac{1}{2}\left[I_{2zz} + I_{2a}\left(\frac{dF_2}{d\phi_2}\right)^2\right]\dot{\phi}_2^2 + \\ &+\frac{1}{2}\dot{\phi}_1^2 I_{2xx}\cos^2\phi_2 + \frac{1}{2}\dot{\phi}_1^2 I_{2yy}\sin^2\phi_2 + \frac{1}{2}\left[I_{3zz} + I_{3a}\left(\frac{dF_3}{d\phi_3}\right)^2\right]\dot{\phi}_3^2 + \\ &+\frac{1}{2}I_{3xx}\dot{\phi}_1^2\sin^2\phi_3 + \frac{1}{2}I_{3yy}\dot{\phi}_1^2\cos^2\phi_3 + \frac{1}{2}\left(l_2^2\dot{\phi}_2^2 + l_2^2\dot{\phi}_1^2\sin^2\phi_2\right)m_3 + \\ &+m_3c_{3x}\left[-l_2\dot{\phi}_2\dot{\phi}_3\sin(\phi_2+\phi_3) + l_2\dot{\phi}_1^2\sin\phi_2\cos\phi_3\right]. \end{aligned} \tag{3.142}$$

The total kinetic energy given by Eq.(3.142) depends on the generalised coordinates ϕ_1, ϕ_2, ϕ_3, which refer to Fig. 3.5. These coordinates are chosen in such a way that each motor drives only one link (compare Figs. 3.3 and 3.4). In addition in the expression for the total energy components associated with dynamic parameters I_{2xx} and I_{3xx} were included. In the above expression we did not assume that $I_{2yy} = I_{2zz}$ and $I_{3yy} = I_{3zz}$, what was assumed in reference [73].

Taking into account the quantities defined in Table 3.3 and based on a set of equations (3.130) we can write the following canonical model for the robot depicted in Fig.3.5 (here we write this model in terms of generalised coordinates ϕ_1, ϕ_2, and ϕ_3 which form vector $\boldsymbol{\phi}$, and corresponding derivaties $\dot{\boldsymbol{\phi}}$ and $\ddot{\boldsymbol{\phi}}$)

$$
\begin{aligned}
\boldsymbol{\tau} &= \sum_{i=1}^{13} \boldsymbol{D}^i(\boldsymbol{\phi}, \dot{\boldsymbol{\phi}}, \ddot{\boldsymbol{\phi}}) X_i = \begin{bmatrix} \ddot{\phi}_1 \\ 0 \\ 0 \end{bmatrix} X_1 + \\
&+ \begin{bmatrix} \ddot{\phi}_1 \sin^2\phi_2 + \dot{\phi}_1 \dot{\phi}_2 \sin 2\phi_2 \\ -\frac{1}{2}\dot{\phi}_1^2 \sin 2\phi_2 \\ 0 \end{bmatrix} X_2 + \begin{bmatrix} 0 \\ \ddot{\phi}_2 \\ 0 \end{bmatrix} X_3 + \begin{bmatrix} 0 \\ 0 \\ \ddot{\phi}_3 \end{bmatrix} X_4 \\
&+ \begin{bmatrix} -\ddot{\phi}_1 \sin^2\phi_3 - \dot{\phi}_1 \dot{\phi}_3 \sin 2\phi_3 \\ 0 \\ \dot{\phi}_1^2 \sin 2\phi_3 \end{bmatrix} X_5 + \\
&+ \begin{bmatrix} 2l_2 \ddot{\phi}_1 \sin\phi_2 \cos\phi_3 - 2l_2 \dot{\phi}_1 \dot{\phi}_3 \sin\phi_2 \sin\phi_3 + \\ +2l_2 \dot{\phi}_1 \dot{\phi}_2 \cos\phi_2 \cos\phi_3 \\ \\ -l_2 \ddot{\phi}_3 \sin(\phi_2+\phi_3) - l_2 \dot{\phi}_3^2 \cos(\phi_2+\phi_3) - \\ -l_2 \dot{\phi}_1^2 \cos\phi_2 \cos\phi_3 \\ \\ -l_2 \ddot{\phi}_2 \sin(\phi_2+\phi_3) - l_2 \dot{\phi}_2^2 \cos(\phi_2+\phi_3) + \\ +l_2 \dot{\phi}_1^2 \sin\phi_2 \sin\phi_3 - g_0 \cos\phi_3 \end{bmatrix} X_6 + \\
&+ \begin{bmatrix} 0 \\ -g_0 \sin\phi_2 \\ 0 \end{bmatrix} X_7 + \begin{bmatrix} \operatorname{sgn} \dot{\phi}_1 \\ 0 \\ 0 \end{bmatrix} X_8 + \begin{bmatrix} 0 \\ \operatorname{sgn} \dot{\phi}_2 \\ 0 \end{bmatrix} X_9 + \\
&+ \begin{bmatrix} 0 \\ 0 \\ \operatorname{sgn} \dot{\phi}_3 \end{bmatrix} X_{10} + \begin{bmatrix} \dot{\phi}_1 \\ 0 \\ 0 \end{bmatrix} X_{11} + \begin{bmatrix} 0 \\ \dot{\phi}_2 \\ 0 \end{bmatrix} X_{12} + \\
&+ \begin{bmatrix} 0 \\ 0 \\ \dot{\phi}_3 \end{bmatrix} X_{13}.
\end{aligned}
\tag{3.143}
$$

In Eq.(3.143) we use the following set of canonical parameters

$$
\begin{aligned}
X_1 &= I_{1zz} + I_{1a}n_1^2 + I_{3zz}, \\
X_2 &= I_{2zz} + m_3 l_2^2, \\
X_3 &= I_{2zz} + I_{2a}\bar{n}_2^2 + l_2^2 m_3, \\
X_4 &= I_{3zz} + I_{3a}\bar{n}_3^2, \\
X_5 &= I_{3zz}, \\
X_6 &= m_3 c_{3x}, \\
X_7 &= m_2 c_{2x} + l_2 m_3, \\
X_8 &= F_{1s}, \\
X_9 &= F_{2s}, \\
X_{10} &= F_{3s}, \\
X_{11} &= F_{1v}, \\
X_{12} &= F_{2v}, \\
X_{13} &= F_{3v} \; .
\end{aligned}
\tag{3.144}
$$

In deriving Eq.(3.143) and Eq.(3.144) we here assumed that $I_{2yy} = I_{2zz}$ and $I_{3yy} = I_{3zz}$, and $I_{2xx} = I_{3xx} = 0$ (compare the total kinetic energy given by Eq.(3.142) with the simplified assumptions). In addition we assumed that $\frac{dF_2}{d\phi_2} = \bar{n}_2$ and $\frac{dF_3}{d\phi_3} = \bar{n}_3$ and both gear ratios are constant. In order to derive equivalent integral model we calcultate the product $\int \boldsymbol{\tau}^T \dot{\boldsymbol{\phi}} dt$ which result in the following equation

$$
\begin{aligned}
\int \boldsymbol{\tau}^T \dot{\boldsymbol{\phi}} dt &= \frac{1}{2}\dot{\phi}_1^2 X_1 + \frac{1}{2}\dot{\phi}_1^2 \sin^2\phi_2 X_2 + \frac{1}{2}\dot{\phi}_2^2 X_3 + \frac{1}{2}\dot{\phi}_3^2 X_4 - \\
&-\frac{1}{2}\dot{\phi}_1^2 \sin^2\phi_3 X_5 - [l_2\dot{\phi}_3\dot{\phi}_2 \sin(\phi_2+\phi_3) - l_2\dot{\phi}_1^{\,2} \sin\phi_2 \cos\phi_3] X_6 + \\
&+g_0 \cos\phi_2 X_7 + X_8 \int |\dot{\phi}_1| dt + X_9 \int |\dot{\phi}_2| dt + X_{10} \int |\dot{\phi}_3| dt + \\
&+X_{11} \int \dot{\phi}_1^2 dt + X_{12} \int \dot{\phi}_2^2 dt + X_{13} \int \dot{\phi}_3^2 dt \; .
\end{aligned}
\tag{3.145}
$$

Note that both the differential and integral models are expressed in term of the same set of canonical parameters given by Eq.(3.144). It is easy to verify that the matrix $\dot{\boldsymbol{A}}(\boldsymbol{\phi}) - [\boldsymbol{C}(\boldsymbol{\phi}, \dot{\boldsymbol{\phi}}) + \boldsymbol{C}^T(\boldsymbol{\phi}, \dot{\boldsymbol{\phi}})]$ satisfies the skew–symmetry property.

In order to clarify the interpretation of the joint variables ϕ_2 and ϕ_3 consider Fig.3.7. Different angles which are indicated in Fig. 3.7 refer also to angles from Figs.3.3 and 3.4. Note that the joint variable ϕ_2 in Fig.3.7 has a negative sign assuming that the same variable in Fig.3.3 has positive sign. We also recognise the sign of ϕ_3 variable which is positive on both Figs.3.4 and 3.7. Joint variables ϕ_2 and ϕ_3 are driven independently by the second and third motor, respectively. Now we explain this situation. Taking into account geometrical considerations we observe from Fig.3.7 that,

$$
x_2 = g_2(\phi_3) = n_3 \Psi_3 \; , \tag{3.146}
$$

as quoted before, Ψ_3 denotes a rotor variable of the third motor, n_3 is a gear ratio of the screw drive, and x_2 denotes a length of the screw of the drive.

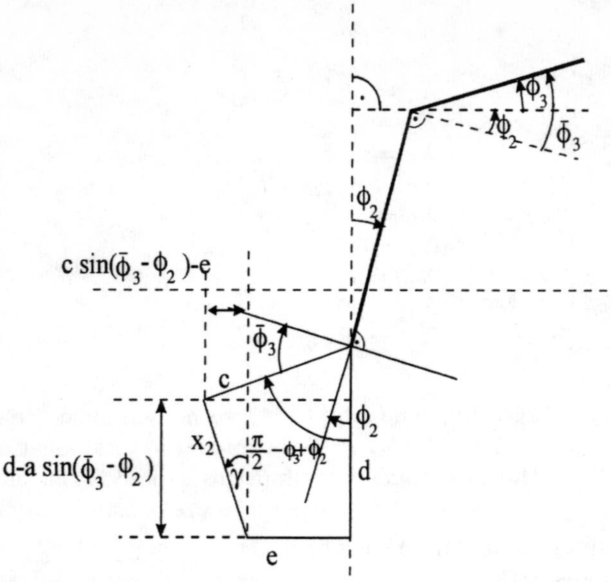

Fig. 3.7. Geometrical interpretation of joint variables ϕ_2 and ϕ_3

Note that the angle ϕ_3 is measured from a line which is perpendicular to the gravity direction. Based on Fig. 3.7 we can write

$$x_2 = \sqrt{(c\cos\phi_3 - e)^2 + (d - c\sin\phi_3)^2} \tag{3.147}$$

and

$$\gamma = g_3(\phi_3) = \arctan\frac{c\cos\phi_3 - e}{d - c\sin\phi_3} \tag{3.148}$$

where γ denotes the angle between the gravity direction and the axis of the stator of the third motor (note that this angle is not constant). Expanding quadratic terms in Eq.(3.147) and after some algebraic manipulations we can rewrite Eq.(3.147) in the following form

$$\begin{aligned} x_2 = g_2(\phi_3) &= \sqrt{d^2 + c^2 + e^2 - 2c\sqrt{d^2+e^2}\sin(\phi_3 + \beta)} = \\ &= \sqrt{B^2 - A^2\sin(\phi_3 + \beta)}\,, \end{aligned} \tag{3.149}$$

where $\beta = \arctan\frac{e}{d}$ is a constant calculated from geometrical distances of the robot and $c^2 = a^2 + b^2$. Differentiation of Eq.(3.146) with respect to time leads to

$$\frac{d\Psi_3}{dt} = \frac{dF_3}{d\phi_3}\cdot\frac{d\phi_3}{dt}\,, \tag{3.150}$$

where

$$\frac{dF_3}{d\phi_3} = -\frac{2\pi}{SK}\cdot\frac{dg_2}{d\phi_3}\,. \tag{3.151}$$

In the last equation $\frac{dF_3}{d\phi_3}$ represents a nonlinear function which is a gear ratio between the joint velocity of the third link and the velocity of the third motor and SK denotes the slip of the screw drive. Minus sign which appears in Eq.(3.151) is due to the oposite changes of the joint variable ϕ_3 and motor variable Ψ_3. Based on Eq.(3.149) we can rewrite Eq.(3.150) as follows (compare also Eq.(3.133))

$$\dot{\Psi}_3 = \frac{\pi}{SK}\frac{A^2\cos(\phi_3+\beta)}{\sqrt{B^2-A^2\sin(\phi_3+\beta)}}\dot{\phi}_3 \,. \tag{3.152}$$

Integration of Eq.(3.152) results in

$$\Psi_3 = -\frac{2\pi}{SK}\sqrt{B^2-A^2\sin(\phi_3+\beta)}+C \tag{3.153}$$

where the integration constant is calculated from the synchronisation position of the robot for the joint variable ϕ_3 namely, $\phi_3 = 0$ for $\Psi_3 = 0$. Hence we can write

$$C = \frac{2\pi}{SK}\sqrt{B^2-A^2\sin\beta}\,. \tag{3.154}$$

From Eq.(3.153) one can calculate the joint variable ψ_3 as follows

$$\phi_3 = -\beta + \arctan\frac{E}{\sqrt{1-E^2}}\,, \tag{3.155}$$

were E is given by Eg.(3.118) We can write Eq.(3.149) also in terms of $\bar{\phi}_3$ and ϕ_2 variables by substituting into it $\phi_3 = \bar{\phi}_3 - \phi_2$ (from Fig.3.7 it is clear that $\bar{\phi}_3 = \phi_2 + \phi_3$):

$$x_2 = f_4(\phi_2,\bar{\phi}_3) = \sqrt{B^2-A^2\sin(\bar{\phi}_3-\phi_2+\beta)}. \tag{3.156}$$

In this particular situation when the third motor moves the third joint, angles $\bar{\phi}_3$ and ϕ_2 change by the same amount, therefore their difference does not alter. This is due to the parallelogram structure of the IRp-6 robot. Equation (3.156) can easily be obtained from Eq.(3.122) if we change the signs of the joint variables $\bar{\phi}_3$ and ϕ_2 (see also Figs. 3.4b and 3.7) in the last one. As a result, in order to calculate the gear ratio in this case we have to differentiate $\frac{df_4}{d\phi_3}$. Note that this derivative depends on ϕ_2.

Actually the same situation happens when we use the change of coordinates given by Eq.(3.141). Now we can write

$$x_2 = f_5(q_2,q_3) = \sqrt{B^2-A^2\sin(q_2+q_3+\beta)}\,, \tag{3.157}$$

and

$$\frac{d\Psi_3}{dt} = \frac{\partial F_3}{\partial q_2}\frac{dq_2}{dt} + \frac{\partial F_3}{\partial q_3}\frac{dq_3}{dt} = \frac{\partial F_3}{\partial q_2}(\dot{q}_2+\dot{q}_3)\,, \tag{3.158}$$

where

$$\frac{\partial F_3}{\partial q_2} = \frac{\partial F_3}{\partial q_3} = \frac{\pi}{SK} \frac{A^2 \cos(q_2 + q_3 + \beta)}{\sqrt{B^2 - A^2 \sin(q_2 + q_3 + \beta)}}. \tag{3.159}$$

These calculations are quite complicated and they result formally from the application of the modified Denavit-Hartenberg notation. In order to avoid this complexity one can use a different set of joint coordinates namely ϕ_1, ϕ_2, and ϕ_3 instead of using the generalised coordinates q_1, q_2, and q_3.

Finally, we write explicitly a function $\frac{dF_2}{dt}$ which results as a formal substitution of $\phi_2 = \frac{\pi}{2} - q_2$ in Eq.(3.113) and performing differentiation operation with respect to q_2 we get (compare sings of ϕ_2 in Figs. 3.3a and 3.7)

$$\frac{dF_2}{dq_2} = \frac{\pi}{SK} \frac{A^2 \sin(-\beta + \phi_{20} - q_2)}{\sqrt{B^2 + A^2 \cos(-\beta + \phi_{20} - q_2)}}. \tag{3.160}$$

Equations (3.159) and (3.160) are necessary when writing the equations of motion in terms of a set of coordinates q_1, q_2, and q_3 which are used below.

Now we calculate the total potential energy of the manipulator based on Eq.(3.10). It is clear that the potential energy of the first link is zero. Here as from the very beginning, we employ generalised coordinates q_1, q_2, and q_3. The potential energy of the second link is

$$E_{p2} = -g_0 m_2 c_{2x} \sin q_2 \ , \tag{3.161}$$

where we have assumed that the only non–zero element of the first order moment is along x axis, (compare Fig.3.6). The potential energy of the third link is given by

$$E_{p3} = -g_0 l_2 m_3 \sin q_2 - g_0 m_3 c_{3x} \sin(q_2 + q_3). \tag{3.162}$$

Therefore the total potential energy of the manipulator can be written in the following form

$$E_{pc} = -g_0 \left(l_2 m_2 + m_2 c_{2x}\right) \sin q_2 - g_0 m_3 c_{3x} \sin\left(q_2 + q_3\right). \tag{3.163}$$

Based on Eqs.(3.139) and (3.163) we can calculate the Lagrangian function and the equations of motion derived from the Lagrange formalism for the first three degrees of freedom of the IRp-6 robot:

$$\begin{aligned}
\tau_1 &= \left[I_{1zz} + n_1^2 I_{1a} + I_{2xx} \sin^2 q_2 + I_{2yy} \cos^2 q_2 + I_{3xx} \sin^2(q_2 + q_3) + \right.\\
&+ \left. I_{3yy} \cos^2(q_2 + q_3) + m_3 l_2^2 \cos^2 q_2 + 2 l_2 m_3 c_{3x} \cos q_2 \cos(q_2 + q_3)\right] \ddot{q}_1 + \\
&+ \dot{q}_1 \dot{q}_2 \left(I_{2xx} \sin 2q_2 - I_{2yy} \sin 2q_2 + I_{3xx} \sin\left[2(q_2 + q_3)\right] - \right.\\
&- I_{3yy} \sin 2(q_2 + q_3) - m_3 l_2^2 \sin 2q_2 - 2 l_2 m_3 c_{3x} \sin q_2 \cos(q_2 + q_3) - \\
&- \left. 2 l_2 m_3 c_{3x} \cos q_2 \sin(q_2 + q_3)\right) + \dot{q}_1 \dot{q}_3 \left[I_{3xx} \sin 2(q_2 + q_3) - \right.\\
&- \left. I_{3yy} \sin 2(q_2 + q_3) - 2 l_2 m_3 c_{3x} \cos q_2 \sin(q_2 + q_3)\right] + F_{1v} \dot{q}_1 + \\
&+ F_{1a} \mathrm{sgn} \dot{q}_1 \ ,
\end{aligned} \tag{3.164}$$

$$
\begin{aligned}
\tau_2 &= \left(I_{2zz} + I_{2a}\left(\frac{dF_2}{dq_2}\right)^2 + I_{3zz} + I_{3a}\left(\frac{\partial F_3}{\partial q_3}\right)^2 + l_2^2 m_3 + \right.\\
&+ \left. 2m_3 c_{3x} l_2 \cos q_3 \right)\ddot{q}_2 + \left[\frac{1}{2}\left(I_{3zz} + \left(\frac{\partial F_3}{\partial q_3}\right)^2 I_{3a}\right) + \right.\\
&+ \left.\frac{1}{2} m_3 c_{3x} l_2 \cos q_3\right]\ddot{q}_3 + \dot{q}_2^2 \frac{dF_2}{dq_2}\frac{d^2F_2}{dq_2^2} I_{2a} - \frac{1}{2}\dot{q}_1^2\left[I_{2xx}\sin 2q_2 - \right.\\
&- I_{2yy}\sin 2q_2 + I_{3xx}\sin 2(q_2+q_3) - I_{3yy}\sin 2(q_2+q_3) - m_3 l_2^2 \sin 2q_2 - \\
&- \left. 2l_2 m_3 c_{3x}\sin q_2\cos(q_2+q_3) - 2l_2 m_3 c_{3x}\cos q_2 \sin(q_2+q_3)\right] + \\
&+ 2I_{3a}\frac{\partial F_3}{\partial q_3}\frac{\partial^2 F_3}{\partial q_3^2}\dot{q}_3\dot{q}_2 - 2m_3 c_{3x} l_2 \dot{q}_2\dot{q}_3 \sin q_3 + \left(\frac{\partial F_3}{\partial q_3}\frac{\partial^2 F_3}{\partial q_3^2} I_{3a} - \right.\\
&- \left. m_3 c_{3x} l_2 \sin q_3\right)\dot{q}_3^2 + I_{3a}\frac{\partial F_3}{\partial q_3}\frac{\partial^2 F_3}{\partial q_3^2}\dot{q}_2^2 - \\
&- g_0 m_2 c_{2x}\cos q_2 - g_0 l_2 m_3 \cos q_2 - g_0 m_3 c_{3x}\cos(q_2+q_3) + F_{2v}\dot{q}_2 + \\
&+ F_{2a}\mathrm{sgn}\dot{q}_2 \,, \\
\tau_3 &= \left[\frac{1}{2}\left(I_{3zz} + \left(\frac{\partial F_3}{\partial q_3}\right)^2 I_{3a}\right) + \frac{1}{2} m_2 c_{3x} l_2 \cos q_3\right]\ddot{q}_2 + \\
&+ \frac{1}{2}\left(I_{3zz} + \left(\frac{\partial F_3}{\partial q_3}\right)^2 I_{3a}\right)\ddot{q}_3 + 2I_{3a}\frac{\partial F_3}{\partial q_3}\frac{\partial^2 F_3}{\partial q_3^2}\dot{q}_2\dot{q}_3 - \\
&- \frac{1}{2}\dot{q}_1^2\left[I_{3xx}\sin 2(q_2+q_3) - I_{3yy}\sin 2(q_2+q_3) - \right.\\
&- \left. 2l_2 m_3 c_{3x}\cos q_2 \sin(q_2+q_3)\right] + \dot{q}_2^2 m_3 c_{3x} l_2 \sin q_3 + I_{3a}\frac{\partial F_3}{\partial q_3}\frac{\partial^2 F_3}{\partial q_3^2}\dot{q}_2^2 + \\
&+ I_{3a}\frac{\partial F_3}{\partial q_3}\frac{\partial^2 F_3}{\partial q_3^2}\dot{q}_3^2 - g_0 m_3 c_{3x}\cos(q_2+q_3) + F_{3v}(\dot{q}_3+\dot{q}_2) + \\
&+ F_{3a}\mathrm{sgn}(\dot{q}_3+\dot{q}_2).
\end{aligned}
$$

First, we show that the matrix $\dot{\boldsymbol{A}} - 2\boldsymbol{C}$ has a skew-symmetric property. Observe that the time derivative of elements of $\boldsymbol{A}$ matrix are

$$
\begin{aligned}
&\dot{A}_{11} = I_{2xx}\dot{q}_2\sin 2q_2 - I_{2yy}\dot{q}_2\sin 2q_2 + I_{3xx}(\dot{q}_2+\dot{q}_3)\sin 2(q_2+q_3) - \\
&-I_{3yy}(\dot{q}_2+\dot{q}_3)\sin 2(q_2+q_3) - m_3 l_2 \dot{q}_2 \sin 2q_2 - \\
&-2l_2 m_3 c_{3x}\sin q_2 \dot{q}_2\cos(q_2+q_3) - 2l_2 m_3 c_{3x}(\dot{q}_2+\dot{q}_3)\cos q_2\sin(q_2+q_3)\,, \\
&\dot{A}_{22} = 2\frac{dF_2}{dq_2}\frac{d^2F_2}{dq_2^2} I_{2a}\dot{q}_2 - 2m_3 c_{3x} l_2 \dot{q}_3 \sin q_3 + 2I_{3a}\frac{\partial F_3}{\partial q_3}\frac{\partial^2 F_3}{\partial q_3^2}(\dot{q}_3+\dot{q}_2)\,, \\
&\dot{A}_{33} = 2\frac{\partial F_3}{\partial q_3}\frac{\partial^2 F_3}{\partial q_3^2} I_{3a}(\dot{q}_3+\dot{q}_2)\,, \qquad (3.165) \\
&\dot{A}_{23} = \dot{A}_{32} = 2\frac{\partial F_3}{\partial q_3}\frac{\partial^2 F_3}{\partial q_3^2} I_{3a}(\dot{q}_3+\dot{q}_2) - l_2 m_3 c_{3x}\dot{q}_3\sin q_3 \quad \text{, and}
\end{aligned}
$$

$$\dot{A}_{12} = \dot{A}_{13} = \dot{A}_{21} = \dot{A}_{31} = 0\,.$$

Now based on Eq.(3.164) we write the elements of matrix $\boldsymbol{C}(\boldsymbol{q}, \dot{\boldsymbol{q}})$:

$$\begin{aligned}
C_{11} &= \frac{1}{2}\left[I_{2xx}\sin 2q_2 - I_{2yy}\sin 2q_2 + I_{3xx}\sin 2(q_2+q_3) -\right.\\
&- I_{3yy}\sin 2(q_2+q_3) - m_3 l_2 \sin q_2 - 2l_2 m_3 c_{3x}\sin q_2\cos(q_2+q_3) -\\
&- \left. 2l_2 m_3 c_{3x}\cos q_2 \sin(q_2+q_3)\right]\dot{q}_2 + \frac{1}{2}\left[I_{3xx}\sin 2(q_2+q_3) -\right.\\
&- \left. I_{3yy}\sin 2(q_2+q_3) - 2l_2 m_3 c_{3x}\cos q_2\sin(q_2+q_3)\right]\dot{q}_3\,,\\
C_{12} &= \frac{1}{2}\left[I_{2xx}\sin 2q_2 - I_{2yy}\sin 2q_2 + I_{3xx}\sin 2(q_2+q_3) -\right.\\
&- I_{3yy}\sin 2(q_2+q_3) - m_3 l_2 \sin q_2 - 2l_2 m_3 c_{3x}\sin q_2\cos(q_2+q_3) -\\
&- \left. 2l_2 m_3 c_{3x}\cos q_2\sin(q_2+q_3)\right]\dot{q}_1\\
C_{13} &= \frac{1}{2}\left[I_{3xx}\sin 2(q_2+q_3) - I_{3yy}\sin 2(q_2+q_3) -\right.\\
&- \left. 2l_2 m_3 c_{3x}\cos q_2\sin(q_2+q_3)\right]\dot{q}_1\\
C_{21} &= -C_{12},\\
C_{22} &= -l_2 m_3 c_{3x}\dot{q}_3 \sin q_3 + I_{2a}\frac{dF_2}{dq_2}\frac{d^2F_2}{dq_2^2}\dot{q}_2 + I_{3a}\frac{\partial F_3}{\partial q_3}\frac{\partial^2 F_3}{\partial q_3^2}(\dot{q}_3+\dot{q}_2)\,,\\
C_{23} &= -l_2 m_3 c_{3x}\dot{q}_3\sin q_3 - l_2 m_3 c_{3x}\dot{q}_2 \sin q_3 + I_{3a}\frac{\partial F_3}{\partial q_3}\frac{\partial^2 F_3}{\partial q_3^2}(\dot{q}_3+\dot{q}_2)\,,\\
C_{31} &= -C_{13}\,,\\
C_{32} &= l_2 m_3 c_{3x}\dot{q}_2\sin q_3 + I_{3a}\frac{\partial F_3}{\partial q_3}\frac{\partial^2 F_3}{\partial q_3^2}(\dot{q}_2+\dot{q}_3)\,,\\
C_{33} &= I_{3a}\frac{\partial F_3}{\partial q_3}\frac{\partial^2 F_3}{\partial q_3^2}(\dot{q}_3+\dot{q}_2)\,.
\end{aligned} \tag{3.166}$$

Taking into account the above expressions and calculating $\boldsymbol{C}^T + \boldsymbol{C}$, it is easy to verify that $\dot{\boldsymbol{A}} = \boldsymbol{C} + \boldsymbol{C}^T$. This means that a basic property of the differential model is satisfied. In a similar manner, it can be proved that this property also holds for the set of equations (3.130). Note that the differential model given by Eq.(3.164) contains elements which are functions of nonlinear gear ratios $\frac{dF_2}{dq_2}$, $\frac{\partial F_3}{\partial q_3}$ and their second derivatives. At this point, we have to make a simplifying assumption that the gear ratios are constant values around the synchronisation point of the IRp-6 robot. This was verified during the experiments (see Chapter 5). Therefore we can write that

$$\frac{dF_2}{dq_2} = \bar{n}_2 = \text{constant} \quad \text{and} \quad \frac{\partial F_3}{\partial q_3} = \bar{n}_3 = \text{constant}. \tag{3.167}$$

As a consequence the second derivatives of these functions are zero so are the appropriate elements in the differential model.

Taking into account the above assumption we formulate an integral model of the IRp-6 robot. Based on Eqs.(3.139) and (3.163) for the total kinetic and potential energy and using into Eq.(3.38) we can write the following integral model in the canonical form

parameters	***expressions*** $d^i(\boldsymbol{q}, \dot{\boldsymbol{q}})$
$X_1 = I_{1zz} + I_{1a}n_1^2 + I_{2xx} + I_{3yy} - I_{3xx}$	$0.5\dot{q}_1^2,$
$X_2 = I_{2yy} - I_{2xx} + m_3 l_2^2 - I_{3yy} + I_{3xx}$	$0.5\dot{q}_1^2 \cos^2 q_2,$
$X_3 = I_{2zz} + I_{2a}\bar{n}_2^2 + I_{3zz} + I_{3a}\bar{n}_3^2 + l_2^2 m_3$	$0.5\dot{q}_2^2,$
$X_4 = I_{3zz} + I_{3a}\bar{n}_3^2$	$0.5\dot{q}_3^2 + \dot{q}_2\dot{q}_3,$
$X_5 = I_{3yy} - I_{3xx},$	$-0.5\left[\sin^2(q_3)\cos 2q_2 + 0.5\sin 2q_2 \sin 2q_3\right]\dot{q}_1^2,$
$X_6 = m_3 c_{3x}$	$l_2\dot{q}_2(\dot{q}_2 + \dot{q}_3)\cos q_3 + l_2\dot{q}_1^2\cos(q_2 + q_3)\cos q_2 + g_0\sin(q_2 + q_3),$
$X_7 = m_2 c_{2x} + l_2 m_3$	$-g_0 \sin q_2,$
$X_8 = F_{1s}$	$\int \lvert\dot{q}_1\rvert dt,$
$X_9 = F_{2s}$	$\int \lvert\dot{q}_2\rvert dt,$
$X_{10} = F_{3s}$	$\int \lvert\dot{q}_2 + \dot{q}_3\rvert dt,$
$X_{11} = F_{1v}$	$\int \dot{q}_1^2 dt,$
$X_{12} = F_{2v}$	$\int \dot{q}_2^2 dt,$
$X_{13} = F_{3v}$	$\int (\dot{q}_2 + \dot{q}_3)^2 dt .$

(3.168)

In the derivation of the integral model in section 3.4 we have assumed that the inertial parameters are constant. The simplifying assumption given by Eq.(3.167) is necessary in the integral model. Now we reformulate the differential model in terms of a set of canonical parameters (compare Eq.(3.164))

$$
\begin{aligned}
\boldsymbol{\tau} &= \sum_{i=1}^{13} \boldsymbol{D}^i(\boldsymbol{q}, \dot{\boldsymbol{q}}, \ddot{\boldsymbol{q}})X_i = \begin{bmatrix} \ddot{q}_1 \\ 0 \\ 0 \end{bmatrix} X_1 + \begin{bmatrix} \ddot{q}_1\cos^2 q_2 - \dot{q}_1\dot{q}_2\sin 2q_2 \\ \frac{1}{2}\dot{q}_1^2 \sin 2q_2 \\ 0 \end{bmatrix} X_2 + \\
&+ \begin{bmatrix} 0 \\ \ddot{q}_2 \\ 0 \end{bmatrix} X_3 + \begin{bmatrix} 0 \\ 0 \\ \frac{1}{2}(\ddot{q}_2 + \ddot{q}_3) \end{bmatrix} X_4 + \\
&+ \begin{bmatrix} \ddot{q}_1\cos^2(q_2+q_3) - \dot{q}_1\dot{q}_2\sin 2(q_2+q_3) - \dot{q}_1\dot{q}_3\sin 2(q_2+q_3) \\ \frac{1}{2}\dot{q}_1^2\sin 2(q_2+q_3) \\ \frac{1}{2}\dot{q}_1^2\sin 2(q_2+q_3) \end{bmatrix} X_5 + \\
&+ \begin{bmatrix} -\ddot{q}_1 2l_2\cos q_2\cos(q_2+q_3) + 2l_2\dot{q}_1\dot{q}_2\sin(2q_2+q_3) + \\ +\dot{q}_1\dot{q}_3 2l_2\cos q_2\sin(q_2+q_3) \\ \\ \ddot{q}_2 2l_2\cos q_3 + \frac{1}{2}\ddot{q}_3\cos q_3 - \dot{q}_1^2 l_2\sin(2q_2+q_3) - \\ g_0\cos(q_2+q_3) - 2l_2\dot{q}_2\dot{q}_3\sin q_3 - l_2\dot{q}_3^2\sin q_3 \\ \\ \frac{1}{2}l_2\ddot{q}_2\cos q_3 - \dot{q}_1^2 l_2\cos q_2\sin(q_2+q_3) - g_0\cos(q_2+q_3) + \\ +\dot{q}_2^2 l_2\sin q_3 \end{bmatrix} X_6 +
\end{aligned}
$$

$$+ \begin{bmatrix} 0 \\ -g_0 \cos q_2 \\ 0 \end{bmatrix} X_7 + \begin{bmatrix} \mathrm{sgn}\dot{q}_1 \\ 0 \\ 0 \end{bmatrix} X_8 + \begin{bmatrix} 0 \\ \mathrm{sgn}\dot{q}_2 \\ 0 \end{bmatrix} X_9 +$$

$$+ \begin{bmatrix} 0 \\ 0 \\ \mathrm{sgn}(\dot{q}_3 + \dot{q}_2) \end{bmatrix} X_{10} + \begin{bmatrix} \dot{q}_1 \\ 0 \\ 0 \end{bmatrix} X_{11} +$$

$$+ \begin{bmatrix} 0 \\ \dot{q}_2 \\ 0 \end{bmatrix} X_{12} + \begin{bmatrix} 0 \\ 0 \\ (\dot{q}_2 + \dot{q}_3) \end{bmatrix} X_{13}. \tag{3.169}$$

Equation (3.169) represents the differential canonical model of the IRp-6 robot. Note that the function $\boldsymbol{D}^i$ associated with each dynamic parameter (being a linear combination of individual link dynamic parameters) are fairly complicated and involve joint positions, velocities, and accelerations. They are more complicated than d^i functions which appear in the integral model. At the same time, information which is included in the differential model is richer than in the integral model. This observation, previously stated, can be seen clearly from Eqs.(3.168) and (3.169). Notice that a vector of parameters is a set of independent parameters. In both formulations the same set of parameters appears. One can calculate a product $\int \boldsymbol{\tau}^T \dot{\boldsymbol{q}} dt$ based on Eq.(3.169) which results in the integral model given by Eq.(3.168). These tedious but straightforward calculations are omitted in the monograph. We can use parameters identified in the integral model directly in the differential model.

From the considerations presented in this section it is clear that formal implementation of the modified Denavit–Hartenberg notation to the kinematical structure of the IRp-6 robot leads to the most complicated equations of motion. However, its integral model is very simple (see Eq.(3.168)) and suitable for the purpose of identification (see Chapter 5). Another set of generalised coordinates, namely ϕ_1, ϕ_2 and ϕ_3, leads to equations of motion which are less complicated and these coordinates allow to calculate the nonlinear gear ratios in efficient way.

Here we make one comment. The $\boldsymbol{D}^i$ or d^i functions play the most important role in experimental identification. Due to the particularly chosen signals $\boldsymbol{q}$, $\dot{\boldsymbol{q}}$, and $\ddot{\boldsymbol{q}}$ they can be linearly dependent. This problem arises when we want to identify the friction coefficients in the integral model. Usually dynamic parameters cannot be identified all in one movement. Separate movements are necessary in order to avoid these difficulties. This problem will be widely discussed in Chapter 5.

CHAPTER 4
IDENTIFICATION OF ROBOT MODEL PARAMETERS

4.1 Introduction

In the previous chapter we derived the two most important models for the purpose of identification of their dynamic parameters. Both models are shown in their canonical form and are represented by a set of dynamic parameters which are usually linear combinations of individual link dynamic parameters. This simplifies the task of identifying these parameters. This chapter is dedicated to various identification schemes which can be used to identify the set of canonical parameters. The techniques proposed here apply to both geared and direct drive robots. From a practical point of view, in order to carry out experimental identification it is necessary to measure joint positions, velocities, accelerations, and joint torques and/or forces. For the purpose of load identification the robot has to be equipped with force/torque sensors and any experimental set-up in a laboratory environment should satisfy these requirements. However, since these sensors are quite expensive, robots are seldom equipped with them in practice.

Identification in robotics can be generally divided into two separate problems. One is to identity a kinematical structure of the robot. The structure can be described in terms of an arbitrary notation, usually standard or modified Denavit-Hartenberg notation. In addition we assume that the parameters which appear in the kinematical structure are constant. In the real world robot is sometimes in collision with the environment so that after a long period of time these parameters can change. Besides, the robot is mechanically designed with a small tolerance which widens with time. Therefore from time to time it is necessary to identify its kinematical parameters and to recalibrate the robot. The values of these parameters are necessary to solve for any robot the inverse and forward kinematics problems. Most of the industrial robots are kinematically controlled (due to the lack of torque measurements). The calibration of the robot is technically quite a difficult problem and is directly connected with the identification of the kinematical parameters of the robot. This problem has been addressed by many researchers. Here we mention only four references [1, 86, 200, 227] which give experimental results. Other references, for example [101, 201, 208], do not include experimental results. The calibration and identification of robot kinematic parameters, though important, are not addressed in this monograph.

The other identification problem in robotics is a problem of finding the inertial parameters of the links, actuators, and unknown load held by the robot. As discussed in Chapter 3, in this monograph, we consider two types of robots: direct drive (DDA) and geared ones. DDA robots are usually faster and dynamic coupling between links cannot be neglected. It has been verified experimentally that friction effects for these robots are important and must not be overlooked [27, 124, 185]. The situation for geared robots is different. These robots are equipped with transmissions with gear ratios in the range of 150 to 200. Due to the fact that the actual inertia along an axis which rotates (for rotational joint) is divided by the square of gear ratio, its dynamic influence is reduced approximately by this value. The dynamics are practically decoupled for robots of this type. It has been verified experimentally that this is true for robots which move with a relatively small velocity (recall that geared robots move slower than direct drive robots). This will be discussed in Chapter 5. On the other hand, friction plays an important role with geared robots and cannot be neglected even if we assume that the controllers are kinematically driven. Usually static friction torques are precompensated at the lowest level at the motor controller. Besides that, for both classes of robot there is a problem of estimation of the dynamic parameters of the load carried by the robot. From the above considerations it is clear that identification of friction, dynamic parameters of a robot with or without gears, and load are important problems.

There are two approaches to solving the problem of robot dynamics identification. One is to identify the dynamic parameters experimentally, either off–line or on–line, by implementing one of the well–known identification procedures. Thanks to the fact that, as shown in the previous section, the dynamic parameters of the robot and the load appear linearly in the differential and integral models discussed in Chapter 3 identification is simplified. This does not apply to nonlinear friction in general. Therefore we propose to precompute the friction torques and subtract their contribution from the generalised forces. This problem will also be addressed in Chapter 5. For the purpose of identification we assume that the dynamic friction model is linear with respect to coefficients. Then we can use any identification scheme applicable for a linear system. The number of parameters is usually in the range of 15 to 20 and we need at least the same number of equations to solve the identification problem. Usually, they are only 3 to 6 degrees of freedom of a robot and the same number of equations for its differential model. In the case of the integral model we have only one equation. As a consequence we have to sample these equations at several discrete time instants in order to obtain a sufficient number of equations to estimate the unknown dynamic parameters. In practice in the case of the differential model we discretise the dynamic equations for all the links. Usually, the same set of parameters appears in each of the n equations of motion and it suffices to take only one equation to solve the identification problem. Generally we suggest choosing those equations which emphasise the parameters making identification easier. Here we separate the problem of finding an exciting trajectory from the identification technique. This problem will be addressed in Chapter 5. On–line identification is more difficult to implement since the identification algorithm has to be very fast in order to estimate the unknown parameters. A software program to cope with this has to be extremely efficiently coded.

Another approach which leads to estimates of the unknown dynamic parameters and load is found in the model–based adaptive control. This particular control scheme makes it possible to identify the unknown parameters and at the same time to control a robot based on the on–line estimated model. Results of this kind of identification can be found in references [30, 52, 111, 207]. Often the estimated model does not result in parameters which match the real values of the dynamic parameters, because in the adaptive control scheme the goal is trajectory end–point tracking or trajectory tracking in general. In addition in robust adaptive control schemes we can assume inaccurate information about the model, but this information is sufficient to control the manipulator without knowing the model accurately. Modern control schemes require no knowledge about the model at all but these schemes will not be discussed in this monograph.

The majority of the literature on the subject focuses on the identification in the time domain. Some authors suggest using the frequency domain [95, 192, 222]. This idea was originally applied to robot manipulators by Kallenbach [95] and later on by Vandanjon and co–authors [222] and comes from the fact that manipulator joint movements are usually polynomial or sinusoidal trajectories. The dynamic equations of motion involve sine and cosine functions of joint variables and joint velocities and accelerations being of shape described above. Therefore we can easily apply the Fourier transformation to such a system (recall that the dynamic parameters appear linearly in the equations of motion). Besides that, the robot as a mechanical system behaves as a low–pass filter. This observation helps to analyse the system in the frequency domain. In order to identify the dynamic parameters we perform a periodic movement of one or at most two joints, and analyse the resulting equation (which comes from the differential model) in the frequency domain. Next we apply the Fast Fourier Transform (FFT) to the resulting equation. It often happens that in the frequency domain some effects can be eliminated, for example, viscous friction characteristics. As a consequence, some effects can be separated in the frequency domain and the identification problem becomes less difficult. In order to identify several dynamic coefficients which appear in the frequency domain equation, we take a sufficient number of equations for a number of different frequencies and from the resulting set of equations we estimate the unknown parameters.

Kallenbach [95] proposes using a covariance method for identification. The stationary covariance relation, which results from applying of the spectral transformation to the equations of motion, is sampled at different frequencies in order to obtain a number of equations in which the dynamic parameters are linear.

From the above considerations we conclude that regardless of the proposed domain - time or frequency - we obtain a set of nonlinear equations in which the dynamic parameters are linear coefficients. In general we have an overdetermined system of equations. In order to estimate the set of unknown parameters from this set of equations we can use, for example, the standard least squares technique. This technique can be used in a recursive form (which is most suitable for on–line identification) or in batch processing and it was used successfully by many authors [1, 10, 11, 21, 22, 36, 59, 60, 108, 124, 134, 136, 181, 185, 196]. The least squares method can be implemented in different forms, for example, damped least squares method, ridge regression and/or modified exponentially-weighted recursive squares method. The last one is proposed by Canudas de Wit and Carrillo [24]. Sequential identification, based on the least squares method, can also be used [26, 64]. Considerations in the chapter concerning the identification techniques are separated from the problem of finding an optimal exciting trajectory which is discussed in Chapter 5.

The practical measurement data: joint positions, velocities, accelerations, and motor currents (which are proportional to the torques) are corrupted by measurement noise. Position encoders, rate generators, and other sensors are prone to stochastic errors, which can usually be approximated by white noise. These stochastically corrupted signals appear in nonlinear equations of motion. Because of that, we cannot extract additive noise from these equations, which would summarise contributions of each measurement noise. As a consequence, the estimates are biased. This difficult problem is also addressed in this chapter and later on in two subsequent chapters which contain experimental results.

This chapter is organised as follows. In the next section we describe an identification scheme for the differential model. In section 4.3 we repeat the same procedure for the integral model. In section 4.4 we present different identification techniques used for the purpose of identification of dynamic and load parameters. The final section includes simulation results illustrating considerations discussed in the chapter.

4.2 Least squares technique for the differential model

In this section we describe the least squares method for estimating the dynamic parameters of the differential model described in Section 3.2. The derivation presented here is heavily borrowed from references [59, 60]. The approach proposed in those papers was implemented by the author of this monograph and verified later by simulation and experimental results. Other approaches which are discussed in the literature were implemented as well and compared with the one presented here.

First we assume that the differential model is canonical, which is a necessary condition for successful identification. We do not concentrate on exciting trajectory generation problem since, as previously mentioned, it does not influence an identification method, and the problem itself will be discussed in Chapter 5. In order to simplify the friction model we assume that it consists of Coulomb and viscous parts, so it depends linearly on these friction coefficients. We denote by m_c the minimum number of dynamic parameters (which are linearly independent) which constitute the differential model. We take into account equation (3.12), which by making use of Eq.(3.23), can be written in the following form:

$$\boldsymbol{\tau} = \sum_{i=1}^{m_c} \boldsymbol{D}^i(\boldsymbol{q}, \dot{\boldsymbol{q}}, \ddot{\boldsymbol{q}}) X_i. \tag{4.1}$$

In Eq.(4.1) column vector $\boldsymbol{D}^i(\boldsymbol{q}, \dot{\boldsymbol{q}}, \ddot{\boldsymbol{q}})$ represents a function which is associated with the dynamic parameter X_i from a vector of the minimum set of the dynamic parameters $\boldsymbol{x}_c$. Notice that vector $\boldsymbol{D}^i$ in Eq.(4.1) differs from vector $\boldsymbol{D}^i$ in Eq.(3.23) due to the fact that the dynamic model in Eq.(4.1) is canonical. Note also that vector $\boldsymbol{D}^i$ associated with the Coulomb friction coefficient representing Coulomb friction torque F_{is} has the form $\boldsymbol{D}^i = [0, \ldots, \mathrm{sgn}(\dot{q}_i), \ldots 0]^T$ and vector $\boldsymbol{D}^i$ associated with the viscous friction coefficient has the form $\boldsymbol{D}^i = [0, \ldots, \dot{q}_i, \ldots 0]^T$. The model described by Eq.(4.1) is a continuous one. In practice we observe it at discrete points t_k, which in this case are assumed to be equidistant. We use the tilde symbol $\sim$ to indicate a signal observed at discrete time t_k. In order to simplify the notation we denote by $\tilde{\boldsymbol{\tau}}(k)$, $\tilde{\boldsymbol{q}}(k)$, $\tilde{\dot{\boldsymbol{q}}}(k)$, and $\tilde{\ddot{\boldsymbol{q}}}(k)$ discrete signals of continuous vector signals $\boldsymbol{\tau}$, $\boldsymbol{q}$, $\dot{\boldsymbol{q}}$, and $\ddot{\boldsymbol{q}}$, respectively, observed at discrete time t_k. Therefore vector $\tilde{\boldsymbol{\tau}}(k)$ can be written as $\tilde{\boldsymbol{\tau}}(k) = [\tilde{\tau}_1(k), \ldots, \tilde{\tau}_n(k)]^T$. For matrix $\boldsymbol{D}(\boldsymbol{q}, \dot{\boldsymbol{q}}, \ddot{\boldsymbol{q}})$, which consists of m_c columns, we use at discrete time t_k the following notation:

$$\boldsymbol{D}(k) = \boldsymbol{D}(\boldsymbol{q}(k), \dot{\boldsymbol{q}}(k), \ddot{\boldsymbol{q}}(k)) = \left[\tilde{\boldsymbol{d}}_1^T(k), \ldots, \tilde{\boldsymbol{d}}_i^T(k), \ldots, \tilde{\boldsymbol{d}}_n^T(k)\right]^T. \tag{4.2}$$

Here, $\tilde{\boldsymbol{d}}_i(k)^T$ denotes the i-th row of matrix $\tilde{\boldsymbol{D}}(k)$ written as a column vector. The number of parameters in Eq.(4.1) is m_c, thus we need at least m_c equations in order to estimate the vector of unknown parameters. Implementation of the least squares method requires more equations. Denote the number of equations by $L > m_c$. Now we define the following vectors:

$$\boldsymbol{y}_L = \left[\tilde{\boldsymbol{\tau}}^T(1), \tilde{\boldsymbol{\tau}}^T(2), \ldots, \tilde{\boldsymbol{\tau}}^T(k), \tilde{\tau}_1(k+1), \ldots, \tilde{\tau}_m(k+1)\right]^T \tag{4.3}$$

and

$$\boldsymbol{h}_L = \left[\tilde{\boldsymbol{D}}^T(1), \tilde{\boldsymbol{D}}^T(2), \ldots, \tilde{\boldsymbol{D}}^T(k), \tilde{\boldsymbol{d}}_1^T(k+1), \ldots, \tilde{\boldsymbol{d}}_m^T(k+1)\right]^T. \tag{4.4}$$

Using definitions (4.3) and (4.4) we can write

$$\boldsymbol{y}_L = \boldsymbol{h}_L \boldsymbol{x}_c + \boldsymbol{w} \tag{4.5}$$

where $\boldsymbol{w}$ is the observation error and $\boldsymbol{h}_L$ is known as the regression matrix. Applying the least squares method to Eq.(4.5) (assuming that $L > m_c$) results in

$$\hat{\boldsymbol{x}}_c = \left(\boldsymbol{h}_L^T \boldsymbol{h}_L\right)^{-1} \boldsymbol{h}_L^T \boldsymbol{y}_L . \tag{4.6}$$

This equation has a solution under the assumption that matrix $\boldsymbol{P}_L = (\boldsymbol{h}_L^T \boldsymbol{h}_L)$ is positive definite, which means that the columns of matrix $\boldsymbol{h}_L$ must be linearly independent. This assumption is satisfied because the model given by Eq.(4.1) is canonical. Note that the elements of matrix $\boldsymbol{h}_L$ depend on the generalised positions, velocities, and accelerations and the columns of this matrix can become linearly dependent for a particular chosen trajectory. In identification terms this trajectory has to satisfy some of the "persistently exciting" conditions, namely the existence of positive constants α, β, and T_0 such that

$$\beta \boldsymbol{I} \geq \int_t^{t+T_0} \boldsymbol{h}_L^T \boldsymbol{h}_L dt \geq \alpha \boldsymbol{I} , \tag{4.7}$$

where α can be recognised as the degree of excitation and the ratio $\kappa(\boldsymbol{P}_L) = \beta/\alpha$ (the condition number) as the conditioning of the persistent excitation of the matrix $\boldsymbol{P}_L$. These two numbers have an important impact on the quality of estimation. On one hand, α determines the amount of data needed to obtain a reasonably good estimate of $\boldsymbol{x}_c$. As a consequence long records are necessary when the estimation is performed using low–excited trajectories. On the other hand, the advantage of trajectories applying small $\kappa(\boldsymbol{P}_L)$, is to reduce the maximum bound of the bias due to noise and unmodelled dynamics [5]. It is clear that planning the trajectories is not a trivial task and this problem will be discussed in detail in Chapter 5. Matrix $\boldsymbol{P}_L$ is often called the input correlation matrix (or the persistent excitation matrix). Note also that Eq.(4.6) is written in terms of the left pseudo–inverse of matrix $\boldsymbol{h}_L$.

In the neighbourhood of singularity of matrix $\boldsymbol{P}_L$ we can solve the rank deficient problem by using the so–called damped least squares inverse:

$$\hat{\boldsymbol{x}}_c = \left(\boldsymbol{h}_L^T \boldsymbol{h}_L + k^2 \boldsymbol{I}\right)^{-1} \boldsymbol{h}_L^T \boldsymbol{y}_L , \tag{4.8}$$

where k is a damping factor that renders the inversion better conditioned numerically and $\boldsymbol{I}$ is an identity matrix. Another possibility to solve the above rank deficient problem would be to use the ridge regression, which makes $\boldsymbol{h}_L^T \boldsymbol{h}_L$ invertible by adding a small number c to the diagonal elements:

$$\hat{\boldsymbol{x}}_c = \left(\boldsymbol{h}_L^T \boldsymbol{h}_L + c\boldsymbol{I}\right)^{-1} \boldsymbol{h}_L^T \boldsymbol{y}_L . \tag{4.9}$$

In this case the estimates are nearly optimal if $c \ll \lambda_{\min}\left(\boldsymbol{h}_L^T \boldsymbol{h}_L\right)$, where $\lambda_{\min}$ is the smallest non–zero eigenvalue of $\boldsymbol{h}_L^T \boldsymbol{h}_L$.

In practice we write Eq.(4.6) in recurrence form:

$$\hat{\boldsymbol{x}}_c(L+1) = \hat{\boldsymbol{x}}_c(L) + \boldsymbol{P}_{L+1}\tilde{\boldsymbol{d}}_i(j)\left[\tilde{\tau}_i(j) - \tilde{\boldsymbol{d}}_i^T(j)\hat{\boldsymbol{x}}_c(L)\right] , \tag{4.10}$$

$$\boldsymbol{P}_{L+1} = \boldsymbol{P}_L - \left(\boldsymbol{I} + \tilde{\boldsymbol{d}}_i^T(j)\boldsymbol{P}_L\tilde{\boldsymbol{d}}_i(j)\right)^{-1} \boldsymbol{P}_L\tilde{\boldsymbol{d}}_i(j)\tilde{\boldsymbol{d}}_i^T(j)\boldsymbol{P}_L. \tag{4.11}$$

Symbol $\hat{\boldsymbol{x}}_c(L)$ denotes the estimate in L-th recurrence step. Equation (4.10) can be interpreted as follows, solution of the estimate $\hat{\boldsymbol{x}}_c(L+1)$ in next recurrence step is formed by adding one element from the torque vector $\tilde{\tau}_i(j)$ and a new vector $\tilde{\boldsymbol{d}}_i(j)$. Stability of the recurrence solution can be achieved by selecting the appropriate factorisation of matrix $\boldsymbol{P}_L$ during the recurrence process. There are many possible ways to perform factorisation of matrix $\boldsymbol{P}_L$. One of them can be written as $\boldsymbol{P}_L = \boldsymbol{U}_L\boldsymbol{\Sigma}_L{\boldsymbol{U}_L}^T$, where $\boldsymbol{U}_L$ is a square upper triangular matrix with ones on the diagonal and the same dimensions as matrix $\boldsymbol{P}_L$ and $\boldsymbol{\Sigma}_L$ diagonal matrix with positive elements on the diagonal (compare also section 3.5). This particular factorisation is called the Agee–Turner factorisation [15]. It guarantees positive definiteness of matrix $\boldsymbol{P}_L$ during the recurrence process. The Agee–Turner factorisation has been implemented successfully during the identification process both in simulation and experimental phases.

Now we build a stochastic model for the purpose of identification of dynamic parameters. As mentioned in section 4.1, measurement data (the generalised positions, velocities, accelerations, torques and/or forces) are corrupted by noise. For subsequent derivation we assume the following noisy signals:

$$\tilde{\boldsymbol{q}}(k) = \boldsymbol{q}(k) + \boldsymbol{v}_p(k), \quad \tilde{\dot{\boldsymbol{q}}}(k) = \dot{\boldsymbol{q}}(k) + \boldsymbol{v}_s(k), \quad \tilde{\ddot{\boldsymbol{q}}}(k) + \boldsymbol{v}_a(k). \tag{4.12}$$

We assume that $\boldsymbol{v}_p$, $\boldsymbol{v}_s$, and $\boldsymbol{v}_a$ are stochastic processes characterised by the white noise with zero expected value.

In order to illustrate the stochastic model consider Fig.4.1, in which we assume the following notation: $\boldsymbol{\tau}$ represents the vector of the generalised forces applied to the joints of the manipulator, and $\tilde{\boldsymbol{\tau}}$ is the vector of the generalised forces calculated based on the robot dynamics model (by solving the inverse dynamics problem). The block denoted by $Min(\boldsymbol{w}^T\boldsymbol{w})$ is aimed at minimising the observation error in term of the least squares method. Similarly, as for the deterministic case considered above, we write $\tilde{\boldsymbol{D}} = \boldsymbol{D}(\tilde{\boldsymbol{q}}, \tilde{\dot{\boldsymbol{q}}}, \tilde{\ddot{\boldsymbol{q}}})$. From Fig.4.1 we have

$$\boldsymbol{\tau} = \boldsymbol{D}(\boldsymbol{q}, \dot{\boldsymbol{q}}, \ddot{\boldsymbol{q}})\boldsymbol{x}_c + \boldsymbol{v}_m , \tag{4.13}$$

where $\boldsymbol{v}_m$ is a stochastic process describing unmodelled dynamics (for example inaccuracies of the friction modelling and other factors which are difficult to be described analytically).

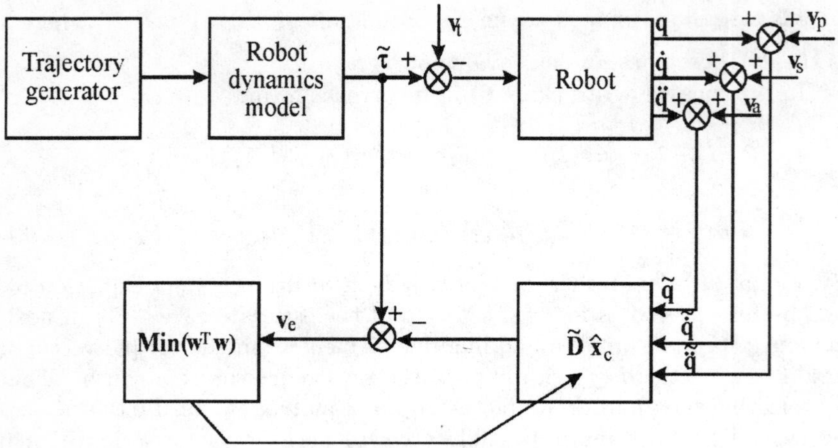

Fig. 4.1. The identification scheme for the differential model

In addition we denote by $\boldsymbol{v}_t$ a stochastic process which characterises a corruptive signal of the torques generated by the motors. Taking into account notation introduced in Fig.4.1 we write

$$\begin{aligned}\tilde{\boldsymbol{\tau}} &= \boldsymbol{\tau} + \boldsymbol{v}_t = \boldsymbol{D}\boldsymbol{x}_c + \boldsymbol{v}_m + \boldsymbol{v}_t = \tilde{\boldsymbol{D}}\boldsymbol{x}_c + \boldsymbol{D}\boldsymbol{x}_c - \tilde{\boldsymbol{D}}\boldsymbol{x}_c + \boldsymbol{v}_m + \boldsymbol{v}_t = \\ &= \boldsymbol{D}\left(\boldsymbol{q} + \boldsymbol{v}_p, \dot{\boldsymbol{q}} + \boldsymbol{v}_s, \ddot{\boldsymbol{q}} + \boldsymbol{v}_a\right)\boldsymbol{x}_c + \boldsymbol{v}_e = \tilde{\boldsymbol{D}}\boldsymbol{x}_c + \boldsymbol{v}_e \ , \end{aligned} \tag{4.14}$$

where $\boldsymbol{v}_e$ is a random vector defined as follows:

$$\boldsymbol{v}_e = \left[\boldsymbol{D}(\boldsymbol{q}, \dot{\boldsymbol{q}}, \ddot{\boldsymbol{q}}) - \boldsymbol{D}(\boldsymbol{q} + \boldsymbol{v}_p, \dot{\boldsymbol{q}} + \boldsymbol{v}_s, \ddot{\boldsymbol{q}} + \boldsymbol{v}_a)\right]\boldsymbol{x}_c + \boldsymbol{v}_m + \boldsymbol{v}_t. \tag{4.15}$$

Note that vector $\boldsymbol{v}_e$ depends on stochastic vectors $\boldsymbol{v}_p$, $\boldsymbol{v}_s$, $\boldsymbol{v}_a$, and $\boldsymbol{v}_t$, but its explicit form is not known. Taking Eq.(4.14) and Eqs.(4.2)–(4.4) we can write the following equation describing the stochastic case:

$$\bar{\boldsymbol{y}}_L = \bar{\boldsymbol{h}}_L \boldsymbol{x}_c + \bar{\boldsymbol{w}}. \tag{4.16}$$

This equation has a different interpretation to the equation (4.5). Here $\bar{\boldsymbol{y}}_L$ and $\bar{\boldsymbol{w}}$ are realisations of stochastic vectors $\boldsymbol{Y}_L$ and $\boldsymbol{W}$, and $\bar{\boldsymbol{h}}_L$ is a realisation of stochastic matrix $\boldsymbol{H}_L$. Generally $\boldsymbol{H}_L$ and $\boldsymbol{W}$ are correlated due to nonlinearities which appear in the $\boldsymbol{H}_L$ matrix. As before, $\boldsymbol{x}_c$ is the deterministic vector of the minimum set of dynamic parameters. Due to the fact that matrix $\bar{\boldsymbol{h}}_L$ is not deterministic we cannot apply to Eq.(4.16) the standard Markov estimation theory. In order to find the optimal estimate of vector $\boldsymbol{x}_c$ we propose to find an estimate $\hat{\boldsymbol{x}}_c$ which has realisation χ_{cL} calculated as a result of the minimisation

$$\chi_{cL} = \min_{\boldsymbol{x}_c} \left[\left(\bar{\boldsymbol{y}}_L - \bar{\boldsymbol{h}}_L \boldsymbol{x}_c\right)^T \left(\bar{\boldsymbol{y}}_L - \bar{\boldsymbol{h}}_L \boldsymbol{x}_c\right)\right]. \tag{4.17}$$

The estimate calculated from Eq.(4.17) is biased and its bias is evaluated as follows:

$$E(\boldsymbol{x}_c - \hat{\boldsymbol{x}}_c) = E\left[(\bar{\boldsymbol{h}}_L{}^T \bar{\boldsymbol{h}}_L)^{-1} \bar{\boldsymbol{h}}_L{}^T \bar{\boldsymbol{w}}\right] ; \tag{4.18}$$

where E is the expectation operator. Due to the fact that $\boldsymbol{W}$ and $\boldsymbol{H}_L$ are not independent stochastic processes this bias cannot be calculated explicitly. An effort was made in this direction by Armstrong [4, 5] and the reader is referred to these references.

Here we suggest investigating the quality of the estimation by simulation. The simulation results will be described in Section 4.5. A block diagram for simulation purposes is presented in Fig.4.2, and is used for simulation

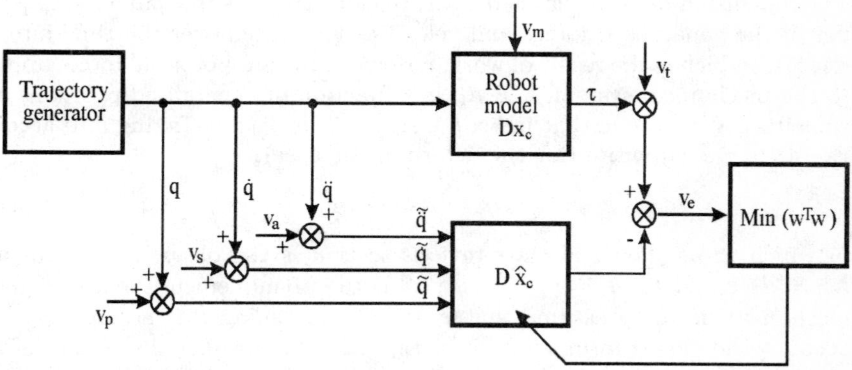

Fig. 4.2. The simulation scheme for the identification of the differential model

experiments for identification of robot dynamic parameters of the differential model. In Fig.4.2 we are not concerned with a real robot as in Fig.4.1 but rather with its model given by $\boldsymbol{D}\boldsymbol{x}_c$. For the purpose of simulation we assume that vector $\boldsymbol{x}_c$ is obtained from CAD modelling or its values are given by the manufacturer. We assume that each signal is corrupted by the white noise with known statistical characteristics, which can be changed in a wide range. We also assume that the vector of generalised forces can be corrupted by the white noise $\boldsymbol{v}_t$ with given characteristics. A block called a trajectory generator can generate an arbitrary trajectory. Usually we assume that the trajectory is a fifth–order polynomial whose coefficients are calculated from the initial conditions (see Craig [31] and Section 2). This trajectory is chosen because its first and second derivatives are smooth functions of time. The simulation results for the PUMA 560 robot are discussed in Section 4.5.

4.3 Identification scheme for the integral model

In this section we propose an identification scheme for the integral model described by Eq.(3.40). Results presented in this section parallel results in the previous section and are borrowed from reference [60]. Recall that Eq.(3.40) is a scalar equation. As before, we assume that we are dealing with the canonical model which is characterised by the minimum set of dynamic parameters. Recall that the integral model is represented by one scalar equation.

In order to make it suitable for the purpose of the identification we sample it, and for the k-th time interval, $(t_1, t_2)_k$, we can write

$$y_k = \boldsymbol{d}_k^T \boldsymbol{x}_c \, , \tag{4.19}$$

where

$$\boldsymbol{d}_k^T = \left[\boldsymbol{dl}_k^T, \boldsymbol{df}_{ks}{}^T, \boldsymbol{df}_{kv}{}^T \right] . \tag{4.20}$$

The dimension of vector $\boldsymbol{x}_c$ is $m_c \times 1$. Recall that y_k is the value of the product of the generalised forces and velocities integrated over the time interval $(t_1, t_2)_k$, which is the result of work performed by non-potential forces applied to the mechanical system, and $\boldsymbol{d}_k^T$ is a function of generalised positions and velocities. We assume that index k changes from 1 to r. Taking into account Eq.(4.19) we can write (for the deterministic case)

$$\boldsymbol{y}_r = \boldsymbol{h}_r \boldsymbol{x}_c + \boldsymbol{w}, \tag{4.21}$$

in which $\boldsymbol{w}$ as shown in the previous section is the observation error and $\boldsymbol{h}_r = [\boldsymbol{d}_1, \ldots, \boldsymbol{d}_r]^T$, $\boldsymbol{y}_r = [y_1, \ldots, y_r]^T$. In order to implement the least squares method we have to assume that $r > m_c$ and taking this assumption into account the closed form solution of Eq.(4.21) minimising the square of the observation error has the following form:

$$\hat{\boldsymbol{x}}_c = \left(\boldsymbol{h}_r^T \boldsymbol{h}_r \right)^{-1} \boldsymbol{h}_r^T \boldsymbol{y}_r . \tag{4.22}$$

As before, matrix $\boldsymbol{P}_r = \boldsymbol{h}_r^T \boldsymbol{h}_r$ has to be nonsingular and this assumption is satisfied when the input trajectory in terms of joint positions and velocities is persistently exciting. The recurrence form of solution (4.22) is written as

$$\hat{\boldsymbol{x}}_c(L+1) = \hat{\boldsymbol{x}}_c(L) + \boldsymbol{P}_{L+1} \boldsymbol{d}_L (y_L - \boldsymbol{d}_L^T \hat{\boldsymbol{x}}_c(L)) \tag{4.23}$$

and

$$\boldsymbol{P}_{L+1} = \boldsymbol{P}_L - (1 + \boldsymbol{d}_L^T \boldsymbol{P}_L \boldsymbol{d}_L)^{-1} \boldsymbol{P}_L \boldsymbol{d}_L \boldsymbol{d}_L^T \boldsymbol{P}_L . \tag{4.24}$$

In order to guarantee the positive definiteness of the $\boldsymbol{P}_L$ matrix during the recurrence we can also apply the Agee–Turner factorisation. The problem of finding the exciting trajectory is discussed in Chapter 5. All comments presented in the previous section also apply in the case of the integral model.

Now we present an identification scheme for the stochastic case [59, 60, 123]. We assume that the joint positions, velocities, and generalised forces are corrupted by additive white noise with known characteristics. As before we assume the following notation $\tilde{\boldsymbol{q}} = \boldsymbol{q} + \boldsymbol{v}_p$, $\tilde{\dot{\boldsymbol{q}}} = \dot{\boldsymbol{q}} + \boldsymbol{v}_s$, and $\tilde{\boldsymbol{\tau}} = \boldsymbol{\tau} + \boldsymbol{v}_t$, in which $\boldsymbol{v}_p$, $\boldsymbol{v}_s$, and $\boldsymbol{v}_t$ denote stochastic processes. Therefore equation (4.19) for the canonical integral model can be rewritten in the following form:

$$y = \boldsymbol{d}^T \boldsymbol{x}_c + v_m, \tag{4.25}$$

where v_m is the modelling error. Integrating the product $\tilde{\boldsymbol{\tau}}^T \tilde{\dot{\boldsymbol{q}}}$ results in

$$\tilde{y} = \int_{t_1}^{t_2} \tilde{\boldsymbol{\tau}}^T \tilde{\dot{\boldsymbol{q}}} dt \tag{4.26}$$

and

$$\tilde{y} = y + v_i \ . \tag{4.27}$$

Substitution of Eq.(4.25) into Eq.(4.27) results in

$$\tilde{y} = \boldsymbol{d}^T \boldsymbol{x}_c + v_m + v_i = \tilde{\boldsymbol{d}}^T \boldsymbol{x}_c + \boldsymbol{d}^T \boldsymbol{x}_c - \tilde{\boldsymbol{d}}^T \boldsymbol{x}_c + v_m + v_i \ . \tag{4.28}$$

and can be rewritten as

$$\tilde{y} = \tilde{\boldsymbol{d}}^T \boldsymbol{x}_c + v_e \ , \tag{4.29}$$

where

$$v_e = \boldsymbol{d}^T \boldsymbol{x}_c - \tilde{\boldsymbol{d}}^T \boldsymbol{x}_c + v_m + v_i \ . \tag{4.30}$$

Note that v_e is a stochastic process which depends on stochastic processes v_p, v_s, v_m, and v_i, but its explicit form is not known. In addition stochastic processes v_e and $\boldsymbol{d}_k$ are correlated.

Analogously to Eq.(4.21) we can write

$$\bar{\boldsymbol{y}}_r = \bar{\boldsymbol{H}}_r \boldsymbol{x}_c + \bar{\boldsymbol{w}} \ . \tag{4.31}$$

In the last equation $\bar{\boldsymbol{y}}_r$ and $\bar{\boldsymbol{w}}$ are realisations of random vectors $\tilde{\boldsymbol{y}}_r$ and $\tilde{\boldsymbol{w}}$, while $\bar{\boldsymbol{H}}_r$ is a realisation of random matrix $\tilde{\boldsymbol{H}}_r$. Note that matrix $\tilde{\boldsymbol{H}}_r$ and vector $\tilde{\boldsymbol{w}}$ are correlated due to the fact that matrix $\tilde{\boldsymbol{H}}_r$ is random, but we do not know its explicit form in terms of the corrupted stochastic processes. As before, we cannot use the standard Markov estimation theory. We propose using the least squares method, which is formulated as follows:

$$\chi_{cr} = \min_{\boldsymbol{x}_c} \left[(\bar{\boldsymbol{y}}_r - \bar{\boldsymbol{H}}_r \boldsymbol{x}_c)^T (\bar{\boldsymbol{y}}_r - \bar{\boldsymbol{H}}_r \boldsymbol{x}_c) \right] \ . \tag{4.32}$$

This means that we are looking for an estimate $\hat{\boldsymbol{x}}$ of the vector $\boldsymbol{x}_c$, the realisation of which is denoted by χ_{cr} and the estimate is biased with the bias calculated from the following equation:

$$E(\boldsymbol{x}_c - \hat{\boldsymbol{x}}_c) = E\left[(\tilde{\boldsymbol{H}}_r^T \tilde{\boldsymbol{H}}_r)^{-1} \tilde{\boldsymbol{H}}_r^T \tilde{\boldsymbol{w}} \right] \ . \tag{4.33}$$

Since $\tilde{\boldsymbol{w}}$ and $\tilde{\boldsymbol{H}}_r$ are correlated the estimate is biased. It means that the expected value of the estimate (which is a vector in this case) is not exactly the same as the estimated vector of the dynamic parameters. This bias is mainly due to the presence of the additive corrupted signals in the nonlinear expressions $\boldsymbol{d}_k$.

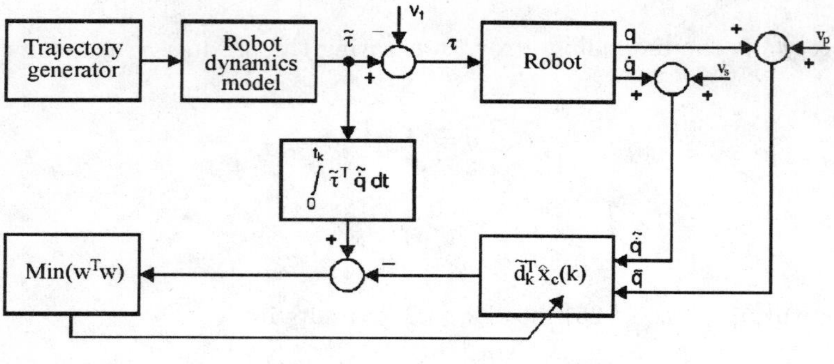

Fig. 4.3. The identification scheme for the integral model

Fig. 4.3 illustrates an identification scheme for the integral model which can be implemented in the real environment. The scheme from Fig.4.3 incorporates all the noisy signals discussed here. An identification scheme which is useful for the purpose of simulation is presented in Fig.4.4. Here we assume that the parameters of the dynamic model are known from CAD modelling. Note that in both schemes we need the robot dynamics equations (namely the differential model). As proved in Section 3.5 both the integral and the differential models are described in terms of the same set of dynamic parameters. The problem of finding the optimal trajectory is nontrivial and is discussed in Chapter 5.

Here we make one comment. In both Figures 4.3 and 4.4 the integral has been calculated over the range of t_{1k}, t_{2k}, and we have assumed equidistant time intervals for each k. In this case we are dealing with a so-called "short integral". If we fix $t_1 = t_0$ as the starting point and t_2 is the total time of the calculations we refer to this integral as a "long integral". Both integrals are

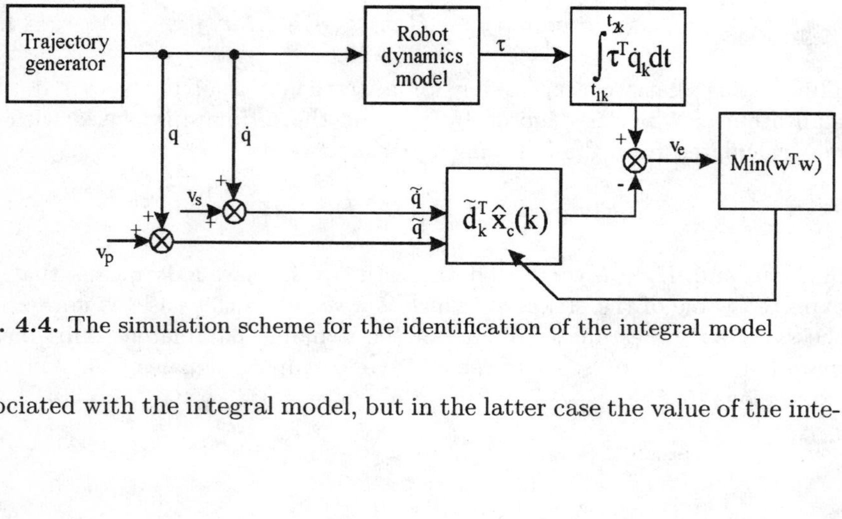

Fig. 4.4. The simulation scheme for the identification of the integral model

associated with the integral model, but in the latter case the value of the integral is accumulated over the whole period of the calculations. Both integrals were implemented in the identification scheme. In references [124, 183, 185] experimental results using these two integrals are compared. The "short integral" gave good results for fast moving robots, for example, for direct drive robots. The "long integral" gave good results for robots which move relatively slowly, as in case of the IRp-6 robot. This situation is discussed in Chapter 5.

4.4 Further comments on identification techniques used for estimation of robot dynamic parameters

This section is dedicated to identification techniques which are different from those presented in the previous sections. The most conceptually interesting technique is the spectral analysis which essentially uses the least squares method. The idea of using spectral analysis to mechanical system is well–known. A robot, as a mechanical system, behaves like a low pass filter in the frequency domain. The gross frequencies are low frequencies mostly due to slow movement of an industrial robot. This is not necessarily true for fast mechanical systems, for example, direct drive robots. Regardless of the type of robot, the spectral analysis can be used.

The suggestion using spectral analysis for the purpose of identification was introduced by Schiehlen and Kallenbach [192]. They reported that the Fast Fourier Transform (FFT) is very suitable for linear systems. In the case of nonlinear systems this method shows however some serious restrictions. They implemented the Fast Fourier Transform to ideal ordinary linear time–invariant multibody systems. This idea was later developed by Vandanjon, and co–authors [222] and successfully applied to the identification of robot dynamic parameters. Since this idea seems to be very interesting we outline it here. Conceptually it consists of several movements of the joints of the manipulator. These movements are periodic trajectories between two points A and B in joint space at frequency f_p. This kind of movement does not apply to the test trajectory which identifies the gravity coefficients. Vandanjon and co-authors [222] propose movements of one or two joints to identify the dynamic parameters of the robots. Before describing this method we consider the following definition.

The spectrum of an arbitrary matrix $\boldsymbol{M}$ of dimensions $m \times n$, normalised by an $m \times 1$ vector $\boldsymbol{a}$ is a $1 \times m$ vector given by

$$Sp(\boldsymbol{M}, \boldsymbol{a}, f) = (\lambda_1, \ldots, \lambda_n) \tag{4.34}$$

with

$$\lambda_k = |M_k(f)| \cos\left(\arg M_k(f) - \arg a(f)\right) / |a(f)| \, , \tag{4.35}$$

where $\boldsymbol{M}_k$ is the k-th column of $\boldsymbol{M}$. $M_k(f)$ and $a(f)$ are the values of the FFT of $\boldsymbol{M}_k$ and $\boldsymbol{a}$ at frequency f, respectively. In this definition we consider each column $\boldsymbol{M}_k$ of matrix $\boldsymbol{M}$ as data which are subject to FFT. At the same time we normalise each column by the value of the FFT of an arbitrary discrete signal consisting of the data forming vector $\boldsymbol{a}$. In Eqs.(4.34) and (4.35) a term $2\pi kf$ does not appear explicitly due to the normalisation of each column of matrix $\boldsymbol{M}$. Both angle arguments depend on a particular frequency. Equations (4.34) and (4.35) are valid for arbitrary frequency f which results from the Fast Fourier Transform.

Now following the idea presented by Vandanjon and co–authors [222] we present four different movements of the manipulator to identify the dynamic parameters of the differential model given by Eq.(4.1).

The first type of movement requires a periodic trajectory of a single joint k whose dynamic model reduces to

$$A_{kk}(q_{k+1},\ldots,q_k)\ddot{q}_k + \frac{\partial E_{pc}}{\partial q_k} + \tau_{kfv}(\dot{q}_k) + \tau_{kfs}(\dot{q}_k) = \tau_k \ , \tag{4.36}$$

where $A_{kk}(q_{k+1},\ldots,q_k)$ denotes a constant momentum as seen from the k-th joint axis of the body consisting of the links numbered $k, k+1, \ldots, n$, $\frac{\partial E_{pc}}{\partial q_k}$ denotes the time derivative of the total potential energy with respect to the generalised coordinate q_k. Finally, τ_{kfv} and τ_{kfs} are the friction torque components due to the velocity and Coulomb friction behaviour. Points A and B are chosen such that joint k achieves the maximum acceleration, in order to increase inertia torque $A_{kk}(\cdot)\ddot{q}_k$, and low velocity, in order to decrease the effect of friction torques τ_{kfv} and τ_{kfs}. Here slow velocity or small velocity is understood as the smallest velocity such there is no friction stick slip phenomenon in the friction model, and the average friction model given by Eqs.(3.29) and (3.30) keeps a physical meaning.

Note also that friction effects in τ_k are decoupled in frequency domain because there is a $\pi/2$ phase shift between fundamental components of velocity and acceleration. Therefore after applying the FFT to Eq.(4.36), friction effects can be eliminated. This is a very nice property of the FFT which cannot be observed in the time domain. We assume that the trajectory is periodic with frequency f_p. In order to apply the FFT we sample Eq.(4.36) at m discrete equidistant points which results in

$$\boldsymbol{Y}_k(\tau_k) = \boldsymbol{D}_k(q_k, \dot{q}_k, \ddot{q}_k)\boldsymbol{x}_c + \boldsymbol{Y}_{kv}(\tau_{kfv}) + \boldsymbol{Y}_{ks}(\tau_{kfs}). \tag{4.37}$$

In this equation $\boldsymbol{Y}_k$ is an $m \times 1$ vector of discrete samples consisting of m discrete values of the generalised torque τ_k, $\boldsymbol{D}_k(q_k, \dot{q}_k, \ddot{q}_k)$ is $m \times m_c$ matrix consisting of m samples of the k-th row of the canonical model of the robot. Note that we prefer using the canonical model in order to satisfy the necessary condition of the identification. Now we apply the FFT to both sides of Eq.(4.37). Using the definition (4.34) we can write

$$Sp(\boldsymbol{D}_k, \ddot{q}_k, f_p)\boldsymbol{x}_c = Sp(\boldsymbol{Y}_k(\tau_k), \ddot{q}_k, f_p). \tag{4.38}$$

Positioning the manipulator at different locked configurations we can obtain a number of equations (4.38) which are linear equations suitable for further least squares identification.

The second type of trajectory movement requires a periodic movement of two joints k and l where we assume that $l > k$. Under this particular conditions we can write the following equation for the k-th generalised torque:

$$\bar{C}_{lk}(q_l,\dots,q_n)\dot{q}_k^2 + A_{lk}(q_l,\dots,q_n)\ddot{q}_k + A_{ll}(q_{l+1},\dots,q_n)\ddot{q}_l + \\ + \frac{\partial E_{pc}}{\partial q_l} + \tau_{lfv} + \tau_{lfs} = \tau_l \ , \tag{4.39}$$

where $\bar{C}_{lk}(q_l,\dots,q_n)$ represents the centrifugal coupling term (see Eqs.(3.19) and (3.21)) and $A_{lk}(q_l,\dots,q_k)$ the inertial coupling term between both links which are moving. Here we write explicitly on which joint generalised coordinates, the elements $\bar{C}_{lk}$, A_{lk}, and A_{ll} depend, in order not to get confused with general equation (3.14). Points A and B are chosen in such a way that joint k achieves maximum velocity in order to increase the centrifugal torque $\bar{C}_{jk}\dot{q}_k^2$. Joint l executes a small movement at "small velocity" in order to decrease other components except gravity.

Next we observe that the centrifugal, inertial and friction effects, can be separated in the frequency domain of $\boldsymbol{Y}_l(\tau_l)$, because the spectrum of $\bar{C}_{lk}(q_l,\dots,q_k)\dot{q}_k^2$ is mainly located in the zero and $2f_p$ frequency while

$$A_{lk}(q_l,\dots,q_l)\ddot{q}_k + A_{ll}(q_{l+1},\dots,q_n)\ddot{q}_l + \tau_{lfv} + \tau_{lfs}$$

has zero component at frequency $2f_p$. Note that in the first and second movements proposed here, the friction gravity torques are not neglected, but in both cases their contribution is small. Now we sample Eq.(4.39) and apply the FFT which gives the following set of linear equation with respect to the vector of dynamic parameters $\boldsymbol{x}_c$:

$$Sp(\boldsymbol{D}_l, \dot{q}_k^2, 2f_p)\boldsymbol{x}_c = Sp(\boldsymbol{Y}_l(\tau_l), \dot{q}_k^2, 2f_p). \tag{4.40}$$

The third type of movement also requires a movement of two joints k and l (as before $l > k$). The reduced model is given by the same equation (4.39). In this movement points A and B are chosen in such a way that joint k achieves maximum acceleration, in order to increase the inertial coupling torque $A_{lk}(q_l,\dots,q_k)\ddot{q}_k$, and low velocity in order to decrease the centrifugal torque $\bar{C}_{lk}(q_l,\dots,q_n)\dot{q}_k^2$. Joint l must execute a small movement at "small velocity" to limit other components which appear in Eq.(4.39). Now we sample Eq.(4.39) and apply to the resulting equation the FFT which gives

$$Sp(\boldsymbol{D}_l, \ddot{q}_k, f_p)\boldsymbol{x}_c = Sp(\boldsymbol{Y}_l(\tau_l), \ddot{q}_k, f_p). \tag{4.41}$$

Note that in both equations (4.38) and (4.41) friction effects are eliminated due to the phase shift, but these equations include a small gravitational contribution. Analysis of gravity torques can be achieved by moving a single joint between two points A and B at a constant and small velocity. In this particular case we obtain the following equation of motion for joint k:

$$\frac{\partial E_{pc}}{\partial q_k} + \tau_{kfv} + \tau_{kfs} = \tau_k. \tag{4.42}$$

This movement has to be large (in space) in order to increase the gravity effect and must be executed at "small velocity" in order to decrease the friction effects. By sampling the above equation along the trajectory we get

$$\boldsymbol{Y}_k(\tau_k) = \boldsymbol{D}_k(q_k)\boldsymbol{x}_c + \boldsymbol{Y}_{kv}(\tau_{kfv}) + \boldsymbol{Y}_{ks}(\tau_{kfs}). \tag{4.43}$$

Friction can be eliminated by centring $\boldsymbol{Y}_k$ and $\boldsymbol{D}_k$. Let $\boldsymbol{D}_k^c$ and $\boldsymbol{Y}_k^c$ denote the centred observation matrix and torque, respectively. Then we can write

$$\boldsymbol{Y}_k^c = \boldsymbol{Y}_k - \bar{\boldsymbol{Y}}_k = (\boldsymbol{D}_k - \bar{\boldsymbol{D}}_k)\boldsymbol{x}_c = \boldsymbol{D}_k^c \boldsymbol{x}_c \ , \tag{4.44}$$

where $\bar{\boldsymbol{D}}_k$ and $\bar{\boldsymbol{Y}}_k$ are the average values of matrices $\boldsymbol{D}_k$ and $\boldsymbol{Y}_k$. By multiplying each term of Eq.(4.44) by $(\boldsymbol{D}_k^c)^T$ we get

$$\boldsymbol{F}^c \boldsymbol{x}_c = (\boldsymbol{D}_k^c)^T \boldsymbol{Y}_k^c \ \text{ with } \ \boldsymbol{F}^c = (\boldsymbol{D}_k^c)^T \boldsymbol{D}_k. \tag{4.45}$$

Note that this equation is suitable for further identification.

Generally, the idea to move one or two joints to follow a prescribed trajectory is not new. Usually in practice we cannot move all the joints due to the difficulties of generating sufficiently exciting trajectories. Mayeda and co–authors [158] introduce this idea for the time domain. Vandanjon and co–authors [222] propose solving the problem in the frequency domain. As can be seen, their method reduces the friction effects and consequently, the measurement noise in the dynamic model.

Note that each test movement results in a single equation (except for the gravity test). This scalar equation can be used to build the rows of a global regression matrix (information matrix). The number of experiments has to be greater than the number of unknown parameters m_c in the canonical model. The goal is to plan experiments so as to ensure the optimal condition number of the information matrix. Vandanjon and co–authors [222] suggest using the Frobenius condition number of the information matrix in order to monitor the above experiment. Optimised experiments are planned in order to obtain a low condition number of the regression matrix and to perform a global identification.

Due to the application of the FFT we only need to measure the position and torque signals. This is also one of the advantages of the proposed method of using the spectrum analysis.

Now we discuss a method closely related to the one presented above, called the covariance method, developed by Schiehlen and Kallenbach [192] and Kallenbach [95]. In order to illustrate this we have to rearrange Eq.(3.12) the differential model. First we rewrite this equation in the following form:

$$\boldsymbol{A}(\boldsymbol{q})\ddot{\boldsymbol{q}} + \boldsymbol{C}(\boldsymbol{q}, \dot{\boldsymbol{q}})\dot{\boldsymbol{q}} + \boldsymbol{G}(\boldsymbol{q}) + \boldsymbol{\tau}_{fv} + \boldsymbol{\tau}_{fs} = \boldsymbol{\tau} \ , \tag{4.46}$$

which can again be rewritten in a more general form as [192]

$$A(q,t)\ddot{q} + k(q,\dot{q},t) = \tau(q,\dot{q},u,t)\,. \tag{4.47}$$

Here vector k represents the vector of generalised gyroscopic forces, which include friction forces, and τ is the vector of generalised applied forces which depends nonlinearly on the position and velocity of the system and on the $m_1 \times 1$ vector of control variables u. The generalised position and velocity vectors have dimension $n \times 1$ as before. Note that the number of control variables does not necessarily have to be equal to n. The vector of generalised friction forces depends on the generalised positions and velocities, therefore it is included in vector $k(q,\dot{q},t)$ in general. The vector of generalised forces depends on control variables; see, for example, the dynamic model of the IRp-6 robot (Section 3.7.2), where the control variables are the DC currents of the motors which are transformed via nonlinear gear ratio functions. The next step is to parametrise Eq.(4.47) with respect to constant parameters such as the dynamic parameters of the links, the spring and damper coefficients and the coefficients in the nonlinear force laws, or their combinations, which results in an $m_c \times 1$ vector x_c of the parameters. As a result Eq.(4.47) can be rewritten in the following form (compare also Eq. (3.23)):

$$A(q,x_c,t)\ddot{q} + k(q,\dot{q},x_c,t) = \tau(q,\dot{q},u,t)\,, \tag{4.48}$$

where

$$A = \sum_{j=1}^{m_c} X_j A_j(q,t), \quad k = \sum_{j=1}^{m_c} X_j k_j(q,\dot{q},t)\,,$$

$$\text{and} \quad \tau = \sum_{j=1}^{m_c} X_j \tau_j(q,\dot{q},u,t) \tag{4.49}$$

are linear with respect to parameters X_j.

Schiehlen and Kallenbach [192] introduces a more general model of the robot dynamics, which is a result of the reformulation of Eqs.(4.48) and (4.49):

$$R(x_c,s,t)\dot{s} = f(x_c,s,u,t). \tag{4.50}$$

In the above s denotes the state vector, $s = \left[q^T, \dot{q}^T\right]^T$, which consists of generalised positions and velocities. As before, vector functions R and f can be parametrised as follows:

$$R = \sum_{j=1}^{m_c} X_j R_j(s,t), \quad \text{and} \quad f = \sum_{j=1}^{m_c} X_j f_j(s,u,t). \tag{4.51}$$

Direct state space representation is not valuable for identification because due to the nonlinear transformation of the model the structure of the system is lost and the transformed model contains the physical parameters explicitly (more clearly because we have to calculate the inverse of the mass matrix).

Therefore, the representation given by Eq.(4.50) is more suitable for identification as it preserves the linearity of the vector of dynamic parameters. Note that matrix $\boldsymbol{R}$ and vector $\boldsymbol{f}$ have the following form:

$$\boldsymbol{R} = \begin{bmatrix} I & 0 \\ 0 & \boldsymbol{A}(\boldsymbol{q}, \boldsymbol{x}_c, t) \end{bmatrix} \quad \text{and} \quad \boldsymbol{f} = \begin{bmatrix} \dot{\boldsymbol{q}} \\ \boldsymbol{\tau}(\boldsymbol{q}, \dot{\boldsymbol{q}}, \boldsymbol{u}, t) - \boldsymbol{k}(\boldsymbol{q}, \dot{\boldsymbol{q}}, \boldsymbol{x}_c, t) \end{bmatrix} . \tag{4.52}$$

In order to identify the vector of dynamic parameters Schiehlen and Kallenbach [192] proposed using the covariance method. This method is based on stationary ergodic coloured noise excitation $\boldsymbol{u}(t)$. For identification, the measured signals $\boldsymbol{u}(t)$ and $\boldsymbol{s}(t)$ are processed by a linear, stable filter of the form

$$\dot{\boldsymbol{y}} = \boldsymbol{F}\boldsymbol{y} + \boldsymbol{E}\boldsymbol{u} + \boldsymbol{H}\boldsymbol{s} \,, \tag{4.53}$$

where $\boldsymbol{y}(t)$ denotes the filter state. The system configuration is presented in Fig.4.5. Note that $\boldsymbol{y}(t)$ is the $l \times 1$ vector which is the output of the filter

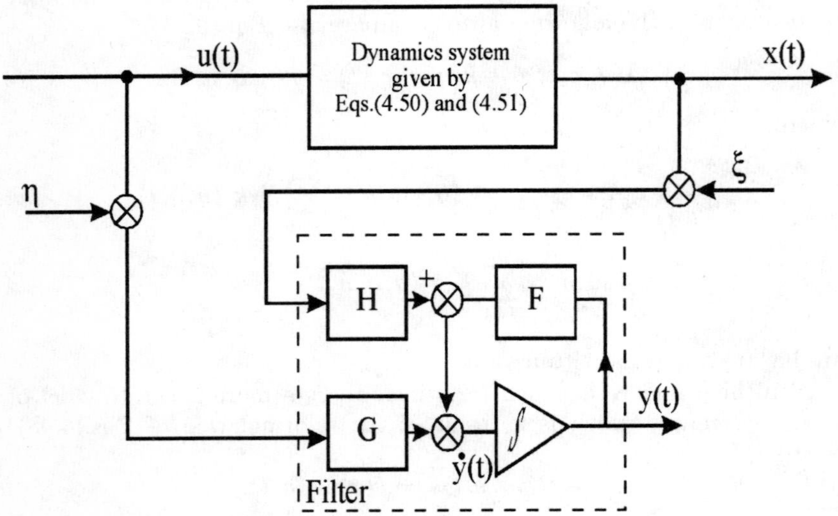

Fig. 4.5. System configuration for the covariance method

indicated in Fig.4.5. It is assumed that $\boldsymbol{s}(t)$, as well as

$$\bar{\boldsymbol{s}} = \boldsymbol{R}\boldsymbol{s}(t) = \sum_{j=1}^{m_c} X_j \bar{\bar{\boldsymbol{s}}}_j(\boldsymbol{s}, t) \tag{4.54}$$

and $\boldsymbol{y}(t)$ are stationary ergodic coloured stochastic processes, which are continuous and differentiable in quadratic mean. Measurement disturbances are taken into account by the $m_1 \times 1$ and $n_1 \times 1$ (here $n_1 = 2n$) processes $\boldsymbol{\eta}(t)$

and $\boldsymbol{\xi}(t)$, which are assumed to be stationary ergodic coloured random signals, mutually uncorrelated with the undisturbed signals. Consider now the following covariance matrix:

$$\boldsymbol{C}_{y\bar{s}}(t_1) = E\left[\boldsymbol{y}(t+t_1) - \boldsymbol{m}_y\right]\left[\bar{\boldsymbol{s}}(t) - \boldsymbol{m}_{\bar{s}}\right]^T , \tag{4.55}$$

where $\boldsymbol{m}_y$ and $\boldsymbol{m}_{\bar{s}}$ denote the constant mean values of the processes $\boldsymbol{y}(t)$ and $\bar{\boldsymbol{s}}(t)$. Note that matrix $\boldsymbol{C}_{y\bar{s}}$ has dimension $l \times n_1$. Now taking into account the stationarity condition we can write

$$\frac{d}{dt}\boldsymbol{C}_{y\bar{s}}(t_1) = \boldsymbol{0}_{l\times n_1} \tag{4.56}$$

where $\boldsymbol{0}_{l\times n_1}$ denotes the zero matrix of dimensions $l\times n_1$. Employing Eq.(4.55) and calculating the stationarity condition [175] we can easily find the following stationary covariance relation:

$$\boldsymbol{C}_{yr}(t_1) + \boldsymbol{C}_{yf}(t_1) + \boldsymbol{F}\boldsymbol{C}_{y\bar{s}}(t_1) + \boldsymbol{E}\boldsymbol{C}_{u\bar{s}}(\tau_1) + \boldsymbol{H}_{s\bar{s}}(t_1) = \boldsymbol{0}_{l\times n_1} , \tag{4.57}$$

where

$$\boldsymbol{r} = \dot{\boldsymbol{R}}\boldsymbol{s}(t) = \sum_{j=1}^{m_c} X_j \bar{\bar{\boldsymbol{r}}}_j(\boldsymbol{s}, t) \tag{4.58}$$

and

$$\dot{\boldsymbol{R}} = \sum_{j=1}^{m_c} X_j \dot{\boldsymbol{R}}_j(\boldsymbol{x}_c, \boldsymbol{s}, t). \tag{4.59}$$

Substituting the linear parametrisation given by Eqs.(4.51), (4.54), (4.58), and (4.59) into Eq.(4.57) we obtain

$$\sum_{j=1}^{m_c} X_j \left[\boldsymbol{C}_{y\bar{r}_j}(t_1) + \boldsymbol{C}_{yf_j}(t_1) + \boldsymbol{F}\boldsymbol{C}_{y\bar{s}_j}(t_1) + \right.$$
$$\left. + \boldsymbol{E}\boldsymbol{C}_{u\bar{s}_j}(t_1) + \boldsymbol{H}\boldsymbol{C}_{s\bar{s}_j}(t_1)\right] = \boldsymbol{0}_{l\times n_1} . \tag{4.60}$$

Equation (4.60) is the fundamental equation of the covariance method. Generally all the covariance matrices in Eq.(4.60) can be estimated from sampled measurement of $\boldsymbol{u}(t)$ and $\boldsymbol{s}(t)$. We sample Eq.(4.60) at equidistant points t_{1i} in order to obtain an overdetermined set of equations which can be solved in the sense of least squares. Note that the covariance method requires knowing state measurements $\boldsymbol{s}(t)$ as well as the control variables. This method can be interpreted as a variant of the instrumental variable method [152] for the differential model. This method is insensitive to measurement noise and can handle unmeasured states.

In Sections 4.2 and 4.3 we presented the batch least squares estimation method for both differential and integral models. At the end of this section we update this information and make several comments on the sequential identification algorithm and the least mean squared algorithm. A good beginning is to take Eq.(3.95) which describes the canonical differential model. This model is described in terms of vector $\boldsymbol{x}_c$ which consists of vectors $\boldsymbol{x}_{1c}, \boldsymbol{x}_{2c}, \ldots, \boldsymbol{x}_{nc}$ (see Eq. (3.95)). Recall that each vector $\boldsymbol{x}_{ic}$ has a dimension less than or equal to the dimension of the vector consisting of the dynamic parameters of link i plus the Coulomb and viscous friction coefficients (namely vector $\boldsymbol{x'}_i$ in Eq.(3.92). Therefore its dimension is less than or equal to 12. It is clear from Eq.(3.95) that we can take the i-th equation from the differential model and use it for the purpose of identification. Nevertheless in this case we are not able to identify the dynamic parameters described by vectors $\boldsymbol{x}_{1c}, \boldsymbol{x}_{2c}, \ldots, \boldsymbol{x}_{(i-1)c}$. Therefore, in general, we cannot identify a set of canonical parameters from only one equation from the set of equations given by (3.95). In order to describe this problem we introduce the following notation:

$$\boldsymbol{x}_c = \hat{\boldsymbol{x}}_c + \tilde{\boldsymbol{x}}_c \ , \tag{4.61}$$

where $\hat{\boldsymbol{x}}_c$ is the estimate of the parameter vector $\boldsymbol{x}_c$ and $\tilde{\boldsymbol{x}}_c$ is the misadjustment vector. Now denote by $\boldsymbol{\Phi}_i^T = \left[\boldsymbol{\Phi}_{i1}^T, \boldsymbol{\Phi}_{i2}^T, \ldots, \boldsymbol{\Phi}_{in}^T\right]$ the concatenate vector of the $\boldsymbol{\Phi}_{if}^T$'s. The parameter update law, which is based on the i-th equation from the set of equations given by (3.95), can be written in the following discrete form (here, as in Sections 4.2 and 4.3, we have omitted the position, velocity, and acceleration signals as formal arguments of $\boldsymbol{\Phi}_{ij}^T$'s):

$$\hat{\boldsymbol{x}}_c(k) = \hat{\boldsymbol{x}}_c(k-1) + \boldsymbol{\Gamma}_i \boldsymbol{\Phi}_i(k) e_i(k) \ , \tag{4.62}$$

where $e_i(k)$ is the model error given by

$$e_i(k) = \tau_i(k) - \boldsymbol{\Phi}_i^T(k)\hat{\boldsymbol{x}}_c(k-1) = \hat{\boldsymbol{\Phi}}_i^T(k)\tilde{\boldsymbol{x}}_c(k-1) + \tilde{\boldsymbol{\Phi}}_i^T(k)\boldsymbol{x}_c(k-1). \tag{4.63}$$

In Eq.(4.62), $\boldsymbol{\Gamma}_i$ is an update gain matrix (also called the adaptation gain matrix). In the least mean squared algorithm this matrix is arbitrary and usually diagonal [30]. The $\boldsymbol{\Gamma}_i$ matrix is subscripted because it is possible that different gains can be applied to the data from different joints, especially if the noise of $||\boldsymbol{\Phi}_i||$ varies markedly between joints. This matrix is time–varying if the least squares algorithm is used and it changes with discretisation step k according to Eq.(4.10). In Eq.(4.63) it is assumed that the regression vector $\boldsymbol{\Phi}_i(k)$ consists of two components (here in order to be consistent in notation used in Eq.(4.61)we use slightly different notation for the measurement noise from that in Section 4.2):

$$\boldsymbol{\Phi}_i(k) = \hat{\boldsymbol{\Phi}}_i(k) + \tilde{\boldsymbol{\Phi}}_i(k) \ , \tag{4.64}$$

where $\hat{\boldsymbol{\Phi}}_i(k)$ is the estimated regression vector and $\tilde{\boldsymbol{\Phi}}(k)$ is the measurement noise, and is zero mean and nicely distributed. Remember that the regression vector is a function of joint positions, velocities, and accelerations, which are noisy signals. Usually, $\tilde{\boldsymbol{\Phi}}_i(k)$ is not, in general, nicely distributed. Using the explicit dynamic model of the robot the standard deviation of $\tilde{\boldsymbol{\Phi}}_i(k)$ can easily be calculated.

A logical consequence of the identification scheme presented above is to combine data from several joints to produce an estimate of $\hat{\boldsymbol{x}}_c$. This idea was suggested by Armstrong [4, 5, 6]. In this case Eq.(4.62) appears as follows:

$$\hat{\boldsymbol{x}}_c(k) = \hat{\boldsymbol{x}}_c(k-1) + \sum_{i=1}^{n} \boldsymbol{\Gamma}_i \boldsymbol{\Phi}_i(k) e_i(k) \ , \tag{4.65}$$

where $e_i(k)$ is given by Eq.(4.63). Note that the adaptation gain is different for each regression vector. In Eq.(4.65) we summarised the data from different links, which is equivalent to data averaging. Now using Eqs.(3.95) and (4.61) the misadjustment vector can easily be calculated from the following equation:

$$\tilde{\boldsymbol{x}}_c(k) = \left(\boldsymbol{I} - \sum_{i=1}^{n} \boldsymbol{\Gamma}_i \boldsymbol{\Phi}_i(k) \boldsymbol{\Phi}_i^T(k) \right) \tilde{\boldsymbol{x}}_c(k-1) \ , \tag{4.66}$$

where $\boldsymbol{I}$ is an identity matrix. The estimate $\hat{\boldsymbol{x}}$ to converge to the true vector of parameters $\boldsymbol{x}_c$ (that is $\tilde{\boldsymbol{x}}_c \to \boldsymbol{0}$) if the matrix $\sum_{i=1}^{n} \boldsymbol{\Gamma}_i \boldsymbol{\Phi}_i(k) \boldsymbol{\Phi}_i^T(k)$ is positive definite over each sufficiently long portion of the trajectory, which leads to the customary persistently exciting condition [30]:

$$\exists \alpha, \beta > 0 \ \text{such that} \ \forall l \ \ \alpha \boldsymbol{I} \le \sum_{i=1}^{n} \sum_{k=l}^{\rho} \boldsymbol{\Gamma}_i \boldsymbol{\Phi}_i(k) \boldsymbol{\Phi}_i^T(k) \le \beta \boldsymbol{I}, \ \ \rho > m_c. \tag{4.67}$$

In this equation the requirement for sufficient excitation is stronger, α must not only be positive, it must also be reasonably large. The matrix in the middle of the above inequality is denoted by $\boldsymbol{P}$ and is called the persistent excitation matrix.

Canudas de Wit and Aubin [26] propose an interesting identification scheme based on the triangular structure of the matrix given by Eq.(3.95). Now we exploit the structure of the upper triangular matrix in Eq.(3.95). Note that link n depends on parameter sub–vector $\boldsymbol{x}_{nc}$. The dynamics of link $n-1$ depends only on vectors $\boldsymbol{x}_{(n-1)c}$ and $\boldsymbol{x}_{nc}$. Similarly, link i depends on sub–vectors $\boldsymbol{x}_{ic}, \boldsymbol{x}_{(i+1)c}, \ldots, \boldsymbol{x}_{nc}$. This suggests identifying sub–vector $\boldsymbol{x}_{nc}$ during the first estimation pass, and then using it in the second pass as a known vector in order to estimate $\boldsymbol{x}_{(n-1)c}$. Successive repetition of this procedure yields the remaining vector subsets. According to [26] the i-th parameter vector is estimated as follows:

$$\hat{\boldsymbol{x}}_{ic}(k) = \hat{\boldsymbol{x}}_{ic}(k-1) + \boldsymbol{P}_i \boldsymbol{\Phi}_{ii}(k) e_i(k) \ \ , \forall i = n, n-1, \ldots, 1 \tag{4.68}$$

and

$$e_i(k) = \left(\tau_i(k) - \sum_{j>i}^{n} \boldsymbol{\Phi}_{ij}^T \boldsymbol{x}_{jc}^0 \right) - \boldsymbol{\Phi}_{ii}^T \hat{\boldsymbol{x}}_{ic}(k-1) \ , \tag{4.69}$$

where $\boldsymbol{x}_j^0$ represents the estimate of $\boldsymbol{x}_j$ previously obtained in the pass corresponding to link j for all $j > i$, and $\boldsymbol{P}_i$ is a constant matrix chosen by the engineer (the $\boldsymbol{P}_i$ matrix plays the same role as the gain adaptation matrix $\boldsymbol{\Gamma}_i$ in Eq.(4.62) and due to the slightly different structure of the update Eq.(4.68) we use this new notation). This matrix is associated with the regression factor $\boldsymbol{\Phi}_{ii}(k)$. Here we use the discrete–time formulation. Note that estimation is performed in the upwards direction: starting with link n which is the one located furthest from the base. Assuming that $\hat{\boldsymbol{x}}_{jc}^0$ tends to $\boldsymbol{x}_{jc}$, the error equation for each link is given by

$$\tilde{\boldsymbol{x}}_{ic}(k) = \left(\boldsymbol{I} - \boldsymbol{P}_i \boldsymbol{\Phi}_{ii}(k)\boldsymbol{\Phi}_{ii}^T(k)\right) \tilde{\boldsymbol{x}}_{ic}(k-1) \, . \tag{4.70}$$

The customary persistently exciting condition in this particular case can be written in the following form

$$\exists \alpha_i, \beta_i > 0 \text{ such that } \forall l \; \alpha_i \boldsymbol{I} \leq \boldsymbol{P}_i \sum_{k=l}^{\rho_i} \boldsymbol{\Phi}_{ii}(k)\boldsymbol{\Phi}_{ii}^T(k) \leq \beta_i \boldsymbol{I}, \tag{4.71}$$

where ρ_i is greater than or equal to the number of parameters in vector $\boldsymbol{x}_{ic}$. Furthermore, due to the fact that $\boldsymbol{\Phi}_{ii}(k)$ depends on the data for the current and proceeding links (it is a function of the joint generalised coordinates $q_1, q_2, \ldots, q_i$ and their first and second derivatives) the estimation of $\boldsymbol{x}_{ic}$ can be performed even if links $i+1, i+2, \ldots, n$ are fixed (no dynamic movements are needed for these links during the estimation).

Canudas de Wit and Aubin [26] formulated on–line and off-line identification schemes for line–by–line estimation. In order to do so we write Eq.(4.69) in a slightly different form,

$$e_i(k) = y_i(k) - \boldsymbol{\Phi}_{ii}^T(k)\hat{\boldsymbol{x}}_{ic}(k-1) \, , \tag{4.72}$$

where

$$y_i(k) = \tau_i(k) - \sum_{j>i}^{n} \boldsymbol{\Phi}_{ij}^T(k)\boldsymbol{x}_{jc}^0. \tag{4.73}$$

In order to avoid measuring the accelerations we can filter the two equations as it was suggested, for example, in [161]. In order to obtain the on-line algorithm formulation, we have to substitute $\boldsymbol{x}_{jc}^0$ by $\hat{\boldsymbol{x}}_{jc}(k)$ in the above equations. Substituting of Eq.(3.95) into (4.73) gives

$$y_i(k) = \boldsymbol{\Phi}_{ii}^T(k)\boldsymbol{x}_{ic} + \varepsilon_i(k) \, , \tag{4.74}$$

where

$$\varepsilon_i(k) = \sum_{j>i}^{n} \boldsymbol{\Phi}_{ij}^T(k)\tilde{\boldsymbol{x}}_{jc}(k) \text{ and } \tilde{\boldsymbol{x}}_{jc}(k) = \boldsymbol{x}_{jc} - \hat{\boldsymbol{x}}_{jc}(k). \tag{4.75}$$

This equation at sample point k, is an unknown value. Taking into account Eq.(4.74) the problem is to estimate $\boldsymbol{x}_{jc}$ in the presence of $\varepsilon_i(k)$. Note that $\varepsilon_i(k)$ can be interpreted as a deterministic noise disturbing the model parametrisation $\boldsymbol{\Phi}_{ii}^T(k)\boldsymbol{x}_{ic}$. We rewrite the error equation (4.72) as follows:

$$e_i(k) = y_i(k) - y_i^n(k) , \tag{4.76}$$

where

$$y_i^m(k) = \boldsymbol{\Phi}_{ii}^T(k)\hat{\boldsymbol{x}}_{ic}(k-1) . \tag{4.77}$$

The objective is to find estimate $\hat{\boldsymbol{x}}_{ic}$ belonging to the set

$$\boldsymbol{\mathcal{M}}(\hat{\boldsymbol{x}}_{ic}) = \left\{ \hat{\boldsymbol{x}}_{ic} : (y_i(k) - y_i^m(k))^2 \leq \sigma_i^2(k) \right\} , \tag{4.78}$$

where $\sigma_i(k)$ defines a known upper bound on $\varepsilon_i(k)$. In order to solve this problem Canudas de Wit and Aubin suggest using the modified exponentially–weighted recursive least squares algorithm (EW–RLS) [26]. The details of this algorithm can be found in reference [26] and are not quoted here because it is of the least squares type. The EW–RLS algorithm and the assumption that the sequence of $\boldsymbol{\Phi}_{ii}(k)$ for sufficient number of sample points satisfies persistently exciting condition guarantee that the following is valid:

$$\lim_{k\to\infty} \sigma_i(k) = 0, \quad \lim_{k\to\infty} \varepsilon_i(k) = 0, \quad \lim_{k\to\infty} e_i(k) = 0 \ \text{ and } \ \lim_{k\to\infty} \hat{\boldsymbol{x}}_{ick} = \boldsymbol{x}_{ic} . \tag{4.79}$$

The proof and all the details can be found in [26]. In the off-line version, the convergence of $\hat{\boldsymbol{x}}_{ic}(k)$ can be ensured by running the EW-RLS algorithm sequentially starting with $i = n$. However, the persistently exciting condition for each link is required. In addition, convergence of vector $\hat{\boldsymbol{x}}_{(i-1)c}(c)$ to $\boldsymbol{x}_{(i-1)c}$ can be assured if the EW-RLS algorithm is used after the proceeding link parameter vector estimate, $\hat{\boldsymbol{x}}_{ic}(k)$, has been converged. Reasoning in the same way for all the links, convergence of the complete parameter set can be established. Note that by using the above algorithm (in either the on–line or off–line version) the burden of the generation of optimal trajectories for identification can be considerable reduced by diminishing the dimension of the persistent excitation matrix (see Eq.(4.71)).

The idea of using sequential estimation was introduced by Mayeda, Osuka, and Kangawa [158]. They applied this idea to the differential model which was not necessarily canonical. Besides that, they did not prove convergence of the set of all the dynamic parameters estimates. Their approach was to move one or two joints in a systematic way such that it was possible to identify all the parameters of the robot. In their identification schemes some of the previously identified estimates were used for subsequent calculations. They also notice that this method is quite effective from the computational point of view, but unfortunately, the results of the estimates are biased. Using the line–by–line estimation scheme proposed by Canudas de Wit and Aubin [26], the noise bias in the estimates is attenuated. It is one of the advantages of the EW-RLS algorithm.

From the considerations presented in this chapter, it is clear that the least squares method is beyond doubt the most popular in the identification of robot dynamics. Different formulations exist for robot dynamics in time and frequency domains. In all formulations robot dynamic parameters appear in a linear form. In a discrete formulation one has to sample equations of motion at a sufficient number of points and then the resulting equations are processed by the last squares algorithm. Different versions of the least squares algorithm can be applied, but this method seems to be the most widely used. It is also clear from the above considerations that part of the story in identification of robot dynamics lies in the planning phase of the experiment. This is evident from Eq.(3.95), written in time domain, which seems to be the most suitable one for planning the identification experiment. This model is canonical and this assumption is a necessary condition for successful identification. It follows from this equation that it is not necessary to move all the joints in order to identify the dynamic parameters. In practice we move one or two joints of the mechanical structure. In order to identify the friction coefficients we only move one joint at a time. It is also important to have an initial knowledge of the mechanical structure. We use physical reasoning to find out which inertias, Coriolis and centrifugal coefficients are dominant and have the greatest impact on the dynamic equations of motion. Then we select the appropriate links and move them in order to estimate the coefficients. Some of the identified coefficients can be used as known parameters in subsequent equations, which are selected for the purpose of identification of other parameters. We believe that intuition in the identification of mechanical structures is very important and in many cases enables successful estimation. As mentioned above, the simplest least squares method can be used in all cases.

A different problem is one of choosing a proper trajectory in the identification procedure. From the above discussion we know that in all cases an input trajectory has to satisfy persistently exciting conditions. In order to identify the dynamic parameters we have to excite the functions which are associated with them. If the level of excitation is too low, is it not possible to identify the corresponding dynamic parameter. Moreover, if we have many parameters in the equations of motion (which is usually the case), we cannot excite only one parameter forgetting about the others because they remain unidentified. This observation is crucial for the planning phase of the experiment. The conclusion is that the mathematical tools which are used for the identification of robot dynamics are rather simple, and intuition plays an important role in planning the experiment. Note also that from the mathematical point of view we choose simple identification techniques since in practice we often want to identify the parameters on-line. All these problems will be discussed in detail in the following chapters.

4.5 Simulation results

In this section we illustrate the identification schemes presented in sections 4.2 and 4.3 by simulation results. Consider the PUMA 560 robot [3]. This robot has 6 degrees of freedom, but for the purpose of simulation we assume that the wrist is an integral part of the third link. This assumption is reasonable since the dynamic interactions between the first three links are dominant when compared to the wrist. The first three links of the robot are presented in Fig.4.6. The coordinate frames are assigned according to the modified Denavit-Hartenberg notation. The kinematical parameters of the first three degrees of freedom are summarised in Table 4.1 The equation of motion

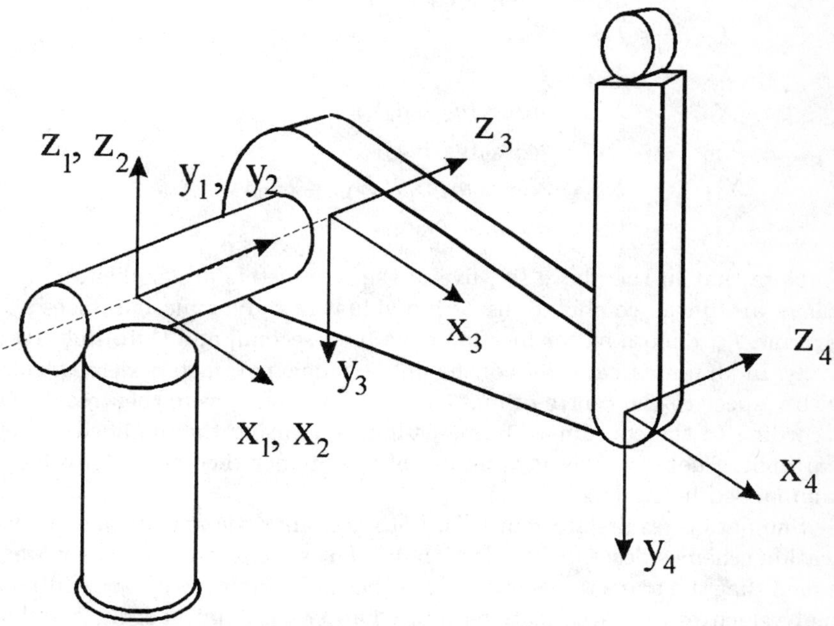

Fig. 4.6. The PUMA 560 robot and its kinematical structure

Table 4.1. Modified Denavit-Hartenberg parameters of the PUMA 560 robot

i	θ_i	d_i	a_i	α_i	σ_i
1	q_1	0	0	0	0
2	q_2	0	a_2	-90°	0
3	q_3	d_3	a_3	0	0

for the PUMA 560 can be found in [3]. Here we present only the dynamic parameters of the canonical model. The differential model is described in terms of 21 parameters and its canonical form contains only 15 parameters (except the friction parameters). They are listed below.

$$\begin{aligned}
X_1 &= I_{1a} + I_{1zz} + m_2 a_2^2 + 2a_2 m_2 c_{2z} + m_3(a_2 + a_3)^2 + \\
&+ 2m_3 c_{3z}(a_2 + a_3) + I_{2xx} + I_{2yy} \ , \\
X_2 &= I_{2xz} - d_3(a_2 + a_3)m_3 - c_2 m_2 c_{2x} - d_3 m_3 c_{3z} \\
X_3 &= I_{2yy} - I_{2xx} + d_3 m_3 \ , \\
X_4 &= I_{2yz} - a_2 m_2 c_{2y} \ , \\
X_5 &= I_{2zz} + I_{3zz} + I_{2a} + I_{3a} + d_3^2 m_3 \ , \\
X_6 &= m_2 c_{2x} + d_3 m_3 \ , \\
X_7 &= I_{3xx} - I_{3yy} \ , \\
X_8 &= I_{3zz} + I_{3a} \ , \\
X_9 &= I_{3xz} - m_3 c_{3x}(a_2 + a_3) \ , \\
X_{10} &= I_{3yz} - m_3 c_{3y}(a_2 + a_3) \ , \\
X_{11} &= I_{2xy}, \ \ X_{12} = m_2 c_{2y}, \ \ X_{13} = I_{3xy}, \\
X_{14} &= m_3 c_{3x}, \ \ X_{15} = m_3 c_{3y}.
\end{aligned}$$

Note that in the above set, five of the parameters are explicit, while the others are linear combinations of individual link dynamic parameters. I_{1a}, I_{2a}, and I_{3a} denote motor inertia of the first, second, and third link, respectively. In [3] numerical values of the link dynamic parameters were calculated with respect to the centre of mass of each link. They were then recalculated according to the coordinate frame assignment (using the modified Denevit–Hartenberg notation) by making use of the Steiner theorem. The values are summarised in Table 4.2.

Simulation results are shown for both deterministic and stochastic identification schemes described in Section 4.2. For simulation purposes it was assumed that the trajectories were fifth–order polynomials with zero initial and final velocities and accelerations (see Chapter 2). Joint coordinates change from their initial values of −2.79rad, −3.925rad, and −0.785rad to their final values of 2.79rad, 0.785rad, and 3.925rad, respectively. Each joint moves in the joint space within the time of 4s. This results in maximum joint velocities of 2.62rad/s for the first joint, and 2.21rad/s for the second and third joints. Maximum joint accelerations are 2.02rad/s for the first joint and 1.7rad/s^2 for the second and third joints. The differential model of the PUMA 560 was discretised at 85 discrete points, which resulted in total 285 equations for the least squares processing.

In Table 4.3 simulation results of the estimates for various cases, both deterministic and stochastic, are included. Based on numerical values from Table 4.2 values of the dynamic parameters are calculated (see column 2 in Table 4.3). Column 3 in Table 4.3 represents estimates which were obtained

Table 4.2. CAD modelled dynamic parameters of the PUMA 560

Parameters	Link 1	Link 2	Link 3
m [kg]	–	17.40	6.04
mc_x [kg · m]	–	1.1832	0.0
mc_y [kg · m]	–	0.1044	-0.864
mc_z [kg · m]	–	-0.2784	0.085
I_{xx} [kg · m^2]	–	0.14	0317
I_{xy} [kg · m^2]	–	0.007	0.0
I_{xz} [kg · m^2]	–	-0.019	0.0
I_{yy} [kg · m^2]	–	0.609	0.0169
I_{yz} [kg · m^2]	–	-0.0017	-0.012
I_{zz} [kg · m^2]	0.35	0.626	0.266

Table 4.3. Parameter estimates of the canonical model of the PUMA 560

Parameters	Estimates			
	Deterministic case	Join torque corrupted by Gaussian noise	D corrupted by Gaussian noise	All signals are noisy
$X_1 = 2.7045$	2.7109	2.7555	2.7108	2.7103
$X_2 = -0.7353$	-0.7387	-0.7253	-0.7387	-0.7279
$X_3 = 1.5952$	1.5830	1.5741	1.5831	1.5833
$X_4 = -0.0271$	-0.0270	-0.0356	-0.0270	-0.0271
$X_5 = 7.555$	7.4700	7.6745	7.4700	7.4786
$X_6 = 3.7912$	3.8000	3.7748	3.8040	3.8028
$X_7 = 0.3001$	0.3003	0.2975	0.3003	0.3004
$X_8 = 1.093$	1.0941	1.1006	1.0941	1.0939
$X_9 = 0.0$	0.0	0.0081	0.0	0.0
$X_{10} = 0.1177$	0.1205	0.1398	0.1200	0.1206
$X_{11} = 0.007$	0.007	-0.0019	0.007	0.007
$X_{12} = 0.1044$	0.1044	0.1113	0.1043	0.1044
$X_{13} = 0.0$	0.0	0.003	0.0	0.0
$X_{14} = 0.0$	0.0	0.0	0.0	0.0
$X_{15} = 0.864$	-0.864	-0.8572	-0.8645	-0.8644

for the differential model of the PUMA 560 by making use of recursive equations (4.10) and (4.11). This case is deterministic, namely we assume that all measurement signals are noise–free. The next column contains estimates for the case when joint moments are corrupted by additive Gaussian noise with mean value of 0.05 and standard deviation of 0.5. The next column represents estimates in situation where each element of the matrix $\boldsymbol{D}$ in Eq.(3.12) is corrupted by the same noise signal with identical characteristics. This case is equivalent to error modelling considered in Section 4.2. Finally the last column in Table 4.3 represents the results of the estimates when all measurement signals, namely the joint positions, velocities, accelerations, and torques are noisy. As before, the noise is Gaussian with zero mean value and standard deviation of 0.005. Note that in the latter case the numerical values of the estimates are very close to the precomputed values given in the second column of Table 4.3. Due to non–zero mean value of the additive noise signals in the third and fourth column, the numerical values of the estimates differ slightly from the given precomputed values and change in a wide range for each measurement signal. The noise characteristics were chosen from references [4, 5, 59]. In order to compare the results joint torques were calculated based on identified dynamic parameters. The predicted torques and the joint torques calculated based on precomputed parameters are identical. This applies to all the cases presented in Table 4.3. The values of the estimates for the deterministic case stabilise after 150 iterations (i.e. after 2.3 s) and their accuracy is about 0.25% in comparison with respect to the precomputed values. It can be observed that the implementation of the Agee–Turner factorisation guarantees numerical stability of the obtained results.

In order to illustrate the numerical results of the identification algorithm described in Section 4.3 several simulation experiments were carried out to show the accuracy and effectiveness of the least squares method for the integral model. We ran experiments for the EDDA robot [185] and the IRp-6 robot [231]. It should be mentioned that for the IRp-6 robot we also tested the identification algorithm for the differential model, with results of similar accuracy presented above. The simulation results for the EDDA robot (recall that this robot has only two degrees of freedom) were similar to those presented above. Nevertheless we encountered some difficulties in identifying of the friction coefficients. This is due mainly to the fact that the functions associated with the Coulomb and viscous friction coefficients are given by Eqs.(3.36) and (3.37). The corresponding functions which are integrated for joint i are $|\dot{q}_i|$ and $\dot{q}_i^2$. These functions, almost regardless of the trajectory shape, are prone to be linearly dependent when integrated over a long time. Therefore in this case we suggest using the short integral. It is generally quite difficult to identify the friction coefficients even in simulation. This was also observed by Prüfer [185]. One possible way to avoid this problem is to precompute the friction torque characteristics and subtract their values from the generalised torques and then identify only the dynamic parameters of the robot. This technique will be discussed in detail in Chapter 5. The estimates of the friction coefficients for the EDDA robot were generally better than those for the IRp-6 robot. This is also due to the joint velocities of the EDDA robot which are almost an order higher in comparison to those for the IRp-6 robot. Nevertheless, when using the short integral the friction coefficients identified for the EDDA robot have an accuracy comparable to that of the differential model presented above.

In order to illustrate the numerical results of the least squares scheme applied to the integral model we show the simulation results for the IRp-6 robot. A set of dynamic parameters for the first three degrees of freedom of this robot is given by Eq.(3.168). These values were estimated initially based on CAD modelling [73]. These values are given in the first column of Table 4.4. The second column in Table 4.4 collects results for the deterministic case.

Table 4.4. Estimates of the canonical model of the IRp-6 robot

Parameters	Deterministic case
$X_1 = 8.25$	8.2458
X_2 not identified	not identified
$X_3 = 26.99$	26.9712
$X_4 = 8.55$	8.5428
$X_5 = 4.02$	4.0285
$X_6 = -0.45$	-0.4428
$X_7 = 8.85$	8.8448

CAD modelling showed the that second parameter X_2 is very small [73]. In this case we suggest not to identify this parameter [110] because it would be poorly identified. The friction parameters $X_8 \div X_{13}$ were not identified either due to the fact that the functions which are associated with them were linearly dependent almost regardless of the type of trajectory used. We tried to identify these parameters but their values were not reproducible when different trajectories were chosen. At the same time, the rest of the dynamic parameters remained the same during these experiments. Therefore we have not presented the values of the friction parameters.

Now observe that the linear dependence of the functions associated with the friction coefficients can be treated in a sense as a local phenomenon which does not effect the rest of the functions associated with the dynamic parameters. This phenomenon was observed by plotting functions associated with different parameters and selecting from them those which were linearly dependent. Surprisingly, poor friction estimates have almost no effect on estimates of other dynamic parameters. Again we use intuition which, as already mentioned in the previous section, is very helpful in identification of mechanical systems.

Various joint trajectories were implemented in order to estimate the dynamic parameters of the IRp-6 robot. One for which the estimates are detailed in Table 4.4 consists of fifth–order polynomial trajectory for the first and third joints and a third–order polynomial trajectory for the second joint. Each movement was within maximal range of joint angles and duration 2s. During the movement, maximum acceleration for each joint was achieved.

Besides that we assumed different noise characteristics imposed on generalised positions, velocities, accelerations, and torques in order to test the identification scheme in the presence of noisy signals. The obtained results were of a similar nature to those presented in the case of the dynamic parameters identification for the differential model. These results are reported in [231]. The experimental results for both EDDA and IRp-6 robot will be discussed in Chapter 5 in order to give more insights of the identification of robot dynamics.

CHAPTER 5

EXPERIMENTAL IDENTIFICATION OF ROBOT DYNAMIC PARAMETERS

5.1 Introduction

In previous Chapters 4 and 5 we considered robot dynamic models and identification techniques which can be used to identify their dynamic parameters. Two classes of robots were considered: geared robots and direct drive robots. As an example of the former we described the differential and integral models of the IRp-6 robot (see Section 3.7.2). A representative robot of the latter was the Experimental Direct Drive Robot (EDDA). For this robot we also developed differential and integral models described in Section 3.7.1. Both formulations were described in terms of a minimum set of dynamic parameters which were linear combinations of individual link parameters (mass, centre of mass, and inertia tensor). We assumed the presence of the friction terms in the dynamic equations of motion, with the friction torque were given by Eq.(3.5). It was also proved that both differential and integral models were described in terms of the same set of the canonical parameters (see Section 3.5).

The experimental identification of the robot model dynamic parameters is a complex procedure, usually involving many dynamic parameters of the canonical model. Some of these parameters have a clear physical meaning while others do not. Some of the parameters have relatively large values while others have small values. This can cause problems if we want to identify all the parameters by inputting one trajectory. The particular trajectory has to be chosen carefully. The trajectory is generally expressed in terms of generalised positions, velocities, and accelerations for the differential model, and in terms of generalised positions and velocities for the integral model. The trajectory has to be smooth in order to avoid jerks in the manipulator movement. It also has to satisfy joint position, velocity, and acceleration constrains. The trajectory is usually designed in joint space, seldom in Cartesian space. Besides that, the trajectory has to satisfy persistently exciting conditions in order to excite each function which is associated with the dynamic parameters. This requirement can easily be explained from a physical point of view. Dynamic parameters appear linearly in differential and integral models. The functions which are associated with them have to be large enough to influence the corresponding torque or energy equation. At the same time, these functions cannot be linearly dependent, otherwise the parameters which are associated with them could not be identified.

Since, in general, these functions are nonlinear and depend on the generalised positions, velocities, and accelerations, our intuition in choosing a good trajectory can be wrong. The robot dynamic equations of motion involve many configuration dependent trigonometric functions which are roughly of the same period: all of the joints start at one extreme and arrive at other extreme roughly together. These functions look sinusoidal and generally have one or two undulations over the trajectory. Nevertheless these functions can easily be linearly dependent. Note also that the trajectory space, in this case the joint space, is huge, and it seems sparsely populated with good choices.

A useful criterion to select good trajectories is the condition number of the regression factor or the condition number of the input correlation matrix (or persistent excitation matrix). The condition number $\kappa(\boldsymbol{A})$ of the nonsingular matrix $\boldsymbol{A}$ with singular values $\sigma_1 \geq \ldots \geq \sigma_n > 0$ is the ratio σ_1/σ_n [72, 110]. If matrix $\boldsymbol{A}$ is normal (i.e. $\boldsymbol{A}\boldsymbol{A}^T = \boldsymbol{A}^T\boldsymbol{A}$) with singular values $\sigma_1 \geq \ldots \geq \sigma_n > 0$, then $\sigma_k = |\lambda_k|$ where λ_k are the eigenvalues of matrix $\boldsymbol{A}$ ($k = 1, \ldots, n$). In particular, the singular and eigenvalues of the matrix $\boldsymbol{A}$, which is nonnegative definite are identical. Note that the input correlation matrix for the canonical differential model and the canonical integral model is nonnegative definite. Therefore, in this particular case, we can define the condition number as the ratio $\lambda_{\max}/\lambda_{\min}$. This definition is used, for example, in reference [156]. A direct consequence of the singular value decomposition theorem [72, 110] is that the singular values of the matrix $\boldsymbol{A}$ are the positive square roots of the eigenvalues (which are nonnegative) of matrix $\boldsymbol{A}^T\boldsymbol{A}$. This justifies the fact that the condition number of the regression factor or the condition number of the input correlation matrix (defined in terms of the singular values or in terms of the eigenvalues) are equivalent. In both cases we look for the smallest condition number.

Gautier and Khalil [65] prefer to calculate the condition number of the regression factor for both differential and integral models. On the other hand, Armstrong [4, 5] calculates the condition number of the input correlation matrix. It seems that in the latter case, the computation load associated with the condition number calculations is greater (this can be drastically reduced when the sequential identification algorithm described in Section 4.4 is implemented). Regardless of the case discussed here the condition number changes with changes of the input trajectory which is chosen for the purpose of identification. A useful criterion of optimal input trajectory design would be the minimisation of the condition number of either the regression factor or the input correlation matrix. This minimisation is difficult and usually leads to suboptimal trajectories. These calculations have to be done off-line because the optimisation of the condition number of the input correlation matrix is a very time consuming procedure. It can take 20 hours on a fast PC computer (IBM PC 486, 133MHz). A very interesting approach towards the design of optimal robot excitation trajectories is presented by Swevers and coauthors [211]. According to their approach the excitation trajectory for each joint is a finite Fourier series. The optimization criterion is the uncertainty on the estimated parameters or a lower bound for it, insted of the often used condition of the parameter estimation problem. As mentioned above, intuition can be wrong and a more systematic approach has to be chosen to solve the difficult problem of optimal trajectory design for the experimental identification of robot dynamics. It is the first problem addressed in this chapter.

Another problem discussed in this chapter is friction analysis and modelling for geared and direct drive robots. This problem plays an important role in experimental identification. It has been observed that a high torque (in general, force or torque) is required to overcome friction, especially for robots with gears. In contrast for direct drive robots, the frictional torque is typically needed to move or to hold the robot link itself, but in general, also cannot be neglected. It is quite difficult to identify the friction coefficients. Usually linear friction coefficients are identified, namely those which appear in Eq.(3.5). The Coulomb and viscous coefficients usually depend on the direction of the movement and are different, but this does not cause a problem since the direction of the movement can easily be detected. Unfortunately the functions which are associated with these two friction coefficients are prone to be linearly dependent particularly, in the integral model. In this chapter we show how to identify the friction coefficients of the integral model by making use of exciting trajectories.

In robotics literature frictional effects are mainly classified with respect to joint velocity. The following effects have to be distinguished: static friction, viscous friction, stiction, breakaway, slip stick, and negative friction. The friction torque is a nonlinear function of these effects and is not described by Eq.(3.5). In order to minimise the effect of unmodelled dynamics in the dynamics equations of motion it is important to take into account the above mentioned phenomena. The simplest way to do this is to measure the friction torque with respect to joint velocity experimentally. References [196, 197] show how to do that for the differential model. In this chapter we propose how to solve this problem for the integral model. Knowing the frictional torques, we can subtract them from the generalised torques. As a result we get only the part of the generalised torques which is affected by dynamic parameters of the canonical model. This idea seems to be more general since additional effects on the frictional torques can be included easily. For example, frictional torques depend on the temperature and this dependence is not linear. In this case, friction precompensation releases the controller from the compensation of the disturbances of nonlinear frictional effects and contributes to a more accurate experimental identification. This is the second main problem addressed in this chapter.

Finally this chapter presents the results of dynamic parameter identification experiments for a class of geared and direct drive robots. The results of the experiments are shown for three robots: an IRp-6, a one link geared robot, and EDDA. A complete dynamic model is estimated experimentally. A discussion concerning both differential and integral models is presented with comparison of the results. Finally, the dynamic models are validated by calculating the input trajectories based on the forward dynamics solution which assumes the identified parameters as input values.

The chapter is organised as follows. In Section 5.2 a general theory of optimal input trajectories is presented. This theory is illustrated by choosing the optimal trajectories for identification of the dynamic parameters of the IRp-6 robot. Different optimisation techniques are discussed and compared. In Section 5.3 a friction analysis and modelling is discussed. A general concept of the Tustin model is presented. Next, a concept of friction characteristics measurements for the integral model is addressed. These considerations are illustrated by friction characteristics measurements for the integral model of the IRp-6 robot. Section 5.4 is dedicated to experimental identification results of the dynamic model of the IRp-6 robot. A complete dynamic model is obtained and verified in terms of reconstruction of the input exciting trajectory based on the forward dynamics solution and identified dynamic model of this robot. Identification of dynamic parameters of the one link geared robot described in Chapter 2 is the subject of Section 5.5. In Section 5.6 identification experimental results for the EDDA robot are presented. Finally, in Section 5.7 we present futher comments on the experimental identification of robot dynamics. The experimental results discussed in this chapter show the specific problems which are important for a class of geared and direct drive robots as focus the identification is concerned.

5.2 Optimal trajectories for robot dynamics identification

5.2.1 Introduction: different optimisation techniques

In this section we present results of the optimisation procedure of the input trajectory for the purpose of identification of robot dynamics. The problem of finding a proper input trajectory is not a trivial one due to the fact that the regression factor, which is a matrix $\boldsymbol{D}(\boldsymbol{q}, \dot{\boldsymbol{q}}, \ddot{\boldsymbol{q}})$ in Eq.(3.12), is a nonlinear function of joint positions, velocities, and accelerations. If the linear model of the friction effect applies the regression factor in the canonical model has a form of the matrix of $\boldsymbol{\Phi}_{ij}^T$ functions defined by Eq.(3.95). In the case of the integral model the regression factor has a form of a vector $\boldsymbol{d}^T(\boldsymbol{q}, \dot{\boldsymbol{q}})$ (see Eq.(3.40)) which is a nonlinear function of joint positions and velocities. In both cases the condition number of the regression factor has to be minimised.

The reasoning for this is very simple and follows from the results given in reference [65]. Note that in this reference the friction coefficients are not included in the regression factor. This is probably due to the fact that it is quite difficult to find exciting trajectories when one has to identify these coefficients.

In this monograph we prefer to precompute the friction torques and subtract them from the generalised forces. This will be discussed later on along with the experimental identification results of the dynamic parameters of the IRp-6 robot. The discussion presented in reference [65] applies to the deterministic case for the integral model presented in Section 4.3. In practical applications function $\boldsymbol{y}_r$ and $\boldsymbol{h}_r$ which appear in Eq.(4.21) are perturbed by noisy measurements of $\boldsymbol{\tau}$, $\boldsymbol{q}$, $\dot{\boldsymbol{q}}$ and error modelling. Suppose that we use the least squares solution to the canonical integral model and study the effect of these perturbation on it. Denote by $\hat{\boldsymbol{x}} + \tilde{\boldsymbol{x}}$ the least squares solution of the perturbed system

$$\boldsymbol{y}_r + \delta \boldsymbol{y}_r = (\boldsymbol{h}_r + \delta \boldsymbol{h}_r)\, \boldsymbol{x}_c + \boldsymbol{w}. \tag{5.1}$$

In the case $r = m_c$ simple bounds for $\tilde{\boldsymbol{x}}_c$ are given by the following equations:

$$\frac{||\tilde{\boldsymbol{x}}_c||}{||\hat{\boldsymbol{x}}_c||} \leq \kappa(\boldsymbol{h}_r) \frac{||\delta \boldsymbol{y}_r||}{||\boldsymbol{y}_r||} \quad \text{with} \quad \delta \boldsymbol{h}_r = \boldsymbol{0} \tag{5.2}$$

and

$$\frac{||\tilde{\boldsymbol{x}}_c||}{||\hat{\boldsymbol{x}}_c + \tilde{\boldsymbol{x}}_c||} \leq \kappa(\boldsymbol{h}_r) \frac{||\delta \boldsymbol{h}_r||}{||\boldsymbol{h}_r||} \quad \text{with} \quad \delta \boldsymbol{y}_r = \boldsymbol{0} \,, \tag{5.3}$$

where $||\cdot||$ is a p vector norm or its subordinate p matrix norm. $\kappa(\boldsymbol{h}_r)$ denotes the condition number with respect to the p norm. From Eqs.(5.2) and (5.3) it is clear that the condition number is a quantity which measures the sensitivity of the solution $\hat{\boldsymbol{x}}_c$ to errors in $\boldsymbol{y}_r$ and $\boldsymbol{h}_r$. As a consequence it is very important that the condition number of $\boldsymbol{h}_r$ is small as possible before computing $\hat{\boldsymbol{x}}_c$. $\kappa(\boldsymbol{h}_r)$ in the p norm is given by

$$\kappa_p(\boldsymbol{h}_r) = ||\boldsymbol{h}_r||_p ||\boldsymbol{h}_r^+||_p \,, \tag{5.4}$$

where $\boldsymbol{h}_r^+$ is the pseudo-inverse of $\boldsymbol{h}_r$. Any two condition numbers in two different norms are equivalent [72] and we omit subscript p. The 2-norm condition can easily be calculated using the singular value decomposition of $\boldsymbol{h}_r$ [72, 110] and results in is the ratio of the largest singular value to the smallest singular value of the matrix $\boldsymbol{h}_r$. Note that the bounds given by Eqs.(5.2) and (5.3) are valid only when errors δ_{hrij} of elements of matrix $\boldsymbol{h}_r$ are all about the same size. Therefore matrix $\boldsymbol{h}_r$ has to be scaled before computing $\hat{\boldsymbol{x}}_c$. This observation comes from reference [65] and can be understood intuitively. We prefer each element of matrix $\boldsymbol{h}_r$ to be of similar size because we want to excite each element of vector $\boldsymbol{x}_c$ by a function of about the same magnitude. Therefore it is clear that matrix $\boldsymbol{h}_r$ has to be well–conditioned and well–equilibrated. Gautier and Khalil [65] pose the following optimisation problem. Find a trajectory in terms of joint positions and velocities $\boldsymbol{q}$ and $\dot{\boldsymbol{q}}$ which minimises the following cost function:

$$f(\boldsymbol{q},\dot{\boldsymbol{q}}) = c_1\kappa(\boldsymbol{h}_r(\boldsymbol{q},\dot{\boldsymbol{q}})) + c_2 S(\boldsymbol{q},\dot{\boldsymbol{q}}) , \tag{5.5}$$

where c_1 and c_2 are two weighing scalars, and S is the measure of equilibrium. When $\boldsymbol{h}_r$ is computed accurately, the absolute error in its element is proportional to the element's size, and the strategy of equilibrium reduces to scale $\boldsymbol{h}_r$ so that all its elements are of the same order of magnitude. Thus S can be defined as

$$S = \frac{|h_{rij}|_{\max}}{|h_{rij}|_{\min}} ,$$

where $|h_{rij}|_{\max}$ and $|h_{rij}|_{\min}$ are the maximum and minimum absolute values of the elements h_{rij} of matrix $\boldsymbol{h}_r$, and $|h_{rij}|_{\min} \neq 0$. Note that the criterion $f(\boldsymbol{q},\dot{\boldsymbol{q}})$ is a nonlinear function of joint positions and velocities. Now the problem is to find the trajectory $(\boldsymbol{q},\dot{\boldsymbol{q}})$ which is defined by a sequence of points which minimise the cost function given by Eq.(5.5).

This problem is a nonlinear optimisation problem with constraints and can be solved by making use of one of the methods quoted by Gautier and Khalil [65]. These are heuristic, quasi–Newton and a gradient conjugate methods. Khalil and Gautier tested three measures of the condition number:

1. The 2–norm condition number using singular value decomposition,
2. An estimate of the 2–norm condition from the QR decomposition [72] of $\boldsymbol{h}_r$ with column pivoting,
3. A condition of the input correlation matrix $\boldsymbol{P} = \boldsymbol{h}^T\boldsymbol{h}$ in the Frobenius norm, given by

$$\kappa_F(\boldsymbol{P}) = ||\boldsymbol{P}||_F||\boldsymbol{P}^{-1}||_F , \tag{5.6}$$

where

$$||\boldsymbol{P}|| = \left(\sum_i\sum_j P_{ij}^2\right)^{\frac{1}{2}} . \tag{5.7}$$

According to Gautier and Khalil [65] the latter measure gives the best convergence rate. Note that the last norm is also used by Vandanjon and co-authors [222] for the purpose of minimisation of the Frobenius condition number of the regression factor in identification of robot dynamics by means of spectrum analysis. This norm seems to be applicable in many situations. SVD and QR decompositions are well–known tools and they are often used in identification of robot dynamics, see for example [1, 63, 66, 203, 205]. As a result of the optimisation procedure the set of optimum points $(\boldsymbol{q},\dot{\boldsymbol{q}})$ is obtained. A continuous and smooth trajectory is calculated by interpolating a line between these points, assuming zero initial and final accelerations and using a fifth–order polynomial. The calculated position trajectory has to be checked to see whether the obtained positions, velocities, and accelerations do not violate the constraints. The procedure outlined above seems to be attractive and not computationally intensive. It shows that a problem of choosing the optimal trajectory is not a trivial one, but intuition can be very helpful. Here the aim was to prepare matrix $\boldsymbol{h}_r$ which is well–conditioned and well–equilibrated. Intuition says, that if the identified parameters are expected to vary drastically (in the range of three to five orders) it is better to scale the vector of dynamic parameters. This hint is very useful and has been verified by many experiments. This practical judgement is equivalent to scaling to some extent, making the regression factor well–equilibrated. Obviously the first requirement saying that matrix $\boldsymbol{h}_r$ has to be well–conditioned is still valid. These considerations are specific to the integral model, but can be applied to the differential model although being more computationally intensive.

Khalil and Gautier [102] assumed that the integral model is perturbed. Presse and Gautier [179] took into account statistical information, namely the expected value and the covariance matrix of the solution vector, when they are available. In their considerations the regression factor (or the input correlation matrix) is supposed to be deterministic. In practice $\boldsymbol{y}_r$ and $\boldsymbol{h}_r$ are random matrices because of noisy measurements and unmodelled dynamics. Therefore Presse and Gautier [179] considered additive perturbations $\boldsymbol{w}$ which appear in Eq.(5.1). This is a simple case which makes it possible to characterise the sensitivity of the method to any perturbation. We assume that $\boldsymbol{w}$ is a zero mean gaussian noise with standard deviation δ_w such that

$$\boldsymbol{C}_w = E(\boldsymbol{w}\boldsymbol{w}^T) = \delta_w^2 \boldsymbol{I}_r \,, \tag{5.8}$$

where $\boldsymbol{I}_r$ is an identity matrix with dimensions $r \times r$. Then the relative norm error on the solution vector assuming that $\boldsymbol{h}_r$ is deterministic is given by Eq.(5.2) in which we have to substitute $\boldsymbol{w}$ instead of $\delta \boldsymbol{y}_r$. Assuming that $\boldsymbol{w}$ is a zero mean gaussian noise we calculate

$$E(\boldsymbol{x}_c - \hat{\boldsymbol{x}}_c) = \boldsymbol{0} \,,$$

$$\boldsymbol{C}_{\hat{\boldsymbol{x}}_c} = E[(\boldsymbol{x}_c - \hat{\boldsymbol{x}}_c)(\boldsymbol{x}_c - \hat{\boldsymbol{x}}_c)^T] = \delta_w^2 \left[\boldsymbol{h}_r^T \boldsymbol{h}_r\right]^{-1} . \tag{5.9}$$

This can easily be verified by taking solution $\hat{\boldsymbol{x}}$ of the overdetermined linear system given by Eq.(4.22) and substituting of it into Eq.(5.9).

Note that the estimate $\hat{\boldsymbol{x}}$ is not biased and the covariance matrix of the estimation error is expressed in terms of the variance δ_w^2 and the inverse of the input correlation matrix. Therefore, as a measure of the covariance matrix of the estimation error we can take the condition number of the input correlation matrix $\kappa(\boldsymbol{h}_r^T \boldsymbol{h}_r)$. Since the eigenvalues of the inverse of the input correlation matrix are the same as the eigenvalues of the regression matrix, the condition number of the input correlation matrix and its inverse are exactly the same. Considering the fact that the condition number of the input correlation matrix and the condition number of the regression factor (which forms the appropriate input correlation matrix) are equivalent, in the sense that they are related by a square root, we use the latter as a criterion for the optimisation problem. The minimum value of the condition number of the regresor factor $\kappa(\boldsymbol{h}_r) = 1$ if and only if $\boldsymbol{h}_r^T\boldsymbol{h}_r = \delta_1^2\boldsymbol{I}_{m_c}$, where $\boldsymbol{I}_{m_c}$ is an identity matrix with appropriate dimensions. Therefore $\boldsymbol{h}_r$ is an orthogonal matrix.

In this particular case $\sigma_1(\boldsymbol{h}_r) = \sigma_2(\boldsymbol{h}_r) = \ldots = \sigma_{m_c}(\boldsymbol{h}_r)$ and the covariance matrix of the estimation error is expressed as follows:

$$\boldsymbol{C}_{\hat{x}} = \left[\frac{\delta_w}{\sigma_1(\boldsymbol{h}_r)}\right]^2 \boldsymbol{I}_{m_c}. \tag{5.10}$$

It can be shown that the singular value $\sigma_1(\boldsymbol{h}_r)$ has to be large to ensure a small value of the standard deviation of each parameter. One possible way to scale the vector of unknown parameters would be to take into account *a prior* knowledge of parameter $\boldsymbol{x}_c$, denoted by $\boldsymbol{x}_a$. Then Eq.(4.21) can be rewritten in the following form:

$$\boldsymbol{y}_r = \boldsymbol{h}_r \mathrm{diag}(\boldsymbol{x}_a)\mathrm{diag}(\boldsymbol{x}_a)^{-1}\boldsymbol{x}_c + \boldsymbol{w} = \bar{\boldsymbol{h}}_r\bar{\boldsymbol{x}}_c + \boldsymbol{w} , \tag{5.11}$$

where $\bar{\boldsymbol{h}}_r = \boldsymbol{h}_r\mathrm{diag}(\boldsymbol{x}_a)$, $\bar{\boldsymbol{x}}_c = [\mathrm{diag}(\boldsymbol{x}_a)]^{-1}\boldsymbol{x}_c$, and $\mathrm{diag}(\cdot)$ denotes diagonal matrix of $(\cdot)$. Note that the transformed variable $\bar{\boldsymbol{x}}_c$ is better scaled than $\boldsymbol{x}_c$. Using the above definitions one can write

$$\boldsymbol{C}_{\hat{\bar{x}}} = [\mathrm{diag}\boldsymbol{x}_a]^{-1}\boldsymbol{C}_{\hat{x}_c}[\mathrm{diag}(\boldsymbol{x}_a)]^{-1T} = \delta_w^2\left[\bar{\boldsymbol{h}}_r\bar{\boldsymbol{h}}_r\right]^{-1}. \tag{5.12}$$

If $\bar{\boldsymbol{h}}_r$ is well conditioned, the efficiency of the identification procedure is the same for each parameter.

Now we formulate a criterion for exciting trajectories in terms of the condition number of matrix $\bar{\boldsymbol{y}}_r$, which must be as small as possible while its smallest singular value must be as large as possible. At the same time, the norm of the measurement vector $\bar{\boldsymbol{h}}_r$ must be as large as possible in order to reduce the relative norm error of the solution vector. This can be taken into account by defining *a prior* measurement $y_{rm} = \min_r|\boldsymbol{y}_{ra}|$ where $\boldsymbol{y}_{ra}$ is the *a prior* knowledge about vector $\boldsymbol{y}_r$. Therefore a more general criterion than that given by Eq.(5.5) can be formulated as follows:

$$f_1(\boldsymbol{q},\dot{\boldsymbol{q}}) = c_1\kappa(\bar{\boldsymbol{h}}_r) + c_2\frac{1}{\sigma_m(\bar{\boldsymbol{h}}_r)} + c_3\frac{1}{y_{rm}} , \tag{5.13}$$

where c_1, c_2, c_3 are three weighted scalars. Minimisation of the criterion given by Eq.(5.13) can be achieved by any of the nonlinear optimisation techniques under the constraints on the joint positions and velocities [65]. The criterion given by Eq.(5.13) is more general than the criterion given by Eq.(5.5), since it takes into account *a prior* information about the vector of the estimated parameters and its statistics. Each term of Eq.(5.13) has clear physical interpretation. The results can be extended to the differential model given by Eq.(3.12).

From the above considerations it follows, that estimate $\hat{\boldsymbol{x}}$ is an unbiased estimate of the vector of dynamic parameters. This assumption is generally far from verified in practice. Usually the estimates are biased due to measurement noise which is present in the regression factor and to unmodelled dynamics. This difficult problem has been addressed by many researchers in robotics. As mentioned in Chapter 4, the measurement signals are corrupted by additive noise with known characteristics. The results of linear estimation theory (for example Markov estimation theory) are not applicable due to nonlinear functions which are present in the regression factor. A comprehensive discussion of a problem is given in Armstrong [4]. This discussion is originally applicable to the differential model but may be applied to the integral model as well [37, 43, 132, 134, 136, 139]. Armstrong presents an interesting noise analysis of the least mean square and least squares identification algorithms. Here we quote these results because we believe them to be the most advanced as far as error analysis is concerned. The batch least squares identification discussed in Section 4.2 is considered first. Rewrite the model given by Eq.(4.5) in the following form:

$$\boldsymbol{y}_L = \boldsymbol{h}_L \boldsymbol{x}_c + \boldsymbol{w} = \boldsymbol{h}_L(\hat{\boldsymbol{x}} + \tilde{\boldsymbol{x}}) + \boldsymbol{w} \ . \tag{5.14}$$

Here we assume that $\boldsymbol{w}$ represents the vector of torques produced by systematic errors (which is a vector of unmodelled dynamics). Then the expected value of the parameter error can be written as follows:

$$E(\tilde{\boldsymbol{x}}_c) = -\left[\boldsymbol{h}_L^T \boldsymbol{W} \boldsymbol{h}_L\right]^{-1} \boldsymbol{h}_L^T \boldsymbol{W} \boldsymbol{w} \ , \tag{5.15}$$

where $\boldsymbol{W}$ is a weighting matrix that reflects varying degrees of confidence in the elements of $\boldsymbol{h}_L$ and $\boldsymbol{y}_L$. If we assume that the experiment is corrupted by additive noise we are dealing with results similar to those given by Eq.(5.9) where we assumed the same noise characteristics. The equation for the covariance vector of the estimation error (5.9) does not contain the weighting matrix $\boldsymbol{W}$.

Now we assume that the regression matrix consists of the two terms given by Eq.(4.64). Recall that $\hat{\boldsymbol{h}}_L$ denotes the estimated regression matrix and $\tilde{\boldsymbol{h}}_L$ is the noise introduced by sensors, due to the noisy measurements. A complete description of the effect of regression noise is very complex and is not available in robotics literature. The same applies to the exact calculations of the variance error. Here we quote after Armstrong [4] the bias of the estimate which is introduced by sensor noise. In order to do that we rewrite Eq.(5.14) (in which we neglected unmodelled dynamics) in the following form:

$$\boldsymbol{y}_L = \hat{\boldsymbol{h}}_L \boldsymbol{x}_c + \tilde{\boldsymbol{h}}_L \boldsymbol{x}_c \ , \tag{5.16}$$

which results in the following normal equation:

$$\hat{\boldsymbol{h}}_L^T \boldsymbol{W} \hat{\boldsymbol{h}}_L \hat{\boldsymbol{x}} = \hat{\boldsymbol{h}}_L^T \boldsymbol{W} \boldsymbol{y}_L \tag{5.17}$$

from which estimate $\hat{\boldsymbol{x}}_c$ can easily be calculated. If in addition, we assume that infinite data length are available, the bias of the estimate can be calculated as follows:

$$E(\tilde{\boldsymbol{x}}) = (\boldsymbol{P} + \boldsymbol{C}_{\tilde{h}_L})^{-1} \boldsymbol{C}_{\tilde{h}_L} \boldsymbol{x}_c, \tag{5.18}$$

where $\boldsymbol{P} = E\left[\boldsymbol{h}_L^T \boldsymbol{W} \boldsymbol{h}_L\right]$ and $\boldsymbol{C}_{\tilde{h}_L} = E\left[\tilde{\boldsymbol{h}}_L^T \boldsymbol{W} \tilde{\boldsymbol{h}}_L\right]$. A complete derivation of this result can be found in [4]. The least mean squares identification scheme was discussed in Section 4.4. In this case the input correlation matrix and regression covariance matrix are, respectively,

$$\boldsymbol{P} = \frac{1}{K} \sum_{k=0}^{K} \boldsymbol{\Phi}_k \boldsymbol{\Phi}_k^T \text{ and } \boldsymbol{C}_{\tilde{\Phi}} = \frac{1}{K} E\left\{ \sum_{k=0}^{K} \tilde{\boldsymbol{\Phi}}_k \tilde{\boldsymbol{\Phi}}_k^T \right\}. \tag{5.19}$$

Armstrong [4] proved that the bias due to random sensor noise is given by the same equation (5.18) in which definitions (5.19) are used. He noticed that in the least mean squares algorithm, which minimises the expected value of the square error, the error signal due to the signal noise bucks the error signal due to $\tilde{\boldsymbol{x}}_c$ and tends to cause the least mean squares algorithm to underestimate the parameters. Considering only the effect of the systematic error $w(k)$, which appears in k-th equation given by Eq.(5.14) the bias is given by

$$E(\tilde{\boldsymbol{x}}_c) = -\boldsymbol{P}^{-1} \frac{1}{K} \sum_{k=0}^{K} \boldsymbol{\Phi}_k^T w(k) , \tag{5.20}$$

which is exactly Eq.(5.15) when $\boldsymbol{W} = \boldsymbol{I}$. Notice that in Eqs.(5.15), (5.18), and (5.20) the inverse of the input correlation matrix appears. If this matrix is ill-conditioned, moderate noise contributions may have a large effect on the estimated parameters. In the following section two methods are described to improve the condition number of the input correlation matrix.

5.2.2 Optimisation procedures

In this section two optimisation procedures are discussed in detail. These two methods can be used for both differential and integral models in the process of identification of their parameters. Here we concentrate on optimisation schemes which minimise the condition number of the input correlation matrix. Marrels and co-authors [156] propose also other optimisation criteria. They consider the inverse of the smallest singular value of the input correlation matrix or the minimum of the inverse of its determinant. The input correlation matrix is a nonlinear function of the joint positions, velocities, and accelerations and therefore the optimisation techniques are usually computationally time consuming. Following the considerations presented in the preceding section we have chosen a method which optimises the condition number of the input correlation matrix. As a result of the optimisation procedure the optimal input trajectory is generated such that the condition number of the input correlation matrix has the smallest value. Two methods which were verified experimentally are presented.

The simplest method, proposed by Van Der Linden and Van Der Weiden [221], is essentially a trial and error search through the space of allowable $\boldsymbol{q}$, $\dot{\boldsymbol{q}}$, and $\ddot{\boldsymbol{q}}$. During the search the condition number $\kappa(\boldsymbol{h}_L^T \boldsymbol{h}_L)$ is calculated based on a number of selected points. Then the points of the best guess are connected to a smooth trajectory using an approach identical to that of Gautier and Khalil [65]. For better efficiency, one can repetitively zoom it in on the neighbourhood of a good estimate. This method is actually a Monte Carlo method which in this particular case can be summarised in the following steps.

1. Initiate search: set the extremes of the search space to the allowable extreme values of the robot. This observation is rather trivial and allows to avoid points which are located at the limits of the range of each joint of the manipulator.
2. Take a certain number of draws (e.g. 1000), calculate the condition number of the input correlation matrix for each draw and remember the best set of points $\boldsymbol{q}$, $\dot{\boldsymbol{q}}$, and $\ddot{\boldsymbol{q}}$.
3. If the search space is too large, calculate the new extremes by taking, for example, half of the previous search centred around the best guess so far (making sure that the allowable extremes are not exceeded); go back to step 2.
4. Connect the best guess to the trajectory.

The set of points resulting from the procedure can be connected by pairing two points using polynomials. The polynomial has to satisfy the joint displacements at the start and end points with zero initial conditions for velocities and accelerations at these points. During the movement, velocity and acceleration should be kept as low as possible to avoid saturation. This problem can be solved by making use of the least squares fit for each axis assuming the initial conditions mentioned above. This method seems to be very effective, is based on the engineer's intuition and gives satisfactory results in practice. It was successfully implemented for searching the joint space of the IRp-6 robot for its differential model. Note that this method leads to suboptimal solutions. It verifies our intuition that in the working space of the manipulator there are good suboptimal (namely exciting) trajectories.

The other method which is mathematically more sophisticated is largely borrowed from Armstrong [4, 5]. Armstrong applied his method to the differential model using the batch least squares and sequential least squares processing (see Section 4.4). Here we extend his results to the integral model discussed in Section 3.3. The integral model is expressed by Eq.(3.38) which for our purposes we rewrite in the following form:

$$\int_{t_1}^{t_2} \boldsymbol{\tau}^T \dot{\boldsymbol{q}} dt = \boldsymbol{dl}^T \boldsymbol{x} + \int_{t_1}^{t_2} \boldsymbol{\tau_f}^T \dot{\boldsymbol{q}} dt , \tag{5.21}$$

where t_1 and t_2 denote two different time instants. In Eq.(5.21) the energy which is lost due to friction phenomena is separated. Recall that vector $\boldsymbol{dl}$ depends on joint positions and velocities. Note that in Eq.(5.21) the friction torques are not restricted to the form given by Eq.(3.5). Here we assume a full friction model representation. This particular form is very useful due to the fact that it is quite difficult to identify the friction Coulomb and velocity coefficients directly from the integral model. The functions associated with them are prone to be linearly dependent even for carefully selected exciting trajectories. It was mentioned before that functions $|\dot{q}_i|$ and $\dot{q}_i^2$ which are associated with the Coulomb and velocity friction coefficients are ill–suited for the purpose of identification of the integral model. This problem can be solved by precomputing the friction torques and subtracting them from the generalised torques. The method will be discussed in Section 5.3. Applying this observation to the differential model we get the following equation for generalised torque i:

$$\tau_i - \tau_{if} = \boldsymbol{d}_i^T(\boldsymbol{q}, \dot{\boldsymbol{q}}, \ddot{\boldsymbol{q}})\boldsymbol{x}_c + w_i \ . \tag{5.22}$$

Here friction torque τ_{if} is not treated as a noise signal, but rather is a deterministic function precomputed before the identification experiments. Recall that $\boldsymbol{d}_i$ denote i-th row of the matrix $\boldsymbol{D}$ written as a column vector. This idea of separating the friction torques is not new. It was used by Seeger and Leonhard [196] and later by Seeger [197] with the differential model in the identification scheme for the Manutec R3 robot. In the case of the differential model precomputation of the friction torques is not necessary (by making use of the exciting trajectories we are able to identify the friction coefficients easily) but when applied it results in very good estimates of the dynamic parameters of the robot.

The minimisation of the observation error according to the least squares fit leads to the following equation governing estimates $\hat{\boldsymbol{x}}_c$:

$$\hat{\boldsymbol{x}}_c = (\boldsymbol{d}_i \boldsymbol{d}_i^T)^{-1} \boldsymbol{d}_i (\tau_i - \tau_{if}). \tag{5.23}$$

It is assumed that the input correlation matrix $\boldsymbol{d}_i \boldsymbol{d}_i^T$ is invertible, which is the case if the so-called persistently exciting conditions are satisfied. We also assume that the input correlation matrix in Eq.(5.23) is invertible. We write this matrix in terms of discrete time observations since in practice the problem is discritised and we have

$$\boldsymbol{P}_k = \sum_{j=1}^{k} \boldsymbol{d}_i(j\Delta T)\boldsymbol{d}_i^T(j\Delta T). \tag{5.24}$$

Vector $\boldsymbol{d}_i$ is a function of joint positions, velocities, and accelerations which due to the simplicity of notation are omitted here. Subscript k can be treated as a number of discrete time points and usually this number is large. Note that the input correlation matrix has the form of Eq.(5.19). It is clear that the existence of estimate $\hat{\boldsymbol{x}}$ depends on the input trajectory which is used for identification. Recall that the necessary condition, namely that the model is canonical is satisfied and can be obtained as a result of the procedure described in Section 3.5.

Now define the following cost function:

$$J = F(\boldsymbol{P}_K). \tag{5.25}$$

F can be either the condition number of the input correlation matrix or the inverse of its smallest singular value or the inverse of the determinant value of the input correlation matrix. Capital K denotes the maximum number of iteration steps, thus the total observation time is $t = K\Delta T$. Usually this time is equal to the duration of the input trajectory.

Now we calculate the differential of the cost function with respect to the elements of the input correlation matrix:

$$dJ = \sum_m \sum_n \frac{\partial J}{\partial P_{mn}} \sum_k \sum_r \frac{\partial P_{mn}}{\partial d_{ir}(k\Delta T)} dd_{ir}. \tag{5.26}$$

P_{mn} denotes element mn of matrix $\boldsymbol{P}_K$ (here superscript K is omitted to simplify the notation), r denotes a sequential number of the function which all together form the vector $\boldsymbol{d}_i(\boldsymbol{q}, \dot{\boldsymbol{q}}, \ddot{\boldsymbol{q}})$ in Eq.(5.22) (these functions are known as the basis functions, and i-th equation has m_c basis functions, which are at the same time the number of the minimum dynamic parameters in the canonical model) and finally, dd_{ir} denotes differential of the function d_{ir}. Note that the last differential can be calculated explicitly since the knowledge about the canonical model is assumed to be known. The other derivatives which appear in Eq.(5.26) are partial derivatives. The differential of the dd_{ir} is calculated at the point $k\Delta T$ where k changes from 0 to K. Summation indices m and n change depending on the number of elements of matrix $\boldsymbol{P}_K$. The dimensions of this matrix vary according to the number of equations which are involved in the optimisation procedure.

The partial derivatives $\frac{\partial P_{mn}}{\partial d_{ir}(k\Delta T)}$ can be calculated explicitly according to the equation

$$\frac{\partial P_{mn}}{\partial d_{ir}(k\Delta T)} = \begin{cases} 0 & \text{for } r \neq m, n, \\ d_{in}(k\Delta T) & \text{for } r = m, \\ d_{im}(k\Delta T) & \text{for } r = n, \\ 2d_{im}(k\Delta T) & \text{for } r = m = n. \end{cases} \tag{5.27}$$

The first partial derivative which appears in Eq.(5.26) cannot be calculated explicitly since we usually calculate the condition number or the inverse of the minimum singular value or the inverse of the determinant numerically, except for very simple cases when these functions can be calculated symbolically. Therefore, the first partial derivative is also calculated numerically. These calculations are straightforward.

Note that the calculations of the cost function are the most time consuming. In the case of the sequential estimation algorithm proposed by Canudas de Wit and Aubin [26], the dimensions of matrix $\boldsymbol{P}_K$ are as small as possible and the computational load is less. In the case of batch least squares identification we exploit the structure of the regression matrix which appears in Eq.(3.12) in order to reduce the number of arithmetic operations. Moving one joint at a time results in smaller matrix $\boldsymbol{P}_K$ which will be discussed later along with the experimental results.

The optimisation of the condition number which uses the cost function defined by Eq.(5.25) can be solved by making use of the Lagrange multipliers [19]. In order to do that, define the following Hamiltonian function at discrete time $k\Delta T$:

$$H(k\Delta T) = J(k\Delta T) + \boldsymbol{\lambda}^T[(k+1)\Delta T]\boldsymbol{f}(k\Delta T) , \tag{5.28}$$

where $\boldsymbol{\lambda}$ is a vector of Lagrange multipliers and $\boldsymbol{f}$ is a function which describes the dynamics of the system. This function depends on the generalised positions, velocities, and accelerations calculated at time instant $k\Delta T$. This function can be chosen arbitrarily by the engineer and many functions defined by the designer can fill this role. In order to be more specific, denote by state $\boldsymbol{s} = [\boldsymbol{q}^T, \dot{\boldsymbol{q}}^T]^T$ the vector of joint positions and velocities calculated at time instant $k\Delta T$. For example, for a robot with three degrees of freedom, the state is

$$\boldsymbol{s}(k\Delta T) = \begin{bmatrix} q_1(k\Delta T) \\ q_2(k\Delta T) \\ q_3(k\Delta T) \\ \dot{q}_1(k\Delta T) \\ \dot{q}_2(k\Delta T) \\ \dot{q}_3(k\Delta T) \end{bmatrix} . \tag{5.29}$$

Then the dynamics of the system at discrete time point $k\Delta T$ can be written as

$$\boldsymbol{f}(k\Delta T) = \boldsymbol{f}[\boldsymbol{s}(k\Delta T), \ddot{\boldsymbol{q}}(k\Delta T)]. \tag{5.30}$$

The definition of the dynamics of the system can be seen as being orthodox. The function $\boldsymbol{f}(k\Delta T)$ does not play the role of the dynamics of the manipulator itself. It is a very important component in the definition of the Hamiltonian function, which is the constraint function. It is obvious that this function should be defined so as to make it possible to express joint accelerations in terms of joint positions and velocities. The optimisation of the condition number of the input correlation matrix is equivalent to finding a sequence of joint accelerations at discrete time instants $k\Delta T$. We assume that the initial trajectory of a vector of joint accelerations is known. Here the engineer's intuition is very helpful. We can assume that the initial joint acceleration trajectory vector is specified as a result of the Monte Carlo search described at the beginning of this section.

The optimisation procedure is summarised in the following iterative steps [4, 5].

$$
\begin{aligned}
\boldsymbol{\lambda}(K) &= 0, \text{ initial condition,} \\
\boldsymbol{\lambda}^T(k\Delta T) &= \frac{\partial \boldsymbol{P}_K(\boldsymbol{s},\ddot{\boldsymbol{q}})}{\partial \boldsymbol{s}(k\Delta T)} + \boldsymbol{\lambda}^T[(k+1)\Delta T]\frac{\partial \boldsymbol{f}(\boldsymbol{s},\ddot{\boldsymbol{q}})}{\partial \boldsymbol{s}(k\Delta T)},
\end{aligned}
\tag{5.31}
$$

$$
\begin{aligned}
\ddot{\boldsymbol{q}}(k\Delta T)_{I+1} &= \ddot{\boldsymbol{q}}(k\Delta T)_I - \boldsymbol{\mu}_{cv}\left[\frac{\partial \boldsymbol{P}_K(\boldsymbol{s},\ddot{\boldsymbol{q}})}{\partial \ddot{\boldsymbol{q}}(k\Delta T)} + \right. \\
&\left. \boldsymbol{\lambda}^T(k\Delta T)\frac{\partial \boldsymbol{f}(\boldsymbol{s},\ddot{\boldsymbol{q}})}{\partial \ddot{\boldsymbol{q}}(k\Delta T)}\right].
\end{aligned}
\tag{5.32}
$$

Equations (5.31) and (5.32) constitute two iterations. The first iteration is initiated by the condition $\boldsymbol{\lambda}(K) = 0$ and starting from the end of the trajectory Lagrange multipliers are calculated. Then a new trajectory in terms of joint accelerations is calculated at discrete point $k\Delta T$ (Eq.(5.32)). This is illustrated in Fig.5.1

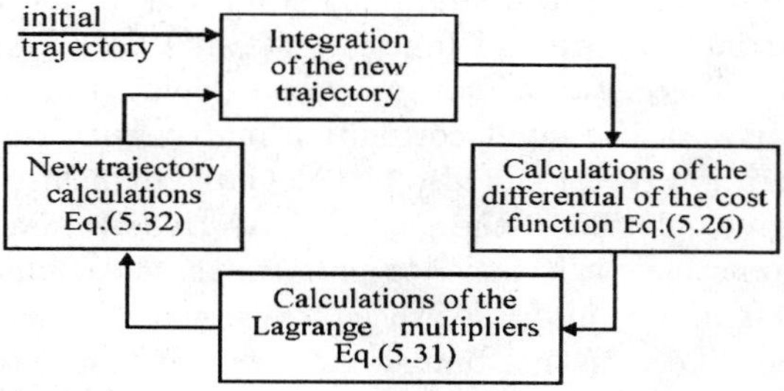

Fig. 5.1. Optimisation procedure of the condition number

The optimisation procedure consists of fours steps described by Eqs.(5.26), (5.31), and (5.32). Equation (5.26) evaluates partial derivatives of the cost function at the trajectory end. Equation (5.31) sweeps the Lagrange multipliers backward. Equation (5.32) is called the control update equation. In order to calculate the joint positions and velocities Eq.(5.32) is integrated twice. The iteration algorithm presented here is also known as a two sweeps algorithm. Matrix $\boldsymbol{\mu}_{cv}$, which appears in Eq.(5.32), controls the speed of the iteration process. This matrix is constant during the process but can be changed in order to make the algorithm converge to the optimal trajectory as fast as possible.

There are several partial derivatives which appear in Eqs.(5.31) and (5.32). First consider the partial derivatives of the dynamic function $\boldsymbol{f}(k\Delta T)$ which depends on the aforementioned state and the joint accelerations. For example, for a manipulator with three degrees of freedom and for the differential model this function can be written as [48, 130, 132, 134]

$$\begin{bmatrix} q_1[(k+1)\Delta T] \\ q_2[(k+1)\Delta T] \\ q_3[(k+1)\Delta T] \\ \dot{q}_1[(k+1)\Delta T] \\ \dot{q}_2[(k+1)\Delta T] \\ \dot{q}_3[(k+1)\Delta T] \end{bmatrix} = \begin{bmatrix} 1 & 0 & 0 & \Delta T & 0 & 0 \\ 0 & 1 & 0 & 0 & \Delta T & 0 \\ 0 & 0 & 1 & 0 & 0 & \Delta T \\ 0 & 0 & 0 & 1 & 0 & 0 \\ 0 & 0 & 0 & 0 & 1 & 0 \\ 0 & 0 & 0 & 0 & 0 & 1 \end{bmatrix} \begin{bmatrix} q_1(k\Delta T) \\ q_2(k\Delta T) \\ q_3(k\Delta T) \\ \dot{q}_1(k\Delta T) \\ \dot{q}_2(k\Delta T) \\ \dot{q}_3(k\Delta T) \end{bmatrix} + \\ + \begin{bmatrix} \frac{1}{2}\Delta T^2 & 0 & 0 \\ 0 & \frac{1}{2}\Delta T^2 & 0 \\ 0 & 0 & \frac{1}{2}\Delta T^2 \\ \Delta T & 0 & 0 \\ 0 & \Delta T & 0 \\ 0 & 0 & \Delta T \end{bmatrix} \begin{bmatrix} \ddot{q}_1(k\Delta T) \\ \ddot{q}_2(k\Delta T) \\ \ddot{q}_3(k\Delta T) \end{bmatrix}. \tag{5.33}$$

Note that the two terms on the right hand side of Eq.(5.33) define function $\boldsymbol{f}(\boldsymbol{s}(k\Delta T), \ddot{\boldsymbol{q}}(k\Delta T))$. At the same time this function describes the state of the system at the next time instant $(k+1)\Delta T$. The definition of function $\boldsymbol{f}(k\Delta T)$ by Eq.(5.33) is clear because it allows calculation of the state at the next discrete time instant from the state and the acceleration at time $k\Delta T$. The partial derivative of function $\boldsymbol{f}(k\Delta T)$ with respect to the state and the vector of acceleration can easily be calculated in a closed form. The partial derivatives of the input correlation matrix with respect to the state and the acceleration vector can also be calculated explicitly, according to the rule governing calculation of the partial derivative of a matrix with respect to a vector. Sometimes it is easier to perform these calculations numerically because the whole optimisation procedure is generally numerical. In order to perform the differentiation numerically, we calculate the ratio of the infinitesimal increment of the function itself to the appropriate increment of the independent variable.

In the batch least squares identification process, the two sweeps algorithm given by Eqs.(5.31) and (5.32) is the same as presented above, except for the fact that the dimensions of the input correlation matrix and other related vectors are much greater because the number of basis functions is also greater.

Now we extend these results to the integral model described by Eq.(5.21). Recall that the friction torque characteristics are precomputed. Function $\boldsymbol{f}(k\Delta T)$, which describes dynamics, depends on generalised positions and velocities. The role of the state is played by the vector of generalised positions. The control vector which is represented by the generalised velocities. Note that the state equation given by Eq.(5.33), which at the same time defines function $\boldsymbol{f}(k\Delta T)$ in the case of the differential model, describes a uniform motion during time period ΔT.

In order to make it more clear, we write the state equation for the integral model of a manipulator with three degrees of freedom in the following form

[48, 130, 132, 134]:

$$\begin{bmatrix} q_1[(k+1)\Delta T] \\ q_2[(k+1)\Delta T] \\ q_3[(k+1)\Delta T] \end{bmatrix} = \begin{bmatrix} q_1(k\Delta T) \\ q_2(k\Delta T) \\ q_3(k\Delta T) \end{bmatrix} + \begin{bmatrix} \Delta T & 0 & 0 \\ 0 & \Delta T & 0 \\ 0 & 0 & \Delta T \end{bmatrix} \begin{bmatrix} \dot{q}_1(k\Delta T) \\ \dot{q}_2(k\Delta T) \\ \dot{q}_3(k\Delta T) \end{bmatrix}. \tag{5.34}$$

The state vector for this particular example has the form

$$\boldsymbol{s}(k\Delta T) = \begin{bmatrix} q_1(k\Delta T) \\ q_2(k\Delta T) \\ q_3(k\Delta T) \end{bmatrix}. \tag{5.35}$$

It is clear from Eq.(5.34) that the state equation represents uniform motion. The optimisation procedure is described by Eqs.(5.31) and (5.32). The latter gets the following form (due to the definition of the function $\boldsymbol{f}(k\Delta T)$:

$$\dot{\boldsymbol{q}}(k\Delta T)_{I+1} = \dot{\boldsymbol{q}}(k\Delta T)_I - \boldsymbol{\mu}_{cv}[\frac{\partial \boldsymbol{P}_K(\boldsymbol{s},\dot{\boldsymbol{q}})}{\partial \dot{\boldsymbol{q}}(k\Delta T)} + \boldsymbol{\lambda}^T(k\Delta T)\frac{\partial \boldsymbol{f}(\boldsymbol{s},\dot{\boldsymbol{q}})}{\partial \dot{\boldsymbol{q}}(k\Delta T)}]. \tag{5.36}$$

Also in this case all the partial derivatives which appear in Eqs.(5.31) and (5.36) can be calculated in a closed form, which simplifies the overall calculations. Note also, that in the case of the integral model, the input correlation matrix $\boldsymbol{P}_K$ has a simpler form than in the case of the differential model because the integral model is always represented by one equation. In this case Eq.(5.21), compared to the differential model represented by Eq.(3.12) contains less information. Generally, the numerical calculation of the optimisation procedure for the integral model is less computationally expensive than for the differential model. Note also that in the case of integral model the generalised positions are calculated by performing numerical integration of both sides of the control equation (5.36). This procedure is usually stable and not prone to errors. During the numerical calculations it is necessary to verify the generalised acceleration, velocity, and position constraints. When these constraints are exceeded the appropriate limits are substituted and the calculation are continued from that point. Another solution to this problem is to choose different initial acceleration and velocity trajectories for the differential and integral models, respectively.

5.2.3 Exciting trajectories for the differential and integral models of the IRp-6 robot

In this section we present simulation results which illustrate the theoretical considerations of the preceding section. These results concern optimal trajectory generation for the purpose of identification of dynamic parameters for the differential and integral models of the IRp-6 robot. Both models are described in Section 3.7.2. A set of dynamic parameters of the canonical model is described by Eq.(3.168). The differential canonical model is described by

Eq.(3.169). In order to design the optimal trajectory based on the input correlation matrix calculated from the differential regression factor we consider two cases.

First we took all three equations (3.169) and assumed the friction torque representation given by Eq.(3.5). In this particular case the input correlation matrix was poorly conditioned assuming that the initial accelerations were spline polynomials of the fifth, third, and fifth order for the first, second, and third joint, respectively. All initial joint positions, velocities, and accelerations were chosen in such a way that they satisfy the constraints for the three degrees of freedom of the IRp-6 robot. The initial value of the condition number was of order 10^{14} and after 90 iterations its value was by only one order reduced. This result was very unsatisfactory mainly due to the friction model represented by Eq.(3.5) and to the fact that the attempt was to find the exciting trajectory for all the three joints in one run of the iteration algorithm given by Eqs.(5.31) and (5.32) with the initial condition $\boldsymbol{\lambda}(K) = \mathbf{0}$. Therefore, we did not continue the simulation experiments involving three joints.

Next, in the optimisation procedure we took only one equation from a set of equations (3.169), namely the first one (with the assumption that the friction forces are expressed by Eq.(3.5)). As the initial trajectory we chose spline polynomials for the first and second joint and a cosine trajectory for the third one. This initial guess for the input trajectory was much better then before because the initial value of the condition number was $19 \cdot 10^5$. After 87 iteration steps this value was reduced to 88. Each iteration of the Eqs.(5.31) and (5.32) was calculated in 300 positions ($K = 300$). We assumed that matrix $\boldsymbol{\mu}_{cv}$ was diagonal with elements equal to the reciprocal of the maximum values of the Lagrange multipliers. The results of the optimisation procedure for the first three joints of the IRp-6 robot are presented in Fig.5.2; only the acceleration signals are depicted. The initial trajectory for each joint is indicated by a continuous line and the optimal trajectory by a dotted line.

Note that the optimal trajectories have approximately the same frequency, except that the one for the first joint is of higher amplitude and third joint is shifted in time. From Eq.(3.169) it is clear that it consists of six basis functions. The optimisation procedure tries to shift the third trajectory in time so as to be able to excite all six parameters. The final accelerations are generally grater than the initial values. This can easily be explained. We have to accelerate the dynamic elements sufficiently in order to be able to identify them. The initial trajectory for the second joint is symmetric. As a result of the optimisation procedure we obtained a trajectory which was no longer symmetric. This observation was verified during experimental identification of the dynamic parameters of the IRp-6 robot. The same observation, most evident, for the second joint is also valid for the two other joints. Results presented here show that although the initial guess about the shape of the trajectory can be correct, it is usually wrong as far as the amplitude and phase shift are concerned. The optimal trajectories were used for the identification

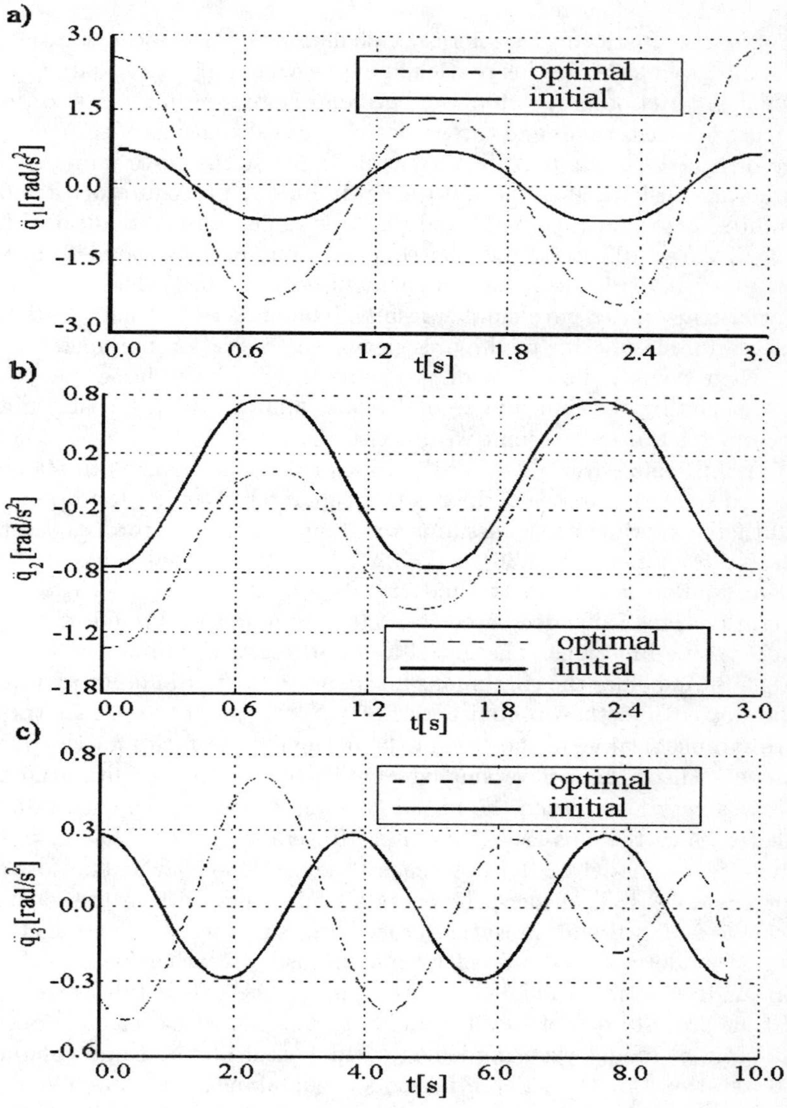

Fig. 5.2. Initial and optimal trajectory for the a) first, b) second, and c) third joint of the IRp-6 robot

experiments and gave satisfactory results. In the optimisation procedure we did not neglect the friction coefficients and when planning single movements of each joint it was possible to design the optimal trajectory which identified the friction coefficients.

Next we performed numerical calculations of the optimal trajectory for the integral model. In this particular case we had only one equation (3.168) in which six friction parameters were present. Note that in the case of the IRp-6 we used three torques and three velocities for calculations of the energy. First, we designed the input trajectory assuming that the three joints are moving simultaneously. Unfortunately the optimisation procedure of the condition number based on Eqs.(5.31) and (5.36) with the initial condition $\boldsymbol{\lambda}(K) = \mathbf{0}$ failed. After 100 iterations the condition number decreased from value of about 10^5 to 10^4 and remained constant with further calculations. Different initial trajectories gave similar results. The numerical calculations were not successful mainly due to the presence of all six friction parameters.

Next we moved only one or two joints. Moving only the second joint while not actuating the first and second joints, and assuming a cosine initial trajectory for the second joint we got the initial value of the condition number of about 3000. After about 23 iterations this value changed to 500. Next we moved only the first and third joints separately keeping the others idle. In all these experiments the assumption about the friction torques described by the vector functions (3.36) and (3.37) held. The initial value of the condition number in the last two movements was about $1.9 \cdot 10^5$. After only ten iterations this value decreased to 1000, assuming that the initial trajectories were cosine functions. The initial and optimal trajectories are illustrated in Fig.5.3. As before the continuous line denotes the initial input trajectory and the dotted line the optimal trajectory. Note, that the optimal trajectories are symmetrical as in the case of the optimal trajectories for the differential model. All of them have smaller amplitude and the condition number decreases very fast. When the initial guess is very good the condition number decreases to the smallest value after 10 iterations. Note that in the case of the integral model we did not neglect the friction coefficients. Most authors, for example [183], neglect the friction coefficients both in the identification scheme and in the optimisation procedure. Numerical calculations for the integral model were performed by making use of Mathematica package [230] on an IBM compatible PC 486 computer. The calculations of the optimisation procedure took several minutes. The numerical calculations for the differential model were implemented in Pascal on the same computer and 90 iterations for the differential model took about 100 hours. Obviously the computational requirements for the integral model are smaller than for the differential model.

The Monte Carlo method described at the beginning of the previous section gave surprisingly good results. This method was applied to the optimisation search in the space of available trajectories for all three joints in the

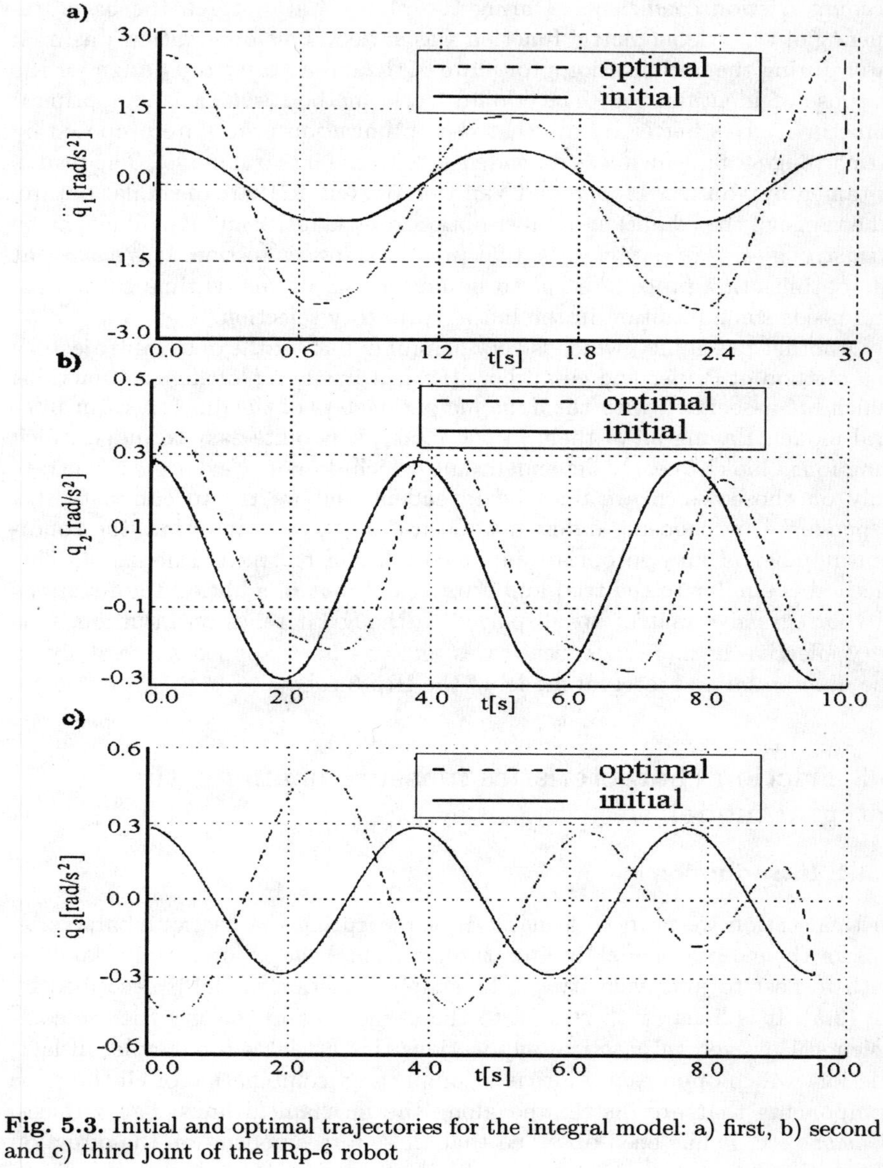

Fig. 5.3. Initial and optimal trajectories for the integral model: a) first, b) second, and c) third joint of the IRp-6 robot

case of the differential model. The optimal trajectories were combinations of trigonometric functions for which the condition number of the input correlation matrix changed within the range of 120 to 150. We also took into account friction coefficients. During the Monte Carlo search the basic frequency of the trigonometric function was varied. This observation was used later during the optimisation procedure of the input trajectory design for the purpose of identification. The Monte Carlo method (which is suboptimal) sometimes gives better results that the optimisation procedure proposed by Armstrong [4, 5], which was extended here to the integral model. The results obtained by Van Der Linden and Van Der Weiden [221] are of similar nature. The range of the condition number obtained by Gautier and Khalil [65] or by Armstrong [4, 5] is very close to that presented in this section. It is clear that the optimisation procedure has to be performed off-line, is time consuming and needs some intuition in the initial trajectory selection.

Another possibility, which is very helpful in finding the optimal trajectory is a method of Prüfer and co-authors [183]. It involves plotting the functions which are associated with the dynamic parameters of the differential or integral model. Having all of them in one figure, it is quite easy to judge which functions "look" linearly dependent and which do not. Next we concentrate only on those which are linearly dependent and we try to eliminate this dependence by choosing different trajectories. At the same time the condition number of an appropriate input correlation matrix is calculated. This method is similar to the trial and error search but in addition the functions of the regressive matrix are displayed. Although it relies on intuition, it is very effective in many practical cases and has been applied successfully to the differential and integral model of the IRp-6 robot.

5.3 Friction characteristics measurements for the integral model

5.3.1 Introduction

In this section we propose a method for measuring the friction characteristics for the integral model. In literature, frictional effects are mainly classified with respect to joint velocity $\dot{q}$. The simplest friction model is described by Eq.(3.5). It is linear with respect to the Coulomb and velocity friction coefficients. However, in practical applications this model is too simple. In fact, friction, which opposes motion, is a complicated combination of all the force components that are distributed along the mechanical links: flat surfaces, bearings etc. It has been observed that these forces depend on the direction of rotation, but the character of the function, in particular for low velocities, gives rise to some disagreements. In addition to well–known frictional models, temperature has to be taken into account. A very detailed analysis of friction torque modelling can be found in reference [6], where a Tustin model is discussed. This model, which seems to be quite a good approximation is described in this section. Canudas de Wit and co-authors [23, 25] analysed the problem of modelling and compensation of friction at low velocities. The need for this type of model is mainly motivated by the instability phenomena that can be caused by over–compensation when simple models are used as a basis for friction compensation.

As mentioned in the previous section, estimation of the friction coefficients is not an easy task, particularly for robots whose dynamics are described in terms of the integral model. This problem was addressed by many researchers. Interesting results of friction coefficient measurements can be found in [182, 184, 196, 208, 240]. All these references are concerned with the presence of friction torques in the differential model, which makes it easier to estimate the friction coefficients. In this section we propose to measure the friction characteristics for the integral model. The method proposed here is very effective and makes it possible to precompute the friction characteristics for the integral model and then to precompensate the friction torques in the integral model in order to identify the dynamic parameters of the model. This technique is discussed later along with the experimental results for the IRp-6 robot. Some preliminary results concerning the friction characteristics measurements for the integral model can be found in [45, 131].

5.3.2 Tustin model

The Tustin model [6] is a more general friction model than that described by Eq.(3.5). It was observed by Canudas de Wit and co-authors [23, 25] that the Tustin model is more appropriate for friction analysis for both geared and non-geared robots. The same assumption was verified earlier by Armstrong [6], who gave a complete friction analysis. Note that even for direct drive robots, friction cannot be neglected, particularly at low velocities, and has to be precompensated [185]. Below we describe the features which are associated with friction phenomena.

1. Asymmetries: It has been observed that imperfection in the motor mechanics and imbalances on the motor shaft yield asymmetric behaviour of the motor dynamics. This applies to the Coulomb friction component, which was verified in [6, 23, 25].
2. Coulomb/Stiction: In classical Coulomb friction models, there is a constant friction torque opposing the motion when the velocity is non–zero. The stiction opposes the motion as long as the torques are smaller in magnitude than the stiction torque. This is valid for zero velocity.
3. Position dependence: In some mechanical systems the friction torques depend on the angular position. In motor drives the position dependence can be considered a consequence of imperfections on the shaft and reductor centres. These imperfections will generate oscillations with a period equal to the reductor ratio. This phenomenon was observed for direct drive robots [185] as well.

4. Friction characteristics also depend on temperature.

In order to take into account all the features described above (except the temperature dependence) one has to use the Tustin friction model which is described by the following equation

$$\tau_f(\dot{q}) = \left(F_s - F_d\left(1 - e^{-\frac{|\dot{q}|}{\dot{q}_c}}\right) + F_v|\dot{q}|\right)\mathrm{sgn}(\dot{q}). \tag{5.37}$$

Here we have omitted the subscript which indicates the joint number in order to keep the consideration general. The joint velocity is denoted by $\dot{q}$. There are several constant coefficients which appear in Eq.(5.37). The first one, F_s, is a static friction coefficient. Formally, this coefficient is associated with static friction, also called stiction. It is the amount of torque (or force) necessary to initiate motion from rest. Static friction is often greater than kinetic friction. Kinetic friction, called Coulomb or dynamic friction, is a friction component that is independent of the magnitude of velocity. Here, in addition, we assume that the stiction does not depend on the position of the motor; although this is not the case. Coefficient F_v, is associated with velocity. The corresponding torque is called the friction viscous torque. This friction component is proportional to velocity, in particular it goes to zero at zero velocity. It is interesting to note a closely related phenomenon, the negative viscous friction, which is reported as the friction which at low velocities decreases with increasing velocity. This friction behaviour, see [23, 25] can in general be a source of instability. Coefficient F_d, is the difference between the static friction and the Coulomb friction. Finally, coefficient $\dot{q}_c$ is a constant which has units of velocity giving the characteristic velocity at which the system transits to kinetic friction. Note also that in Eq.(5.37) the sign function is present, which indicates the asymmetries of the friction torque discussed above.

In spite of the completeness of model (5.37) the nonlinearity of parameter $\dot{q}_c$ restricts its usefulness for on-line identification (linear predictors require a model expression that is linear in parameters). A curve which corresponds to the model given by Eq.(5.37) is shown in Fig.5.4, with coefficients F_s and F_d indicated. This model is more general than that given by Eq.(3.5); however it is nonlinear with respect to one parameter.

In order to give a more complete description of the friction phenomena we quote the following notions from Armstrong [6]. The notion of break-away is the transition from rest (static friction) to motion (kinetic friction). Consequently we introduce break-away torque (or force) as being the amount of torque (force) which is required to overcome static friction. Another notion is a break-away distance which denotes a distance travelled during break-away, that is the distance over which static friction operates, a consequence of the materials used and forces applied. Another term which appears in tribology is the Dahl friction or the Dahl effect. This is a friction phenomenon which arises from the elastic deformation of bonding sites between two surfaces

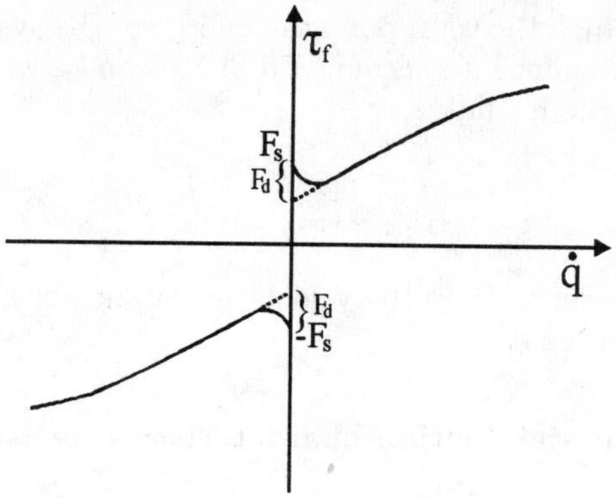

Fig. 5.4. Tustin friction model

which are locked in static friction. The Dahl effect causes a sliding function to behave as a spring for small displacements.

Finally, we introduce a friction phenomenon that arises from the use of lubrication and gives rise to decreasing friction with increasing velocity at low velocities. This friction phenomenon is called the Stribeck friction or the Stribeck effect and is illustrated in Fig.5.4. The Dahl effect, though usually present, is not discussed in this monograph. As we can see from the above description a variety of different friction components are usually presented and the Tustin model is only an approximation of these complicated phenomena. Obviously, the friction model cannot be neglected when we identify the dynamic model of the robot. As already mentioned, it is present in both types of robots.

There are many methods of measuring the friction characteristics for the differential model given by Eq.(3.12). The most representative method is a method proposed by Seeger and Leonhard [196] but it only applies to the differential model. In this section we extend their results to the integral model given by Eq.(3.38). The procedure consists of three steps

1. Step 1. Move the link i with a constant velocity as slow as possible. During this movement measure and store, the joint position and motor current.
2. Step 2. Increase the joint velocity by a constant value $\Delta\dot{q}$ and go back to step 1.
3. Step 3. When the link i achieves the highest joint velocity, end the procedure.

In order to analyse this three–steps procedure consider Eqs.(3.24) and (3.25) and Eq.(3.26). It is clear from these equations that the first component on the right side of Eq.(3.38) is equal to the difference between the total energy of the system calculated at time instant t_1 and t_2. According to the procedure presented above this component is exactly zero if we assume movements in which the total potential energy of the system does not vary. If we take this constraint into account Eq.(3.38) can be written in the following form (which is true for link i):

$$\tau_{fi}\,|_{\dot{q}=\text{constant}} = \frac{\int_{t_1}^{t_2} \tau_i dt}{t_2 - t_1} \tag{5.38}$$

In practice, in order to keep the velocity constant during the movement it is necessary to average it.

5.3.3 Experimental friction characteristics measurements for the IRp-6 robot

In this section we present experimental friction characteristics measurements for the first three links of the IRp-6 robot. These measurements were carried out in the experimental set-up described in Chapter 2. The friction characteristics measurements are associated with friction torques at the joints, which are indirectly influenced by the gear mechanical system between the motor and the appropriate link. Here we consider only the first three links since the experimental identification was carried out only for these links. It is assumed that the wrist is an integral part of the link. First we observed that the friction torques for the first link depend on temperature. This was because the harmonic drive which connects the first motor to the link itself was filled with oil. Temperature dependence was not observed for the second and third links.

Note that for the first link the condition given by Eq.(5.38) is satisfied automatically. There is no gravity for the first link. This situation becomes complicated for the second and third links of the IRp-6 robot; see discussion in Section 3.7.2. These links work in the gravity field. In order to measure the friction torque function versus velocity for the second link, this link is posed in a vertical position with very high accuracy. Next, symmetrical movements in the vicinity of the vertical position of the second link with a constant velocity are performed. Because the movements of the second link were made in both directions, it was necessary to recognise the direction of the joint velocity. Recall that according to Eq.(5.37) the friction torque depends on the direction of the velocity. Due to this particular movement Eq.(5.38) can be used.

For the third link, a different movement was designed. At the very beginning the second link was placed exactly in a vertical position. This was achieved by making use of counterbalances. The horizontal position was measured very accurately using a mason's level and adjusting the counterbalances.

Next, a similar kind of movement was designed for the third link. The link is moved up and down symmetrically around the horizontal position with nine different constant speeds. This special trajectory made it possible to employ

Eq.(5.38), because of gravitational restrictions. As before we recognised the direction of the movement in order to satisfy the requirements of Eq.(5.38).

In order to measure the friction torques for each link of the IRp-6 robot 11 measurement points were calculated. The generalised velocity of each link varied from the minimum, which was chosen as 10% of the maximum, up to the maximum, in increments of 5 % of the maximum velocity. For each joint movement designed according to the method described above, the joint position, its velocity, and the DC current were recorded with a time step of 5ms. The definite integral in Eq.(5.38) was calculated with the time interval 0.5 s for the first joint and 0.3 s for the second and third joints. Following this procedure for each measurement point we were able to record 500 and 300 data points for the first, and second and third links. As a consequence each measurement point of the friction characteristics was averaged from 500 and 300 measurement data. Time intervals 0.5 s and 0.3 s were chosen in order to keep the temperature constant during the experiment. In addition, during each movement the requirement of constant velocity was verified. This verification was based on incremental calculations of the velocity in degrees divided by the constant time increment. As a result, we got so-called predicted velocity and this was compared to the measured velocity; in case of difference the predicted velocity was corrected. By duplicating these calculations, we were able to keep the measured velocity constant at each measurement point. The additional calculations were performed in order to satisfy the very strict requirements concerning the implementation of Eq.(5.38). Moreover, each experiment was repeated a number of times in order to guarantee that these requirements were satisfied. Armstrong [6] notices that perhaps the most fundamental issue in an effort to model any process is repeatability. This concept was applied to the velocity calculations by correcting the predicted velocities and observing the measured velocities.

For each link the stiction friction was measured depending of the movement direction of the link. In order to measure the stiction friction the following method was used. An increasing torque was exerted on a chosen joint, which was measured by the force/torque sensor, while the drive unit for this joint was turned-off. At the instant in which the movement of the corresponding link started, the value of the torque measured by the sensor referred to the stiction friction torque. This experiment was repeated several times in both directions for each joint. Several torque values of each link were averaged in order to filter out measurement errors.

The friction characteristics for the first three links of the IRp-6 robot as functions of joint velocities are presented in Fig. 5.5. Consider the friction torque characteristics for the first link (see Fig. 5.5a). The measured stiction friction is 14.1N·m in both directions of movement. As has been mentioned, friction characteristics of a joint depend on temperature. The temperature of the oil in the harmonic drive was measured in the range from 20°C to 45°C. Different temperatures are indicated in Fig. 5.5a. For the other two joints,

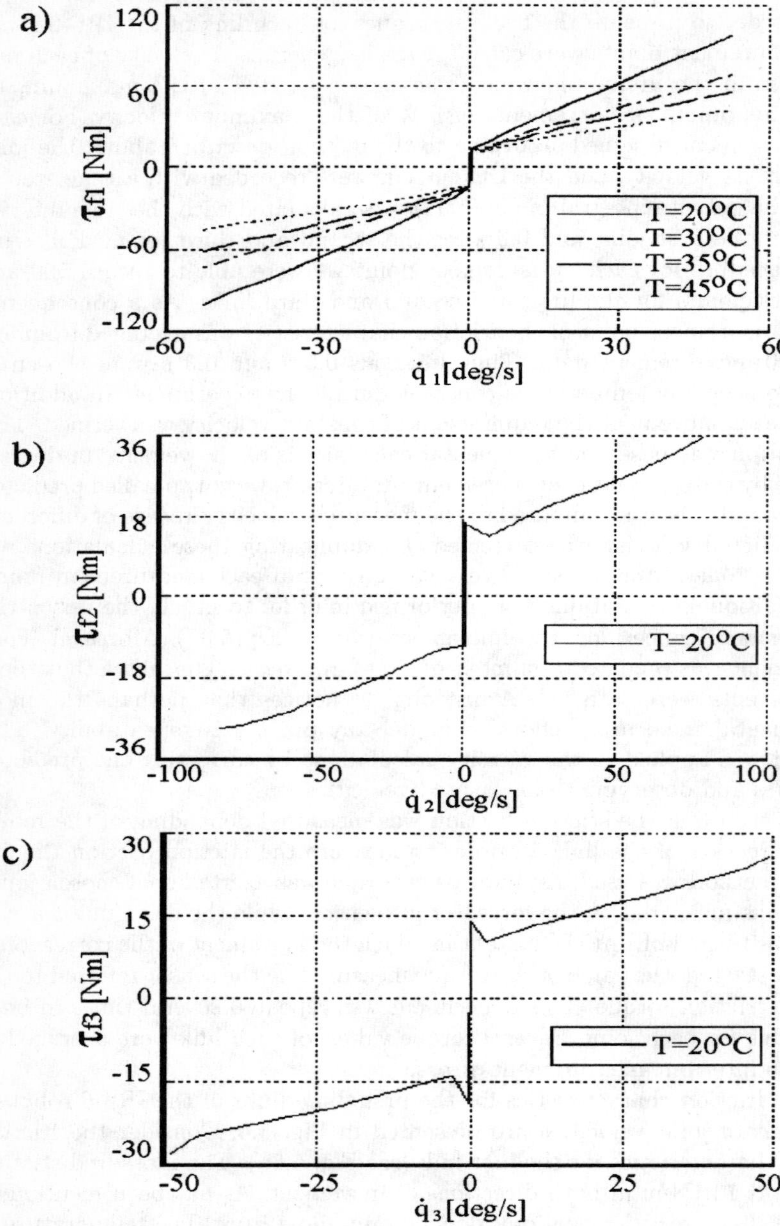

Fig. 5.5. Friction characteristics versus velocities for the first a) second b) third link c) of the IRp-6 robot

this phenomenon was not dominant, and for both curves in Figs. 5.5a and b the temperature was fixed at 20°C.

In Fig.5.5b the friction torque versus velocity for the second link is presented. Note that the stiction friction in the positive direction (here clockwise) is equal to 16.8N·m, while in the negative direction it is 10.6N·m.

In Fig.5.5c the friction torque for the third link is depicted. The stiction friction for the third link in the positive direction is equal to 18N·m and in the negative direction 14.1N·m. For all three movements we neglected the initial part of the movement (about 20% of the total movement) and final part (also about 20% of the total movement). This is possible because the initial and final parts of the velocity satisfy a trapezoidal profile of the velocity. Therefore, it is difficult to guarantee constant velocity in these regions. This assumption was verified in the experiment.

Comparing the curves presented in Figs.5.5a-c all friction phenomena associated with the Tustin model given by Eq.(5.37) can be observed. In addition, friction characteristics depend on temperature, which is not present in the Tustin model. The friction characteristics depicted in Fig.5.5a-c were approximated by making use of spline polynomial functions in Mathematica system [230]. As a result of this approximation we obtained curve which is similar to that in Fig.5.4. The conclusion is that the friction torque curve given by Eq.(5.37) is a more general form of the friction characteristics given by Eqs.(3.29) and (3.30) and is more accurate in practice. This postulate was verified experimentally during identification of dynamic parameters of the IRp-6 robot which is discussed in the next section.

We conclude this section with two remarks. The method proposed in this section applies to the friction characteristics measurements which are valid for integral models. The three–step procedure proposed here can be used for both classes of robot and has been successfully implemented for the IRp-6 robot. It is a generalisation of the methods presented in references [182, 183, 184], which are applicable to the class of geared robots described by the differential model. We believe that the friction phenomenon is very important for robots with gears and that the friction torques are dominant for these robots; therefore the results discussed in this section are complementary to the results given in references [182, 183, 184].

Our second remark concerns the problem of repeatability of friction characteristics measurements. Because the friction characteristics measurements depend on many features and on external conditions such as temperature, direction of rotation and other factors, the measurement data should, as for as possible, be collected under identical circumstances. For example, prior to collecting the measurement data, the robot should be "warmed up" for some time (several minutes) of motion throughout the workspace. The friction torque measurements have to be precomputed and then subtracted from the generalised torques. Next, based on the identified dynamic model of the IRp-6 robot, one has to recover the input trajectory by integrating the equations of motion in order to find an input trajectory described in terms of positions, velocities, and accelerations. The input trajectory should be recovered accurately, which is discussed in detail in the next section. In addition to that we have to perform the experiment many times in order to check the repeatability. One can calculate statistical factors such as the mean square error and the variance of the measurement data to check whether they are within a certain interval of confidence.

Here we can follow the procedure proposed by Armstrong [6]. We believe that it is possible to validate the friction torque measurements by precomputing the friction characteristics and then subtracting these functions as offsets from the total generalised forces measurement. Such elements of friction as stiction and breakaway can be hidden at the software and hardware level as constant offsets which are precompensated values. Depending on the work cycle of the robot, remember that robot can work for hundreds of hours, the chances are that its parameters may change. This is particularly valid for a class of geared robots for which the friction torques are dominant. Therefore, it is advisable to recalculate the precompensated values at regular intervals.

5.4 Experimental identification results for the IRp-6 robot

In Section 3.7.2 we studied the differential and the integral models of the IRp-6 robot. As before we now restrict our considerations to the first three degrees of freedom of this robot. The differential model of the robot is given by Eqs.(3.130)-(3.133); its kinematical structure is presented in Fig.3.6a. Note that the differential model is described in terms of the ϕ_1, ϕ_2, and ϕ_3 generalised coordinates, which are very suitable since each motor drives only one link independently. As can be seen from Fig.3.7 by choosing generalised coordinates ϕ_1, ϕ_2, and $\bar{\phi}_3$ (which is another possible set of coordinates) we are not able to move the third link of this robot independently from the movement of the second link due to the parallelogram structure of the robot. Formally choosing generalised coordinates according to the modified Denavit-Hartenberg notation we get the coordinates which are summarised in Table 3.2. Also in this case we cannot move the third link independently from the first. The resulting equations of motion are given by Eq.(3.169) which yields the differential model described in terms of generalised coordinates q_1, q_2, and q_3. This formal approach is correct, but shows that rigorously following the notation does not always result in the most effective model from a computational point of view. Next we have formulated the integral model in terms of generalised coordinates q_1, q_2, and q_3 which is given by a set of equations (3.168). In the integral model there are 13 dynamic parameters which are combinations of dynamic parameters of individual links. The differential and the integral models are equivalent and are expressed in terms of the same set of canonical parameters. In addition to that, we have assumed that nonlinear functions which describe the gear ratios between the motor velocity and link velocity for the second and third links are constant within the vicinity of the synchronisation position of the robot. These functions given by Eqs.(3.159) and (3.160) were evaluated numerically and their values in the vicinity of the synchronisation position were almost constant. They change by less than 1% of the value at the synchronisation position. Recall that this position is described in terms of generalised coordinates $q_2 = \frac{\pi}{2}$ and $q_3 = \frac{3}{2}\pi$ and in terms of motor coordinates as $\Psi_2 = \Psi_3 = 0$. At the limits Eqs.(3.159) and (3.160) are very inaccurate and change drastically. Surprisingly all dynamic parameters of the canonical model were identified within the vicinity of the synchronisation position and the above assumption was verified.

In this section we present the identification results concerning the integral model given by a set of Eq.(3.168). The experiments for the differential model were carried out as well and, as mentioned in Section 4.5, simulation results for both types of models were also obtained. The experimental results for the integral model are compared to those in [73, 74], which were obtained for the differential model of the IRp-6 robot. This comparison is possible by taking into account Table 3.3, which relates different quantities in the differential model described in [73, 74] to the integral model given by the set of Eq.(3.168). First notice that the equations for the integral model are augmented by two components, $0.5 I_{2xx} \cos^2(q_2)\dot{q}_1^2$ and $0.5 I_{3xx} \sin^2(q_2 + q_3)\dot{q}_1^2$, which according to [73] and [74] were assumed to be zero. Although inertias I_{2xx} and I_{3xx} have indeed small numerical values, in order to get a representative model for the IRp-6 robot we did not neglect them. Functions d^i in Eq.(3.168) depend on generalised positions and velocities. These are the link positions and velocities and at any time of the calculations they have to be transformed to the motor quantities via appropriate gear ratios because these signals are measured by the measurement system described in Chapter 2.

The first identification experiments with the integral model were not successful. It was not possible to identify all 13 parameters of the canonical model in one experiment regardless of whether the integral was short or long and regardless of the type of the input trajectory. Practically it was possible to identify only 10 dynamic parameters in a single experiment. This situation can easily be explained. The IRp-6 robot has a parallelogram structure, which complicates the nonlinear functions $d^i(\boldsymbol{q}, \dot{\boldsymbol{q}})$ even if carefully chosen input trajectories are used. These functions become linearly dependent. They were displayed graphically during the experimental identification. Therefore it was decided to move one or two joints at a time to reduce the dimension of the problem.

Here we propose the following movements of the three links of the IRp-6 robot. By moving only the third link we get the following parameters and functions d^i:

$$
\begin{array}{ll}
\textbf{parameters} & \textbf{functions}\ d^i(\boldsymbol{q},\dot{\boldsymbol{q}}) \\
X_1 = I_{3zz} + I_{3zz}\bar{n}_3^2, & 0.5\dot{q}_3^2, \\
X_2 = m_3 c_{3x}, & -g_0 \sin(q_2+q_3), \\
X_3 = F_{3v}, & \int \dot{q}_3^2 dt, \\
X_4 = F_{3s}, & \int |\dot{q}_3| dt.
\end{array}
\tag{5.39}
$$

By moving only the second link we get the following parameters and functions:

$$
\begin{array}{ll}
\textbf{parameters} & \textbf{functions}\ d^i(\boldsymbol{q},\dot{\boldsymbol{q}}) \\
X_1 = m_2 c_{2x} + m_3 l_2, & -g_0 \cos q_2, \\
X_2 = I_{2zz} + I_{2a}\bar{n}_2^2 + m_3 l_2^2 + I_{3zz} + I_{3a}\bar{n}_3^2, & 0.5\dot{q}_2^2, \\
X_3 = F_{2v}, & \int \dot{q}_2^2 dt, \\
X_4 = F_{2s}, & \int |\dot{q}_2| dt.
\end{array}
\tag{5.40}
$$

By moving only the first link we obtain:

$$
\begin{array}{ll}
\textbf{parameters} & \textbf{functions}\ d^i(\boldsymbol{q},\dot{\boldsymbol{q}}) \\
X_1 = I_{1zz} + I_{a1}n_1^2 + I_{3yy} + I_{2xx} - I_{3xx}, & 0.5\dot{q}_1^2, \\
X_2 = F_{1v}, & \int \dot{q}_1^2 dt, \\
X_3 = F_{1s}, & \int |\dot{q}_1| dt.
\end{array}
\tag{5.41}
$$

In all the above movements we have assumed that the idle links remain in their synchronisation positions. As a consequence, their joint coordinates have the following values: $q_1 = 0$, $q_2 = \frac{\pi}{2}$, $q_3 = \frac{3}{2}\pi$. Links which move can be at an arbitrary position, though the synchronisation position is the most preferable. This assumption was verified in practice.

Finally, we design a movement in which the first and third links are not idle and the second link is in its synchronisation position. Accordingly we get the following parameters and $d^i(\boldsymbol{q},\dot{\boldsymbol{q}})$ functions:

$$
\begin{array}{ll}
\textbf{parameters} & \textbf{functions}\ d^i(\boldsymbol{q},\dot{\boldsymbol{q}}) \\
X_1 = I_{1zz} + I_{a1}n_1^2 + I_{2xx} + I_{3yy} - I_{3xx}, & 0.5\dot{q}_1^2, \\
X_2 = I_{3yy} - I_{3xx}, & 0.5\dot{q}_1^2 \sin^2 q_3, \\
X_3 = m_3 c_{3x}, & -g_0 \sin(q_2+q_3), \\
X_4 = I_{3zz} + \bar{n}_3^2 I_{3a}, & 0.5\dot{q}_3^2, \\
X_5 = F_{1s}, & \int |\dot{q}_1| dt, \\
X_6 = F_{3s}, & \int |\dot{q}_2| dt, \\
X_7 = F_{1v}, & \int \dot{q}_1^2 dt, \\
X_8 = F_{3v}, & \int \dot{q}_2^2 dt.
\end{array}
\tag{5.42}
$$

Note that in Eqs.(5.39)÷(5.42) we have numbered the parameters in the order in which they appear in Eq.(3.168) and the indices of the parameters are not same as in Eq.(3.168). We did it just for simplicity. It is important to clarify that in the above equations whenever the third link is idle we have $q_2 + q_3 = 2\pi$ and $\dot{q}_2 + \dot{q}_3 = 0$. This follows from Eq.(3.141) and the appropriate relationship between generalised positions q_1, q_2, and q_3 and generalised coordinates ϕ_1, ϕ_2, and ϕ_3 which are indicated in Fig.3.5. This is equivalent to the situation when $\phi_3 = 0$ and its time derivative $\dot{\phi}_3 = 0$. Recall that ϕ_2 and ϕ_3 are driven independently. It follows from a very simple analysis of Eqs.(5.39)÷(5.42) that the set of test motions described above make it possible to identify all canonical parameters which appear in Eq.(3.168). In addition we assumed that parameter X_2 which appears in Eq.(3.168) was not identified because that it is very small. This situation was discussed in Section 4.5. This assumption was verified by calculating the dynamic parameters which appear in parameter X_2 in Eq.(3.168) (see [73]).

It has to be pointed out that by performing movements of the three links independently or designing a combination of test movements of two links, it is possible to identify all the dynamic parameters which appear in Eq.(3.168). In addition to that, four sets of parameters which appear in Eqs.(5.39)-(5.42) are independent in the sense that none of the identified parameters in one set appears in the other sets as a parameter whose value is necessary for further identification. Because of this we do not observe accumulation error as in the case of a sequential identification scheme. This problem was discussed by Canudas de Wit and Aubin in [26]. In a hybrid sequential algorithm designed according to their procedure, the bias accumulated errors can easily be avoided by properly choosing the test movements. As mentioned in Chapter 3, the kinematical structure of the robot determines which dynamic parameters appear in the differential and integral models. For the IRp-6 robot it is not possible to position the third link to be perpendicular to the second link due to the parallelogram structure of the robot. At the same time positioning those links to be mutually perpendicular and moving the first link it is possible to identify parameter X_2 in Eq.(3.168). Due to this fact parameter X_2 is poorly identified. This complicates the identification in this particular case. As discussed in Chapter 4.5 experimental identification requires a lot of intuition, which cannot be neglected, to solve this problem successfully for the complicated mechanical systems.

Now we describe the numerical results of the experimental identification of the four designed movements. Consider first the movement of the third link. The dynamic parameters which can be identified during this movement are given by Eq.(5.39). The movement of the third link is characterised by the curves presented in Fig.5.6. In Fig.5.6a we can observe the generalised coordinate q_3; its derivative is shown in Fig.5.6b. Generalised moment τ_3 is depicted in Fig.5.6c. Finally, in Fig.5.6d we observe two curves: one is calculated based on the left hand side of Eq.(3.38) (continuous line) and the other calculated based on the right hand side of Eq.(3.38) (dotted line). The input trajectory for the third link is a fifth-order spline polynomial. The results of the experiment are presented in Fig.5.7a-d. The estimates of the parameters have the following numerical values: $\hat{X}_1 = 9\text{kg}\cdot\text{m}^2$, $\hat{X}_2 = -0.36\text{kg}\cdot\text{m}$, $\hat{X}_3 = 5.2\text{N}\cdot\text{m}\cdot\text{s}$, and $\hat{X}_4 = 18.5\text{N}\cdot\text{m}$. It can be seen that the estimates are becoming stable after 1 to 1.5s of the total observation time.

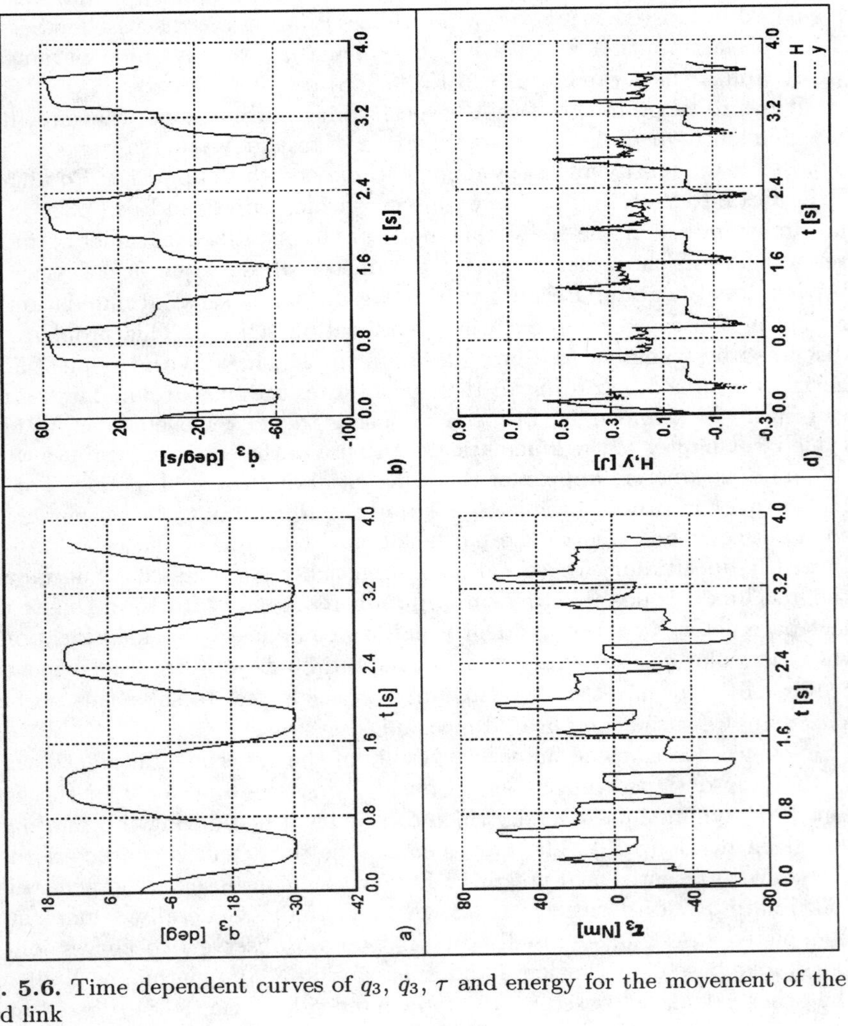

Fig. 5.6. Time dependent curves of q_3, $\dot{q}_3$, τ and energy for the movement of the third link

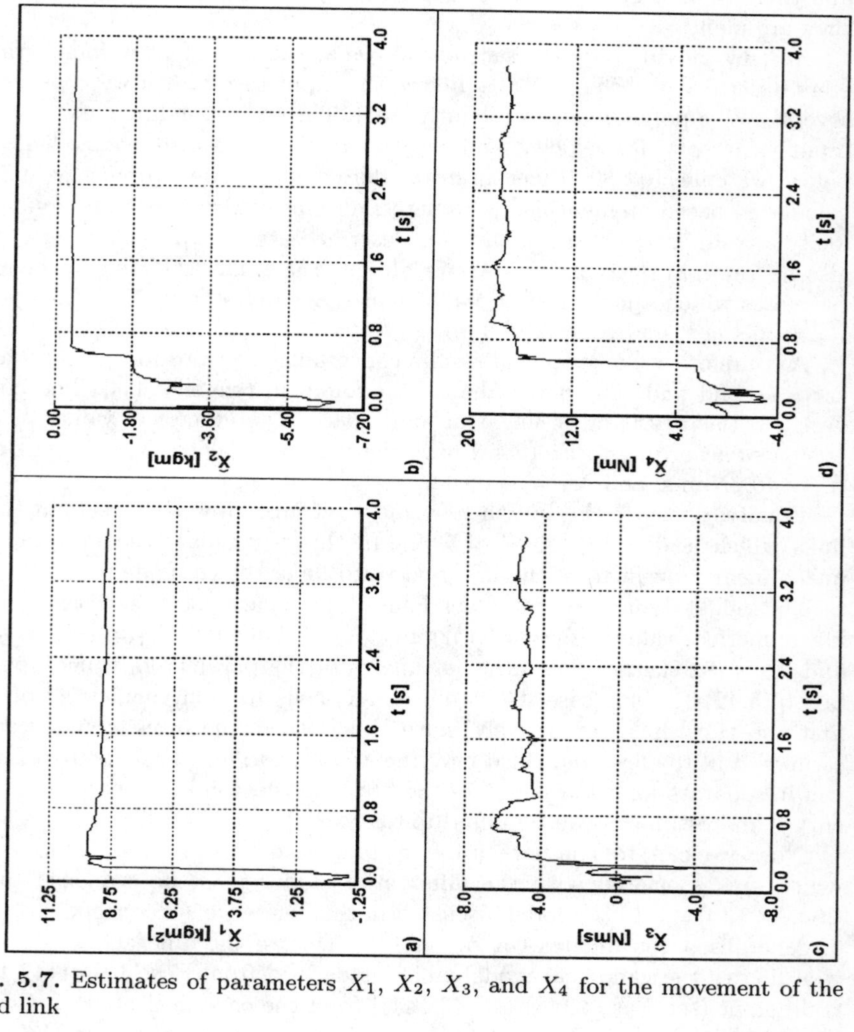

Fig. 5.7. Estimates of parameters X_1, X_2, X_3, and X_4 for the movement of the third link

For this particular movement we collected 8000 measurement points for the input trajectory and corresponding generalised torque. The total observation time was 4s. Note that the two energy curves in Fig.5.6d are very close to each other. The measurement torque τ_3 is noisy (see Fig.5.6c) and this noise is, to some extent, filtered out by the integral. At the peak values and their vicinity two integrals in Fig.5.6d differ, but except for these points they are identical.

Now by moving only the second link we are able to identify four parameters described by Eq.(5.40). As before the input test trajectory consists of several spline polynomials of the fifth order. The corresponding curves of the input trajectory, its velocity, and torque can be found in reference [34]. As before we collected 8000 measurement data points. The two energy curves calculated based on identified parameters and measured torque and joint velocity are similar to those in the case described above. This comparison can also be found in reference [34]. We recall only the numerical value of the four estimates which appear in Eq.(5.40), namely $\hat{X}_1 = 8.1$kg·m, $\hat{X}_2 = 5.5$kg·m^2, $\hat{X}_3 = 26$N·m·s, and $\hat{X}_4 = 28.7$N·m.

A complete set of graphical results concerning the third movement which keeps second and third links idle can be found in reference [34]. The input design trajectory is quasi-linear in joint space. The numerical values of the estimates according to Eq.(5.41) have the following values: $\hat{X}_1 = 11.8$kg·m^2, $\hat{X}_2 = 40.2$N·m·s, and $\hat{X}_3 = 36.06$N·m.

Finally, consider the fourth movement which moves the first and third links which is described by Eq.(5.42). In this particular case we assumed quasi-linear movement of the first and third links, which made it possible to estimate eight dynamic parameters. Four of these parameters have the following numerical values: $\hat{X}_1 = 11.15$kg·m^2, $\hat{X}_2 = 4.3$kg·m^2, $\hat{X}_3 = -0.34$kg·m, and $\hat{X}_4 = 8.98$kg·m^2. The other estimates of the parameters which appear in Eq.(5.42) (in this case, Coulomb and viscous friction coefficients of the first and third link, respectively) have almost the same value as parameters estimated in the first and third movement. This verifies the estimates identified in separate joint movement of the first and third link. All corresponding curves and estimates can be found in reference [34].

Now we can to compare these results with results in reference [73], which are associated with the differential model given be sequence equations (3.130)÷(3.133). Model discussed in this reference is described in terms of generalised coordinates ϕ_1, ϕ_2, and ϕ_3. Due to the different set of joint coordinates the canonical model which is derived from Eqs.(3.130)÷(3.133) is different (see Eqs.(3.143) and (3.144)) from the canonical model given by Eq.(3.169) in terms of the set of canonical parameters. Although the individual link parameters which appear in both representations are the same as far as the dynamics are concerned, they are regrouped differently, and due to the specific kinematic representation the regrouping procedure does not give exactly the same parameters. Nevertheless, some of the parameters from the canonical set of parameters are the same in both representations. Generally, this comparison is often difficult because in practice we do not know the exact values of the dynamic parameters.

The results presented here are more detailed and more accurate than those in reference [73]. First of all in [73] there are no results which are concerning the estimation of the Coulomb friction coefficients. Besides, here we identify parameter X_6 from a set of parameters given by Eq.(3.168), which in our case is close to zero (−0.39kg·m); in reference [73] it is quoted to be about 5kg·m. The latter is obviously wrong, the reason being the counterbalance located at the third link was not included in the calculations. This element of the first moment should be close to zero which has a clear physical interpretation. The counterbalance is mounted on the third link because it keeps the link in a stable position when the control system of the robot is turned off. Otherwise the link would fall down when the control system is not active. In addition to that, in reference [73] there is no comparison between the predicted and measured torques which is the simplest test of the identification quality. This verification was done in this monograph.

The above results for the integral model were obtained by making use of the short integral which was calculated every 5ms. Employing of the long integral instead generally gave worse results due to the fact that joint positions, velocities, and torques are noisy signals and these signals were integrated over a long period of time. This was obviously a source of inaccuracy for estimates. The conclusion is that in the implementation of the integral model the short integral performs better. The experimental identification was also carried out for the differential model represented by Eq.(3.12). The results obtained were similar and are not presented here.

In Sections 5.2 and 5.3 we described techniques in which the optimal test trajectories were introduced and a special technique to measure friction characteristics. We applied these techniques to the experimental identification of the integral model and now discuss the results.

We found that friction characteristics can be measured accurately for the integral model. The experiments carried out for the IRp-6 robot shown that identification of the friction coefficients from the integral model is always difficult because the functions which are associated with the Coulomb and viscous coefficients (Eqs.(3.36) and (3.37)) are usually linearly dependent for a wide class of input trajectories. This is due to the function of the type $|\dot{q}_i|$ and $\dot{q}_i^2$ which appear in Eqs.(3.36) and (3.37). Decreasing the limits of the integrals $\Delta T = t_2 - t_1$ to let's say 5 to 10ms, improves the situation, but even in this case the values of the integrals can still be proportional. An alternative solution suggested in Section 5.3 is to precompute the friction characteristics for the differential and integral models which are presented in Eqs.(3.12) and (3.38), respectively. Following the procedure described in Section 5.3 we can eliminate the friction characteristics from Eq.(3.38). Note that the friction model is represented by the Tustin model, rather than the Coulomb and viscous terms which we identified in movements of a single link.

Now analyse the situation when we only more the first link. The only parameter from the set of parameters represented by Eq.(5.41) which has to be estimated is X_1 The friction characteristics were precomputed and subtracted from the left hand side of Eq.(3.38). Estimate $\hat{X}_1$ of the first mass parameter is plotted in Fig.5.8. The numerical value of the estimate is $\hat{X}_1 \approx 8\text{kg}\cdot\text{m}^2$. This value is different from the value discussed above before ($11.8\text{kg}\cdot\text{m}^2$) because the gripper was not mounted during the experiment. We can easily weigh the gripper and verify that the difference between the numerical values of this parameter is due to the mass of the gripper. Note that estimate $\hat{X}_1$ is very stable. Usually the estimates of the parameters behave stable when displayed as functions of time. The estimate is stable after 3s of experiment time. The friction coefficients, (Coulomb and viscous) have approximate values $\hat{X}_2 = 15\text{N}\cdot\text{m}$ and $\hat{X}_3 = 8\text{N}\cdot\text{m}\cdot\text{s}$ which are different to those identified before. These values can be found in Fig.5.5a when we assume a working temperature of 20°C.

In order to verify these results we designed the optimal trajectory for the movement of the first link according to the optimisation procedure introduced in Section 5.2. The optimal trajectory for first link is presented in Fig.5.2a. Using this trajectory we obtained the estimates which are almost the same as those discussed above. For instance, the estimate of the mass parameter $\hat{X}_1$ is 8 $\text{kg}\cdot\text{m}^2$. We repeated this procedure for all the links of the IRp-6 robot. To

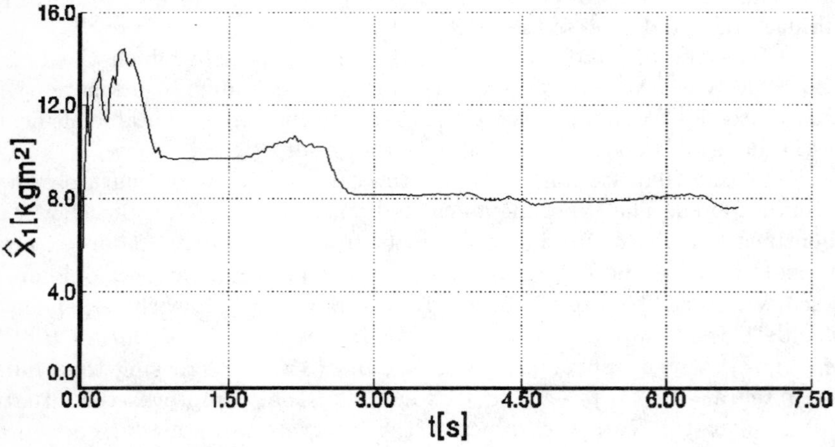

Fig. 5.8. Estimate of the mass parameter of the first link of the IRp-6 robot

sum up, we precomputed and subtracted the friction characteristics and next we identified the mass parameters. In order to verify the estimates we chose the optimal trajectories for the individual joints according to the procedure

Table 5.1. Estimates of the dynamic parameters of the integral model of the IRp-6 robot (see [139])

Parameters	Estimates
$\hat{X}_1$ [kg·m^2]	8.0
$\hat{X}_2$ [kg·m^2]	not identified
$\hat{X}_3$ [kg·m^2]	27
$\hat{X}_4$ [kg·m^2]	9.0
$\hat{X}_5$ [kg·m^2]	4.5
$\hat{X}_6$ [kg·m]	-0.2
$\hat{X}_7$ [kg·m^2]	10
$\hat{X}_8$ [N·m]	15
$\hat{X}_9$ [N·m]	28.5
$\hat{X}_{10}$ [N·m]	35
$\hat{X}_{11}$ [N·m·s]	88
$\hat{X}_{12}$ [N·m·s]	23.5
$\hat{X}_{13}$ [N·m·s]	22

described in Section 5.2 and compared the results. Note that we included the Coulomb and viscous friction coefficients. The results of the estimates are summarised in Table 5.1, in which we assumed that the gripper was not mounted at the wrist of the robot.

The comparison of two integrals: one calculated from the identified parameters and measured generalised positions and velocities expressed as the right–hand side of Eq.(3.40) and the other, called the predicted integral, calculated from the measured velocities and torques (see Eq.(3.38)) is not necessarily a good criterion of the quality of the identified estimates. See, for example, Fig.5.6d in which these two integrals give almost identical values. This observation disagrees with Gautier and Khalil [60] when we use the integral model for the purposes of identification. Minimisation of the mean square error discussed in Chapter 4 in the case of linearly dependent parameters, results in estimates which are not correct and at the same time the two energy functions are identical. Experiments and simulation results confirm this (see Section 4). Intuitively, we can state that the estimates of the parameters are chosen by the least squares method in such a way that the fit of the two integrals is perfect. The quality of the estimation is discussed later in this section.

Now we describe in detail the results of the experimental identification by means of optimal trajectories. The optimal exciting trajectories are described in detail in Section 5.2. Prüfer and co-authors [183] state that the integral model is not suitable for identification mainly due to the difficulties of friction coefficients identification. It is partly true and confirmed by our simulation and experimental practice concerning the IRp-6 robot. On the other hand, if the optimal exciting trajectories and the short integral are employed, the integral model can be very useful for identification purposes.

The results of the estimates for the integral model of the IRp-6 robot using the optimal trajectories (discussed in Section 5.2) are summarised in Figs.5.9, 5.10, and 5.11 for the first, second, and third movements of the joints, respectively. Recall that in these movements the dynamic parameters which appear in Eqs.(5.41), (5.40), and (5.39) are subject to identification. The corresponding estimates are depicted in Figs.5.9(a-c), 5.10(a-d) and 5.11(a-d). In all these figures the continuous line indicates the estimates of the appropriate parameters using non-optimal trajectories. The dotted line indicates estimates obtained using the optimal trajectories designed according to the procedure described in Section 5.2. It happens that the non-optimal trajectories used for the purpose of identification are exactly the same as those used for the identification purposes at the beginning of this section. Note that as a result of applying the optimal trajectories we obtain estimates which are very stable and have numerical values are closer to the real values of the dynamic parameters. Here we refer to the calculations of the mass parameters of the IRp-6 robot by Szkodny [213], also reported in reference [73]. These results were obtained by disassembling the IRp-6 robot and weighing and measuring its links in order to calculate numerically their inertia values. Obviously these calculations are not very precise. Comparing the results of experimental identification and of simulation and using engineering intuition it is possible to evaluate the identification results. It was found that the results of the estimates using the optimal exciting trajectories were close to real values. As mentioned above, the X_1 parameter of the first link was identified with and without friction coefficients using optimal exciting trajectories. Both estimates have similar values. We compared friction coefficient estimates identified from the precomputed friction characteristics and next by estimating them using optimal input trajectories. The results were very close and this again verifies the proposed method of measuring the friction characteristics. Friction plays an important role for the class of geared robots and we are able to identify their coefficients using the integral model. These results disagree with reference [183]. There are not many identification experimental results for the class of geared robots described in terms of the integral model. Results presented here fill this gap in robotics literature.

Now we comment on the quality of the estimates of the integral model. In practice we are not able to implement any control scheme which is based on an identified dynamic model due to the lack of access to the torque loop in the controller of the IRp-6 robot. Recall that the IRp-6 robot is only positionally controlled. Another method which compares both energies calculated from Eq.(3.38) can lead to misunderstanding of the quality of the identified parameters. In order to solve this problem we suggest using two methods. The first method compares the measured torques with the torques calculated on the basis the inverse dynamics problem. This method, described, for example

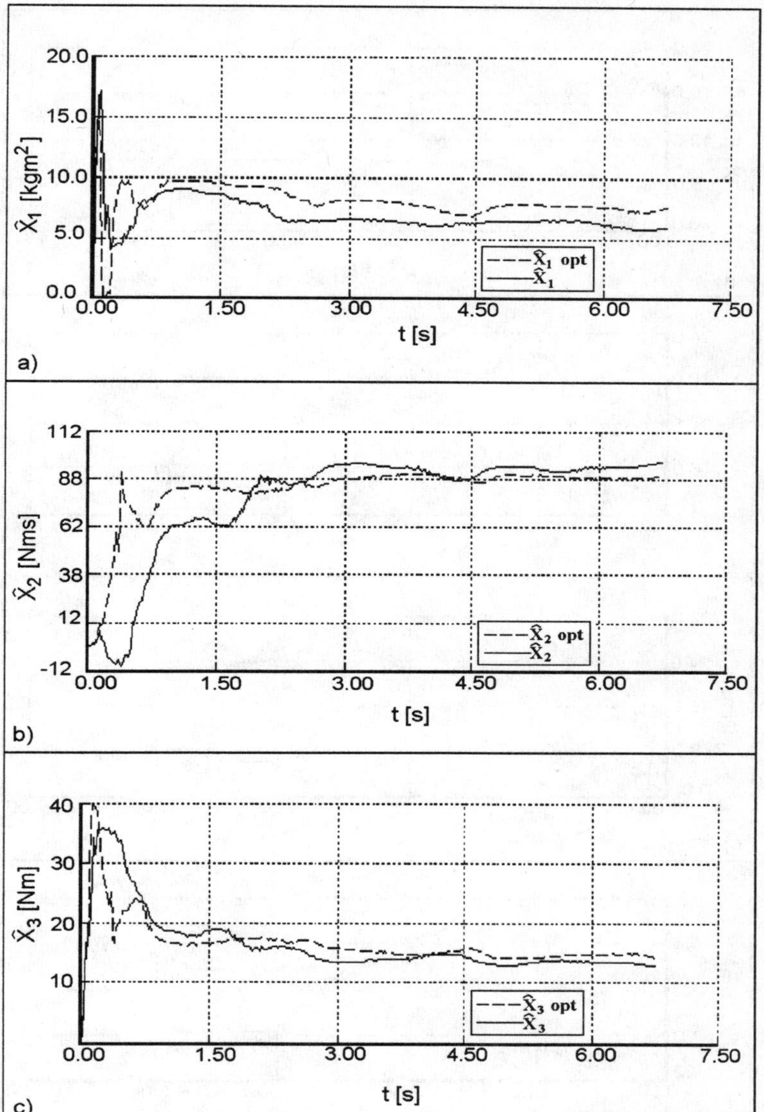

Fig. 5.9. Estimates of the parameters of the first link of the IRp-6 robot: a) $\hat{X}_1$, b) $\hat{X}_2$, c) $\hat{X}_3$

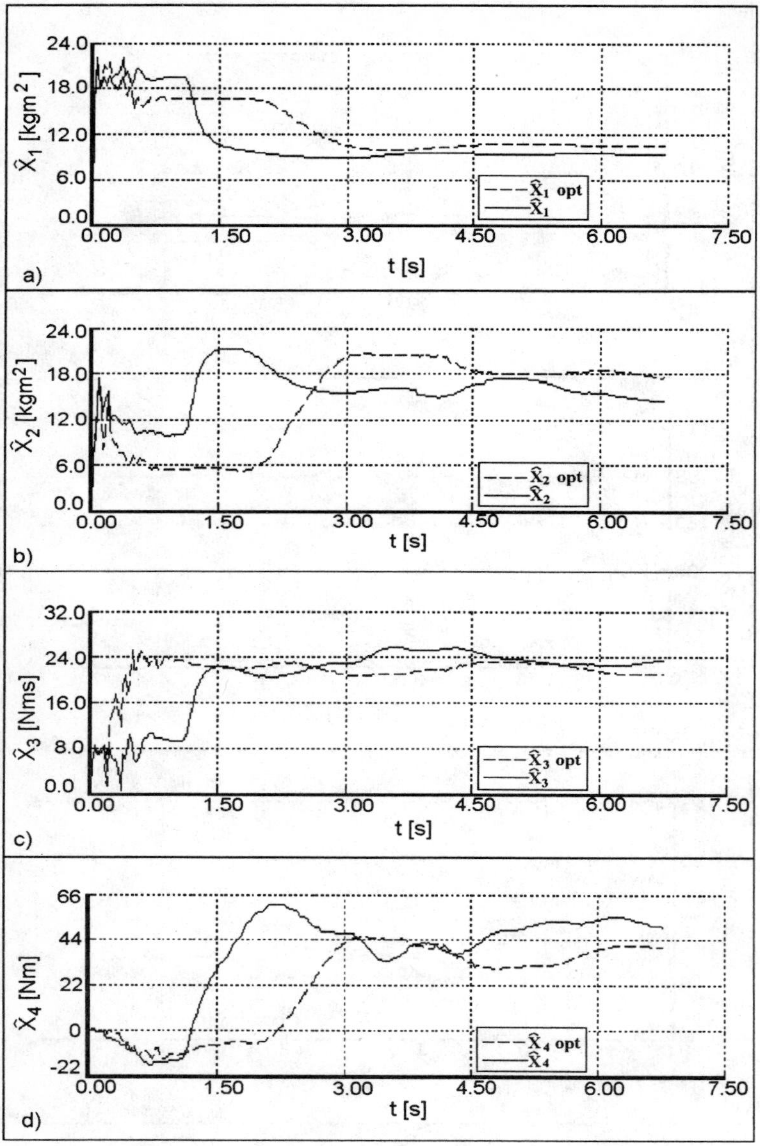

Fig. 5.10. Estimates of the parameters of the second link of the IRp-6 robot: a) $\hat{X}_1$, b) $\hat{X}_2$, c) $\hat{X}_3$, d) $\hat{X}_4$

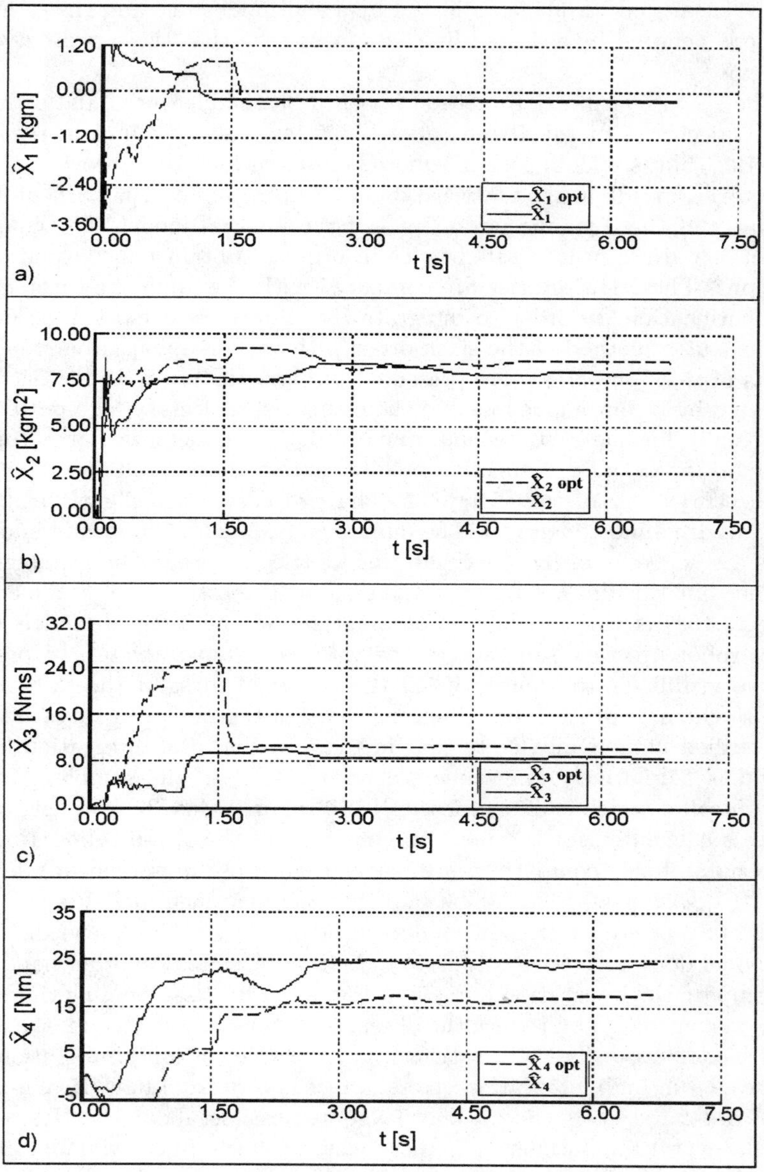

Fig. 5.11. Estimates of the parameters of the third link of the IRp-6 robot: a) $\hat{X}_1$, b) $\hat{X}_2$, c) $\hat{X}_3$, d) $\hat{X}_4$

in reference [1], was applied to the identification results for the IRp-6 robot. The results of this comparison are presented in Figs.5.12a and b. In this figure the measured torque is indicated by a continuous line and the estimated torque is denoted by a dotted line. It can be seen that these two curves are very close.

The second verification method is based on the measured and estimated joint trajectories of the IRp-6 robot. This test can be found in references [177, 192], but is seldom used in robotics. It seems logical to recover the input trajectories by solving the forward dynamics problem on the basis of on the estimates of the dynamic parameters of the integral model. The equations of motions are then integrated twice in order to obtain joint velocities and positions. These trajectories are compared with the input trajectories used for identification. In order to integrate the equations of motion we use the Runge–Kutta method of the sixth order with varying integration steps. The results of this comparison are presented in Fig.5.13. The initial trajectory is indicated by a continuous line and the estimated trajectory by a dotted line. The results for the first, second, and third joint trajectories are presented in Fig.5.13a, b, and c, respectively. Note that the best results were obtained for the trajectory of the first joint. The two curves are slightly shifted in time but are almost identical. The numerical results for the second and third joints are worse. In order to explain this we tried several different integration schemes but this did not improve the results. The estimate of parameter X_6 of Eq.(3.168) is very small. The differential and the integral models of the IRp-6 robot are very sensitive to the value of this parameter. In practice, it is very difficult to identify it, as the centre of mass of the third link is close to the axis of rotation. The sensitivity of the model to this parameter was verified numerically. Note that in Fig.5.13 the total integration time is 6s and is a relatively long time when compared to the sampling time of the controller which is 32ms for the IRp-6 robot. Every 32ms a new control signal is calculated for the position controller of the IRp-6 robot. However, the results of this comparison are better than those presented in references [177, 192]. Anyhow it cannot be said that the identified model is not useful for the purpose of control. In modern control schemes the knowledge about the model does not necessarily have to be accurate. On the other hand when we compare the integration horizon in Fig.5.13 with the sampling time of the position controller and reduce the observation time to one second, the results of this comparison are still satisfactory. Moreover, a long integration time still results in high integration errors, a fact known from numerical analysis.

In order to get as wide a knowledge as possible about the IRp-6 robot model, several simulations end experiments for both differential and integral models were run. The experimental identification results for the differential model are reported in reference [231]. Since in all calculations results of a similar nature were obtained we believe that the identified model including the friction model are reliable and can be used in model based control schemes.

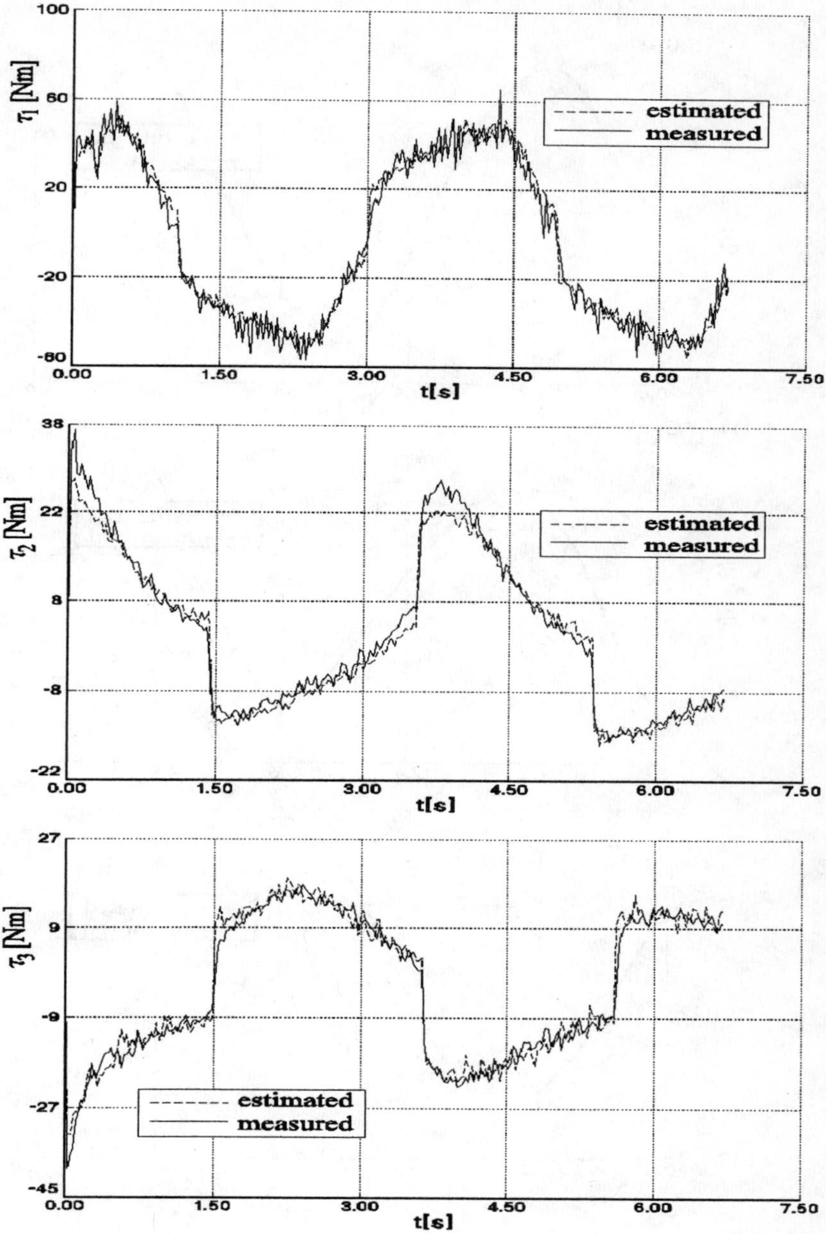

Fig. 5.12. Measured and estimated torque for the first a), second b), third c) joint of the IRp-6 robot

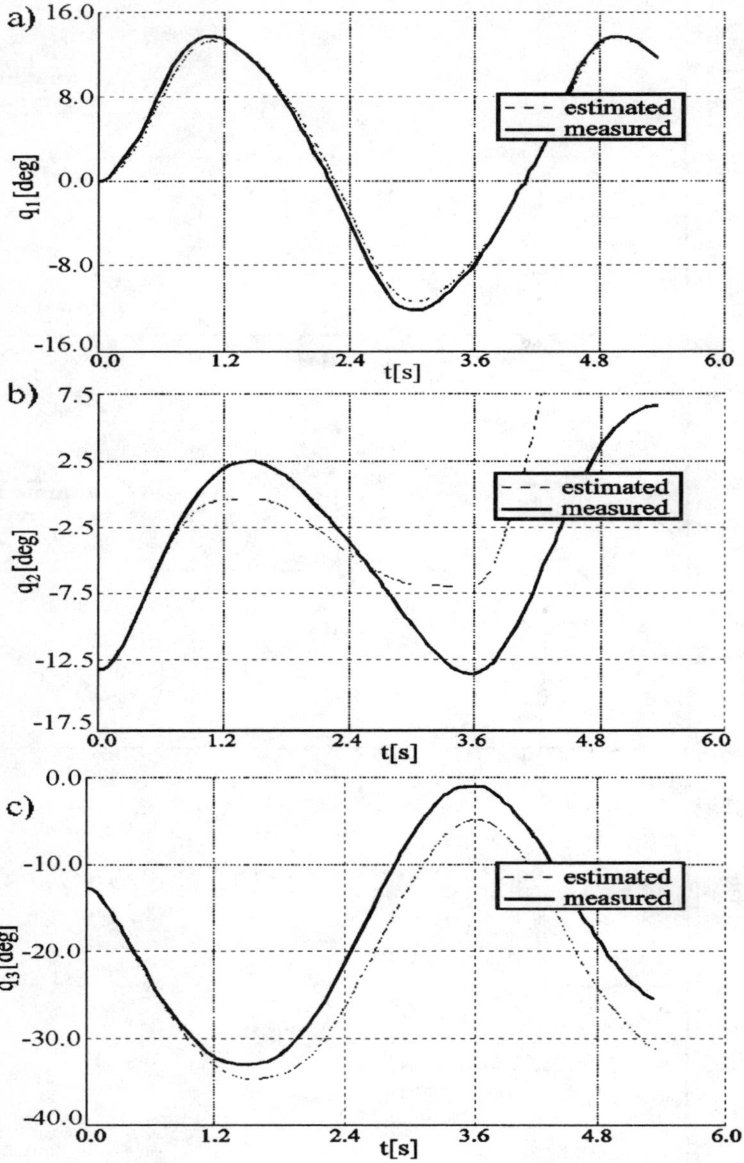

Fig. 5.13. Measured and estimated trajectory for the first a), second b), third c) joint of the IRp-6 robot

5.5 Experimental identification results for a one link geared robot

In this section we present experimental identification results of a one link geared robot described in Section 2.4. The dynamic equations of the system from Fig.2.6 can be presented in the following form

$$A\ddot{q} + B_\nu \dot{q} + B_s sgn(\dot{q}) + N sin(q) = I,$$

$$A = \frac{I_a}{K_\tau}, \quad B_\nu = \frac{F_\nu}{K_\tau}, \quad B_s = \frac{F_s}{K_\tau}, \quad N = \frac{N_0}{K_\tau},$$

where I_a is the rotor inertia plus the inertia of the rotating part of the harmonic drive and load, F_ν is the coefficient of viscous friction of the joint, F_s is the coefficient of static friction of the joint, K_τ is the coefficient which characterizes the electromechanical conversion of armature current to torque, I is the motor armature current and q is the angular motor position (hence the position of the load N_0).

The electrical subsystems dynamics for the permanent magnet brush DC motor is assumed to be

$$L\dot{I} = V_e - RI - K_B\dot{q},$$

where L is the armature inductance, R is the armature resistance, K_B is the back-emf coefficient, and V_e is the input control voltage. For the above electromechanical model we assume that $q, \dot{q}$, and I are all measurable.

In the first experiment we identified the parameters of the dynamical model. For the purpose of identification we used the least squares method in the recursive form with the Agee-Turner factorization. As the input signal for the identification we assumed a rectangular current signal with varying amplitude and frequency. The results of the identification are summarised in Table 5.2.

Table 5.2. Estimates of the mechanical subsystem

Current amplitude	I [kg·m^2]	F_ν [N·m·s]	F_s [N·m]
2	0.000045	0.00175	0.136
2.5	0.000048	0.00193	0.134
3	0.000049	0.00206	0.129
3.5	0.000050	0.00209	0.123
4	0.000050	0.00202	0.123

The results presented in Table 5.2 were obtained assuming that the frequency of the input signal was 1Hz. In the identification process we also tried different frequencies. We ran several identification schemes in order to average the estimates for each current amplitude. Finally, we obtained the following estimates $I_a = 0.00005\text{kg}\cdot\text{m}^2$, $F_\nu = 0.002\text{N}\cdot\text{m}\cdot\text{s}$, and $F_s = 0.127\text{N}\cdot\text{m}$. Based on the geometrical measurements of the load we estimated the load inertia as $0.000015\text{kg}\cdot\text{m}^2$. This result was verified by executing the identification scheme with and without the load.

In a similar fashion we identified the parameters of the electrical subsystem. Here we summarize only the final results, which are as follows: $K_B = 6.2V/1000\text{rpm}$, $R = 1.4\Omega$, and $L = 0.5\text{mH}$.

Next we ran several experiments in order to measure the characteristics of the harmonic drive. More detailed description of the harmonic drive can be found in reference [219]. Here we summarize only some of its characteristics. First we observed input and output velocities as functions of time in order to observe the resonant losses in the harmonic drive transmission. In Figs.5.14 and 5.15 input and output velocities are observed assuming a step function of the current as the input.

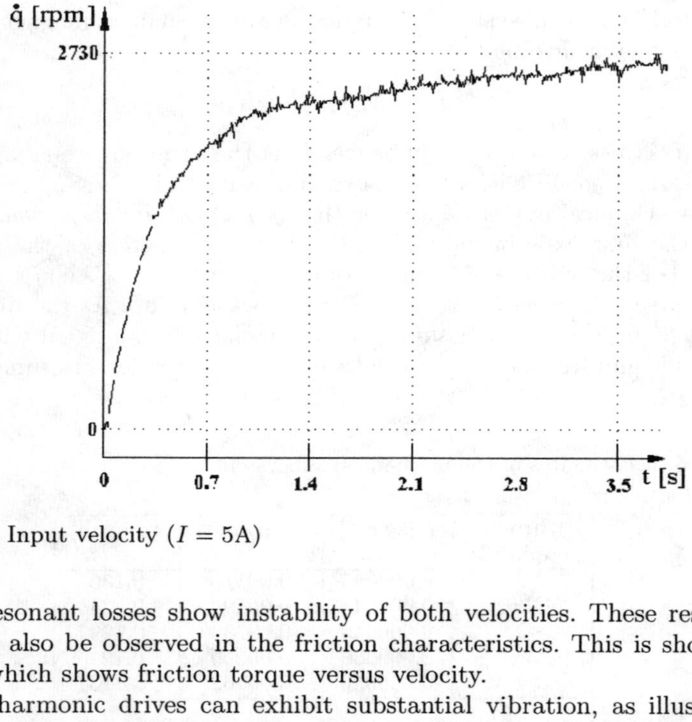

Fig. 5.14. Input velocity ($I = 5\text{A}$)

The resonant losses show instability of both velocities. These resonant losses can also be observed in the friction characteristics. This is shown in Fig.5.16 which shows friction torque versus velocity.

Since harmonic drives can exhibit substantial vibration, as illustrated above, complex behaviour in the transmission has to captured to improve dynamic model. From the experimental stiffness profile, shown in Fig.5.17, the characteristics of stiffering-spring behaviour of harmonic drive flexibility can be observed. This stiffness characteristic is the relationship between the torsional angle and the torque at the output of the harmonic drive.

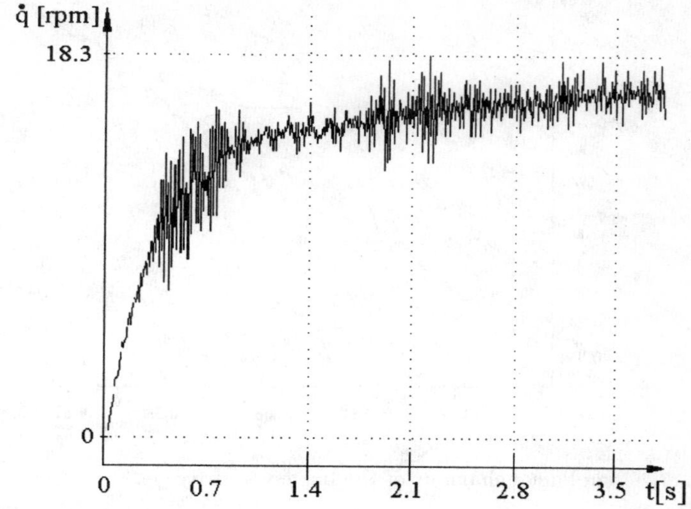

Fig. 5.15. Output velocity ($I = 5$A)

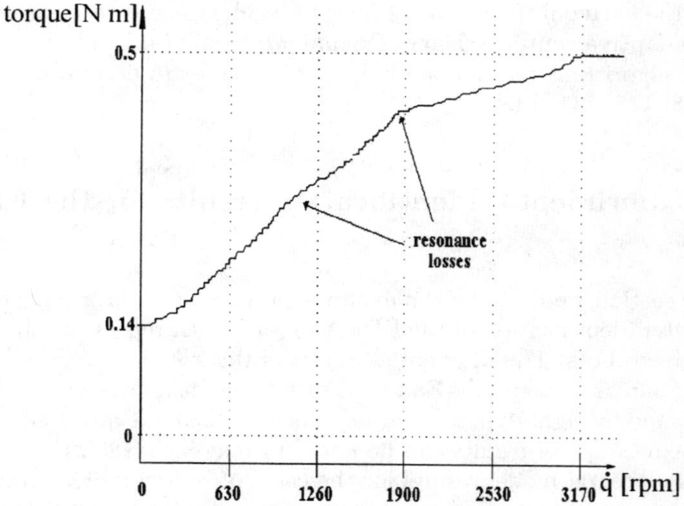

Fig. 5.16. Friction torque of the harmonic drive

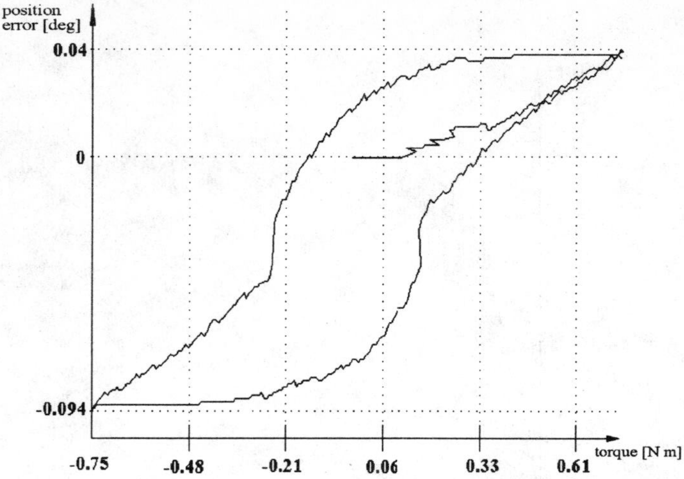

Fig. 5.17. Compliant behaviour of the harmonic drive

Finally, we measured the harmonic drive kinematic error. This error is illustrated in Fig.5.18 which shows position error versus time assuming constant velocity (in our case 150 rpm).

The measured characteristics of the harmonic drive show that its model is quite complicated and has to be included in the control system. The parameters of the harmonic drive do not have to be identified if we assume the model based adaptive control scheme. Preliminary results concerning the adaptive control algorithms for the one link geared robot are described in references [88, 128, 135, 141, 142].

5.6 Experimental identification results for the EDDA robot

In this section we describe the experimental results concerning the dynamic parameter identification of the EDDA robot, which represents the class of direct drive robots. The differential model of this robot is given by Eq.(3.103) and the integral model by Eq.(3.105). Both models are canonical. They are represented by eight dynamic parameters described in Eq.(3.104). A complete set of experimental results can be found in reference [185]. Here we describe only the integral model results for the cases of short and long integrals. The initial values of the dynamic parameters were obtained from CAD modelling. This modelling can be very accurate as the EDDA robot has a very simple mechanical structure and only two degrees of freedom. The structure of the links is very regular and inertia elements can easily be calculated. Each link is

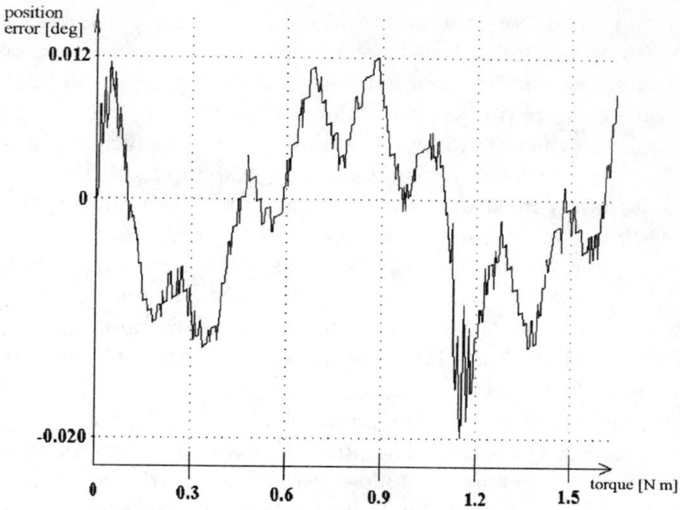

Fig. 5.18. Measured harmonic drive kinematic error

mounted on the megatorque motor, so the inertia characteristics of the motor itself have to be included. The first four dynamic parameters which are numbered according to Eq.(3.104) have the following numerical values obtained from CAD modelling $X_1 = 3.1\text{kg}\cdot\text{m}^2$, $X_2 = 7.5\text{kg}\cdot\text{m}$, $X_3 = 0.53\text{kg}\cdot\text{m}^2$, and $X_4 = 1.06\text{kg}\cdot\text{m}$. The values of the friction coefficients were obtained from the friction characteristics measurements by means of the method introduced by Seeger and Leonhard [196]. It is important to note that this method is applicable to the differential model and is different from the method proposed in Section 5.3. The Coulomb and velocity friction coefficients for the two links of the EDDA robot have the following values: $F_{1s} = 6\text{N}\cdot\text{m}$, $F_{2s} = 3\text{N}\cdot\text{m}$, $F_{1v} = 1\text{N}\cdot\text{m}\cdot\text{s}$, and $F_{2v} = 0.5\text{N}\cdot\text{m}\cdot\text{s}$. These values are different from the appropriate friction characteristics for the differential model.

Next we performed experimental identification using the least squares method discussed in Chapter 4. All identification algorithms were developed at the Chair of Control, Robotics, and Computer Science, Poznań University of Technology. We assumed the knowledge about the Coulomb and viscous friction coefficients in the simplest friction model in the form of Eqs.(3.29) and (3.30) and calculated the second and third terms in Eq.(3.38), which are responsible for the friction torque contribution in the integral model. From now on we assume that the friction characteristics are subtracted from the total energy of the system, which results in the integral model having only four dynamic parameters. In order to identify these parameters we assume the following input trajectory for the two joints of the EDDA robot:

$$q_i = \alpha_{i0} + \alpha_{i1}\cos(2\pi\alpha_{i2}t) + \alpha_{i3}\sin(2\pi\alpha_{i4}t), \quad \text{where } i = 1, 2 \; . \tag{5.43}$$

Here α_{i1} and α_{i2} denote contributions expressed as a percentage of the maximal velocity for each link (measured as a number of revolutions per second), and constants α_{i2} and α_{i4} denote the frequency of the input signal. Note that these frequencies are, in general, different.

In the experiments, different values of the coefficients which appear in Eq.(5.43) were investigated. Choosing the input trajectories given by Eq.(5.43) is mainly motivated by the work of Lu and co-authors [154]. Observe that by varying the frequency of the curves as in Eq.(5.43) we can easily excite different parameters because the peaks of these two curves can be located arbitrarily. Besides that the form of the input trajectories given by Eq.(5.43) guarantees smooth velocities and accelerations. The numerical values of the coefficients shown in Eq.(5.43), were chosen as a result of several experiments. The trial and error method seems to be very effective in this case. Recall that we are dealing with a relatively simple robot structure and with a small number of canonical parameters. The numerical values of the coefficients for the first link displacement are as follows: $\alpha_{10} = 1.25$, $\alpha_{11} = -1.25$, $\alpha_{12} = 0.4$, $\alpha_{13} = 0$, $\alpha_{14} = 0$. The appropriate values of the trajectory parameters for the second joint are $\alpha_{20} = -0.9$, $\alpha_{21} = -0.9$, $\alpha_{22} = 0.4$, $\alpha_{23} = 0$, and $\alpha_{24} = 0$. The joint velocities and accelerations are smooth curves. The position signal can be measured very accurately by the high resolution encoders mounted at the motors. The position signals were differentiated twice in order to obtain the acceleration signals. Because the position signals are almost noise–free repeated differentiation results in signals with very little noisy. Besides that carefully chosen filters were implemented in order to avoid the phase shift usually present in numerical differentiation. Taking several samples symmetrically about the point at which we calculate the derivative, we observed that the measured positions and calculated velocities and accelerations resulted in estimates that were not biased.

The experiments for the EDDA robot were carried out in double pendulum and planar configurations. In order to obtain reasonable data for the motor currents, which are proportional to the generalised torques, we averaged 10 measurements (sample time is 1.1ms). This reduced the measurement noise and ripples which appear in the torques. As in the case of the IRp-6 robot the least squares method in its recursive form, Eqs.(4.10) and (4.11) was used. The initial value of the estimate $\hat{\boldsymbol{x}}_0$ was assumed zero and the initial value of the inverse of the input correlation matrix $\boldsymbol{P}_0$, was an identity matrix with 10^6 on the diagonal. In order to calculate the estimates we took 500 measurements with the time interval of 33ms (because of the averaging procedure of the motor currents); therefore the total observation time was 1.65 s. The values of the estimates were almost exactly the same as the numerical values of the estimates calculated based on CAD modelling. This result is not surprising. As mentioned before CAD modelling is very precise since the EDDA robot is designed and built at the Institut für Robotik und Prozessinformatik, so therefore the full technical specification is known. We recall here that CAD modelling was successfully implemented by authors of the references [1, 155]. In our case the values of the estimates were stable after 120 iterations.

We mentioned in the previous section that repeatability of the estimates is an important factor of the quality of the identification procedure. In order to obtain reproducible results we ran the recursive equations (4.10) and (4.11) and their final values were used as the initial values of the estimates for the second run. We observed that the values of the estimates were exactly the same as the initial input estimates except for a few iterations at the very beginning. This procedure was repeated several times and we found that this procedure always resulted in the same values of the estimates.

In experimental investigations we tested short and long integrals connected with the integral model. The short integral was calculated over the period of 33ms and the long integral over the total integration time, of 1.65s. The results of these two calculations were exactly the same. We observed that when using of the short integral the estimates became stable faster than in the case of the long integral. Generally, the long integral gives more accurate estimates than the short integral. Nevertheless, the estimates resulting from the long integral can contain the bias due to the integration of the low frequency signals, making them less accurate, although it was not the case in the situation considered here. In order to verify these estimates we ran the experiments for the differential model. Next we compared the estimates in both situations. This comparison was satisfactory. Using the differential model we were able to identify all 8 parameters of the canonical model in one run. This is not possible for the integral model. The friction characteristics have to be precomputed first, even if, as in our case, the Coulomb and viscous friction coefficients are smaller than the absolute values of the friction coefficients of the IRp-6 robot. Note that we also recognised the friction coefficients with the direction of motor movement. A more detailed analysis was carried out by Prüfer and reported in reference [185]. In order to verify the experimental results we used the first method described in the previous section, comparing the predicted and measured energy. The friction characteristics were first subtracted then we compared the two energies. We observed that the two curves were almost the same, the maximum absolute difference between them being not bigger than 2.5 %. This result was satisfactory.

The friction torques versus velocity were measured very accurately by the procedure proposed by Seeger and Leonhard [196]. It is known that friction does not play a very important role for direct drive robots. Fiction characteristics measurements for the EDDA robot show that all phenomena associated with friction can also be observed. These characteristics can be found in reference [185]. First we tried to recognise the direction of movement and associated with this the fact that the Coulomb and viscous coefficients had different values in both directions of movement. This was indeed observed in the Tustin model measured for the EDDA. Altogether we have eight friction parameters to be identified plus four dynamic parameters. This results in 12 dynamic parameters which appear in both differential and integral models. Unfortunately the results of the identification of the 12 parameters were not satisfactory. Generally, the results obtained for the differential model were better than for the integral model. In the case of the differential model we observed that after 250 iteration steps the estimates had a tendency to become stable. However these estimates were prone be change when observed over a long period of time. In the case of the integral model the situation was even worse, with no noticeable stabilisation the estimates of the friction parameters. Clearly, a linear relationship between the functions associated with the friction coefficients existed. At the same time we observed that the estimates of the dynamic parameters, $\hat{X}_1 \div \hat{X}_4$, were stable and reproducible in all cases considered. It was found that the dynamic link parameters and the friction parameters constituted two separate groups of parameters and in the process of identification only the former could be identified correctly. Therefore the parameters which are linearly dependent do not have strong a influence on the parameters which are linearly independent. This observation can be very helpful in practical identification. To conclude, the identification of the friction coefficient for both the differential and integral model (the latter regardless of the type of the integral) was not satisfactory.

Prüfer and co-authors [183] try to solve this problem by choosing the optimal test trajectory for identification. They obtain good results calculating the condition number of the input correlation matrix for different input trajectories. This method, which is a combination of engineering intuition and a trial and error approach, gives surprisingly good results. They show that by carefully choosing the test trajectories it is possible to identify the friction coefficients in both directions of movement, particularly for the differential model. In the case of the integral model the results are not so good. The practical considerations carried out by Prüfer and co-authors [183] are not be applicable to the case of the IRp-6 robot. The optimisation procedure discussed in Section 5.2 is necessary for the IRp-6 robot. Therefore, we can state that the experimental results obtained for the IRp-6 robot are, in a sense, richer than those obtained for the EDDA robot. Note that the EDDA robot can move faster than the IRp-6 robot, which simplifies the identification. The inertial parameters are dominant compared to the friction coefficients. In the first approximation the friction coefficients can be neglected, and then only the dynamic parameters are subject to the identification procedure. The experimental results confirm this practical assumption. Moreover, for geared robots, this assumption cannot be verified.

Prüfer in his Ph.D. Thesis [185] verifies the identified dynamic model by using it in various model based control schemes. The results obtained are satisfactory. Unfortunately it is not possible to perform such verification for the IRp-6 robot.

5.7 Further comments on the experimental identification of robot dynamics.

In this section we discuss the results of the experimental identification of robot dynamics and relate them to the results presented in Sections 5.4, 5.5 and 5.6. In Sections 5.4 and 5.5 we presented the results of robot dynamics identification for the class of geared robots and in Section 5.6 for the class of direct drive robots. We do not claim that the approach presented in those sections is general, because identification of robot dynamics is quite a complex problem and the particular structure of a robot determines the steps which should be taken to solve it. Mathematical techniques are very important but engineering intuition plays an almost equal role. Therefore we cannot recommend a general recipe to solve this problem.

It seems reasonable to precompute the friction characteristics of a geared robot regardless of the type of the model used. Another hint is to decompose the model into several submodels having smaller dimensions in order to avoid identification of a set of dynamic parameters of the canonical model using only one test trajectory. Generally, it is better to identify the dynamic parameters step by step using several test trajectories to identify the parameters of submodels. The test movement should be designed is such a way that the submodels are described in terms of the dynamic parameters which appear only once in each submodel. This is possible only if the test trajectory is designed for one, or at most, two joints, but of course this is not always practical. We can observe that the errors of the estimates accumulate. An important motivation to reduce the dimension of the original model the problem comes from of optimisation of the test trajectory, which is usually very time consuming. Besides that, it is important to analyse the kinematical structure of the robot itself and judge how to plan the trajectories for the individual links in order to excite the individual dynamic parameters. Surprisingly, it is our experience that the movements do not have to vary in a wide range. Sometimes periodical movements around a specific point in space are sufficient for successful identification. The manipulator can be positioned in different points in configuration space which are suitable for identification.

As for the identification technique, the least squares method is preferable, especially in its recursive form. This method can be used off–line or on–line since it is not very time consuming in its recursive form.

The results presented in this monograph concern mainly the identification of dynamic parameters of the integral model, for which not many results are available in robotics literature. It has been our objective to show that this model can be useful for identification purposes. The proposed method for friction characteristics measurement in combination with the optimisation procedure with the single or double link movements seems to be the most attractive for the identification of robot dynamics for the integral model.

Now compare the results obtained in Sections 5.4 and 5.6 to similar results found in robotics literature. Consider first the optimisation technique for the minimisation of the condition number of the input correlation matrix. The numerical calculations presented in Section 5.4 show that an intuitively chosen input trajectory results in the input correlation matrix having the condition number of order 10^5. Generally this guess was poor. An and co-authors [1] choose an input trajectory for the MIT/Asada Arm direct drive robot for the differential model, which gives the input correlation matrix condition number of 5710. A similar guess was made by Craig [30] for the identification of the dynamic parameters of the Adept 1 Arm. Their initial trajectories are better than ours. After 87 iteration steps the condition number decreased to 88. Similar optimisation results are reported for the MIT/Asada Arm and Adept 1 Arm. The corresponding optimal condition numbers are 48 and 76, respectively. Caccavale and Chiacchio [22] choose the optimal trajectory for the differential model of the SMART-36.12R industrial robot with the condition number of 115. For the corresponding integral model of the same robot the optimal condition number was 531 [21]. Note that the range of the optimal condition number is the same in the cited references when compared to our results of the integral model of the IRp-6 robot. The optimisation results for the integral model of the SMART-36.12R are even one order worse. Generally, the condition number of the optimal trajectory should be less than 100 and intuitively chosen trajectories may not be appropriate for identification purposes.

There are at least three experimental methods which can validate the identified robot dynamic model (here we are not concerned with model based control schemes which do not necessarily require very precise information about the model). The first one, which in our opinion is the best, compares the input trajectories with the trajectories resulting from the solution of the forward dynamics problem, which is calculated from the identified model. This technique is seldom used by researchers and can only be found in references [158, 192], the results of which are similar to those discussed in this monograph. We believe that this test is the most logical and at the same time the most difficult.

Usually roboticians compare the measured torques with the predicted torques calculated from the identified dynamic model; see the results presented in references [1, 4, 5, 64, 67, 69, 194, 196, 197, 222]. The results described in this monograph are of a similar nature.

The third method compares the measured motor torques with the predicted ones with and without load. Reference [67] shows that the predicted torques match the actual torques closely. It is also possible to compare the energies calculated from the measured velocities and torques and the identified integral model. In general, this kind of comparison is not recommended since it can lead to wrong identification.

The experimental results discussed in this monograph concern mainly the identification of dynamic parameters of the integral model. When using the integral model we do not have to measure the acceleration signals. As reported by Armstrong [4], noisy acceleration signals can result in biased estimates of the dynamic parameters. To avoid this, we have chosen the integral model.

As mentioned in Chapter 4 the measurement signals are usually noisy. In general any noise which is present in position, velocity, and acceleration signals distorts the input correlation matrix and leads to systematic errors in the parameter estimates. Therefore in order to reduce the sensitivity of the least squares method the condition number of the input correlation matrix should be as small as possible, for example, one. If one has to compute the acceleration by numerically differentiating the joint angles, the noise level is unacceptably high. Usually, robot sensors provide discrete joint measurements from encoders or resolvers. Measurement errors can be reduced by making use of high resolution encoders. Resolvers are also accurate position sensors and their noise can be kept very small (refer to the analysis in Armstrong [4]). Gautier [69] analysed various noisy signals and their impact on three dynamic models, the so-called explicit, implicit, and energy.

The first of those is the differential model. The second one is the model discussed in Section 5.2, in which the derivation of the velocity to get the joint accelerations has been replaced by the derivation of function $\boldsymbol{A}(\boldsymbol{q})\dot{\boldsymbol{q}}$ of the velocity. The third one is the well–known integral model discussed in section 3.3. The measurement signals can be pre–filtered if all of them are measurable. If not, the joint velocity and acceleration can be calculated without phase shift by the central difference of signal q.

Another approach is to pre–filter the explicit and implicit dynamics equations by properly choosing the cut off frequency. This method is also proposed by Lu and co-authors [154]. The study conducted by Gautier [69] confirms that there is no advantage in using the dynamic identification model with implicit acceleration, because its symbolic expressions are more complicated then those of the dynamic model with explicit accelerations, while the experimental results are the same.

In the monograph we use mainly the integral model which is the simplest and with properly chosen trajectories can result in accurate estimates of the dynamic parameters. One way to suppress the noise would be to apply spectral analysis to the differential model as suggested by Vandenjon and co-authors [222]. The covariance method proposed by Schiehlen and Kallenbach [192] suppresses the bias errors in the estimates. Seeger [197] also uses the spectral analysis in some parts of the experimental identification procedure because the autocorrelation function of the acceleration error decreases rapidly as the time delay increases and removes most of the bias that was introduced by numerical differentiation. These techniques were not checked, since our method gave satisfactory results.

As mentioned in Section 5.2 Presse and Gautier [179] propose new criteria of exciting trajectories for robot identification. The most general criterion is the minimisation of three components given by Eq.(5.13). They present simulation results concerning this general criterion, which are very promising.

This criterion was not verified in our experimental practice since the results we obtained were satisfactory. Nevertheless it should be pointed out that several approaches to robot dynamics identification have been developed. One can treat them as variety of tools which can be used to solve the difficult identification problem.

CHAPTER 6
LOAD DYNAMICS IDENTIFICATION

6.1 Introduction

In this chapter we present experimental results concerning identification of dynamic parameters of a load hold by a robot. There are ten dynamic parameters of the load namely mass, three parameters of the centre of mass multiplied by its mass, and six parameters of the inertia tensor, all together, ten parameters. These parameters are expressed in local coordinate frame which is associated with the load and therefore they are constant. Experimental identification of this set of parameters is quite difficult because we need very specialised equipment in order to carry out the experiments. Of course we can measure and weigh the load and then calculate all the parameters, but we want to do it experimentally, during the movement of the robot, assuming that the robot grasps the load which is a completely unknown object. The most expensive part is the force/torque sensor which has to be mounted between the wrist and the gripper. As mentioned in Chapter 2 Chair of Control, Robotics, and Computer Science has two force/torque sensors. The sensor allows us to measure forces and torques in the local coordinate frame associated with it. Three components of the forces and torques are measured in three dimensional space. Note, that in case considered here we need measurements of the forces and torques in three dimensions because these forces and torques are not compensated for by the structure of the gripper. In the case of joint generalised forces measurements, we can assume that the sensor measures only the forces and torques along the axis of translation and rotation, respectively. Full sensing is necessary for the load identification. Besides that, in order to carry out the experiments we need measurements of joint positions, velocities, and accelerations. In our experimental set-up we were able to measure these signals [143] except for the acceleration signals which result from the differentiation of the velocity signals. We also need the interface with direct access to the robot controller of the IRp-6 robot which allows us to design an input trajectory in joint space and Cartesian space. The whole measurement set-up is very expensive and usually not available in industrial practice.

In this monograph we design two types of measurements of the load dynamic parameters. The first type is a static measurement of the load parameters which is performed by positioning the manipulator in different configurations in Cartesian space. In each configuration hundreds of measurements are collected (these are force and torque measurements). Based on these measurements we can find the mass and centre of mass of the load. Obviously we cannot measure the inertia tensor [96] elements because the manipulator is at rest. These measurements can be found, for example, in reference [176]. In this reference Paul described two methods to finding the mass of the load when the robot is not moving. Unfortunately none of his methods can be used to find the centre of mass. Olsen and Bekey [173] proposed a method which is similar to that discussed in reference [1]. In the measurements of the first moment the most difficult problem is to extract the first moment components from the gripper and the load, because these two objects constitute one rigid body from which we have to extract the features which are associated with the load. Of course the best situation is when the mass of the gripper is very small in comparison with the load, but unfortunately this is not always the case. An and co-authors [1] propose solving this problem by moving the load along one chosen axis of the local coordinate frame by a known distance. Based on the know distance we can find the relative change of the centre of mass along this axis.

The second type of measurements is called dynamic. These measurements are associated with the movement of the manipulator along known predefined trajectories. During this movement we collect the following data, joint positions, velocities, accelerations, and sensor measurements of forces and torques. Next, based on these measurements, we carry out the identification procedure, usually off–line, although it can be done on–line too and this will be discussed at the end of this chapter. It is clear that from dynamic measurement we can recover all ten inertial parameters of the load. Usually it is not possible to measure the accelerations of the coordinate frame associated with the strain gauge sensor and we have to perform the numerical differentiation of the velocity signal. Recall that the acceleration of the origin of the coordinate frame results from the kinematical considerations which involve all joint positions, velocities, and accelerations. In this situation we are dealing with the differentiation model [1].

In order to avoid numerical differentiation (which is performed on complicated noisy expression involving joint positions and velocities) we can integrate the equations describing the differential model. The differential model is sometimes called the explicit dynamic model when we use the terminology from dynamic models of robots. Integration of the aforementioned differential model results in the so-called implicit dynamic model for load identification. The integration operation is usually performed in the local coordinate frame associated with the strain gauge sensor. This approach was used by Atkeson and co-authors of references [10, 11]. In both these models dynamic parameters of the load appear to be linear. Note, that the last model is also considered to be an integral model for load identification. For the purpose of identification we can use the least squares method in its recursive form [15].

Some other methods such as singular value decomposition can be used too [10, 11]. The last method was used by Yoshikawa in reference [235].

Identification of load dynamic parameters can be performed on–line. In this case we use one of the methods proposed here which is coded very efficiently in machine code. We can also use the adaptive control scheme which is based on a dynamic model not well defined. Unknown parameters of the model are, for example, load, Coulomb and velocity friction coefficients. These unknown parameters can be identified as a result of the adaptation law which is formulated based on the Lyapunov theory. This approach seems to be attractive but is not discussed in this monograph. The results of this approach are discussed in [30, 53, 108].

The chapter is organised as follows. In Section 6.2, basic mathematical relationships which are necessary for load identification are derived. The explicit differential model is derived and the implicit differential model are discussed. Both these models are derived for the IRp-6 robot and load identification. The next section is devoted to optimisation of the condition number of the exciting trajectory for load identification. In Section 6.4 results of the static measurements are presented. Some preliminary results concerning the static measurements can be found in reference [34]. Finally, in Section 6.5 dynamic measurements of load identification are presented.

6.2 Mathematical description of load dynamic models

In order to derive the fundamental mathematical relationship associated with the identification problem of load dynamics, consider the rigid body presented in Fig. 6.1. This body is located in the coordinate frame denoted by x_0, y_0, z_0 which is known as a base coordinate frame. The rigid body has mass m. Point P belongs to the body and constitutes an origin of the coordinate frame which is associated with the body. This local coordinate frame is also a frame in which the forces and torques are measured. All dynamic parameters are expressed in this frame and therefore they are constant. The distance between the origin of the base coordinate frame and the origin of the local coordinate frame is denoted by $\boldsymbol{p}$. Another point, which is associated with the rigid body, is denoted by C and is located exactly in the centre of mass of the body. The distance between points C and P is denoted by $\boldsymbol{c}$.

The distance between point C and the origin of the base coordinate frame is denoted by $\boldsymbol{q}$. Now referring to the notation in Fig.6.1 one can write the following Newton–Euler equations of motion for the rigid body

$$\boldsymbol{f}_c = \boldsymbol{f} + m\boldsymbol{g} = m\ddot{\boldsymbol{q}} \,, \tag{6.1}$$

$$\boldsymbol{m}_c = \boldsymbol{m} - \boldsymbol{c} \times \boldsymbol{f} = \boldsymbol{I}_c\,\dot{\boldsymbol{\omega}} + \boldsymbol{\omega} \times (\boldsymbol{I}_c\,\boldsymbol{\omega}) \,, \tag{6.2}$$

where: $\boldsymbol{f}$ – three dimensional vector of force exerted on the rigid body by strain gauge sensor in point P,

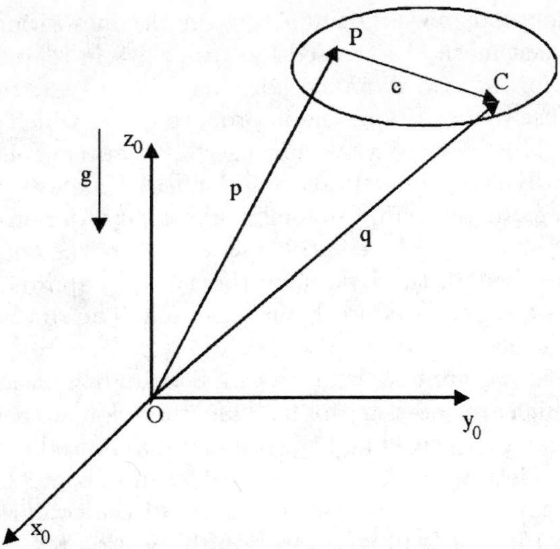

Fig. 6.1. A rigid body in base coordinate frame

$\boldsymbol{g}$ – a vector of gravity $\boldsymbol{g} = [0,\ 0,\ g_0]^T$ where $g_0 = -9,81\ \mathrm{m/s^2}$,
$\ddot{\boldsymbol{q}}$ – an acceleration vector of the centre of mass of the rigid body (this vector is three dimensional and is not a joint acceleration vector),
$\boldsymbol{m}$ – a three dimensional vector of torques exerted by the sensor on the rigid body,
$\boldsymbol{I}_c$ – inertia tensor of the rigid body expressed in coordinate frame with origin point C,
$\boldsymbol{\omega}$ – a three dimensional vector of angular velocities of the rigid body,
$\dot{\boldsymbol{\omega}}$ – a three dimensional vector of angular acceleration of the rigid body.

In Eqs.(6.1) and (6.2) $\boldsymbol{f}_c$ and $\boldsymbol{m}_c$ denote resultant force and torque acting on the rigid body at its centre of mass, and '×' denotes a cross product operation.

The acceleration vectors $\ddot{\boldsymbol{p}}$ and $\ddot{\boldsymbol{q}}$ are related as follows:

$$\ddot{\boldsymbol{q}} = \ddot{\boldsymbol{p}} + \dot{\boldsymbol{\omega}} \times \boldsymbol{c} + \boldsymbol{\omega} \times (\boldsymbol{\omega} \times \boldsymbol{c}) \,. \tag{6.3}$$

Substitution of Eq.(6.3) into Eqs.(6.1) and (6.2) results in:

$$\boldsymbol{f} = m\ddot{\boldsymbol{p}} - m\boldsymbol{g} + \dot{\boldsymbol{\omega}} \times m\boldsymbol{c} + \boldsymbol{\omega} \times (\boldsymbol{\omega} \times m\boldsymbol{c}) \,, \tag{6.4}$$

$$\begin{aligned} \boldsymbol{m} &= \boldsymbol{I}_c\,\dot{\boldsymbol{\omega}} + \boldsymbol{\omega} \times (\boldsymbol{I}_c\,\boldsymbol{\omega}) + m\boldsymbol{c} \times (\dot{\boldsymbol{\omega}} \times \boldsymbol{c}) + \\ &+ m\boldsymbol{c} \times (\boldsymbol{\omega} \times (\boldsymbol{\omega} \times \boldsymbol{c})) + m\boldsymbol{c} \times \ddot{\boldsymbol{p}} - m\boldsymbol{c} \times \boldsymbol{g} \,. \end{aligned} \tag{6.5}$$

Now we assume that the coordinate frames with origins in points P and C are parallel. This assumption allows to calculate the inertia tensor in coordinate frame associated with point P by making use of the Steiner theorem [229] as follows:

$$\boldsymbol{I}_p = \boldsymbol{I}_c + m\left[(\boldsymbol{c}^T\,\boldsymbol{c})\boldsymbol{\mathcal{I}} - (\boldsymbol{c}\,\boldsymbol{c}^T)\right] \,, \tag{6.6}$$

where $\boldsymbol{\mathcal{I}}$ is an identity matrix.

After some algebraic manipulations on double and triple cross products which appear in Eq.(6.5) and using Eq.(6.6) we obtain the following expression:

$$\boldsymbol{m} = \boldsymbol{I}_p\,\dot{\boldsymbol{\omega}} + \boldsymbol{\omega} \times (\boldsymbol{I}_p\,\boldsymbol{\omega}) + m\boldsymbol{c} \times \ddot{\boldsymbol{p}} - m\boldsymbol{c} \times \boldsymbol{g} \,. \tag{6.7}$$

Taking into account Eqs.(6.4) and (6.7) we can state that the mass of the body m, its first moment $m\boldsymbol{c}$ (three elements) and inertia tensor elements $\boldsymbol{I}_p$ (six elements) with respect to the origin P of the coordinate frame, appear linearly in these equations. This statement is very important from the identification point of view because Eqs.(6.4) and (6.7) represent the linear model which is very suitable for the estimation of load dynamic parameters.

In order to write Eqs.(6.4) and (6.7) in a more compact form, we introduce the following notation:

$$\boldsymbol{\omega} \times \boldsymbol{c} = \begin{bmatrix} 0 & -\omega_z & \omega_y \\ \omega_z & 0 & -\omega_x \\ -\omega_y & \omega_x & 0 \end{bmatrix} \begin{bmatrix} c_x \\ c_y \\ c_z \end{bmatrix} \overset{def}{=} [\boldsymbol{\omega}\times]\boldsymbol{c} \,, \tag{6.8}$$

and

$$\boldsymbol{I}\boldsymbol{\omega} = \begin{bmatrix} \omega_x & \omega_y & \omega_z & 0 & 0 & 0 \\ 0 & \omega_x & 0 & \omega_y & \omega_z & 0 \\ 0 & 0 & \omega_x & 0 & \omega_y & \omega_z \end{bmatrix} \begin{bmatrix} I_{xx} \\ I_{xy} \\ I_{xz} \\ I_{yy} \\ I_{yz} \\ I_{zz} \end{bmatrix} \overset{def}{=} [\cdot\boldsymbol{\omega}] \begin{bmatrix} I_{xx} \\ I_{xy} \\ I_{xz} \\ I_{yy} \\ I_{yz} \\ I_{zz} \end{bmatrix} \,, \tag{6.9}$$

where

$$\boldsymbol{I} = \begin{bmatrix} I_{xx} & I_{xy} & I_{xz} \\ I_{xy} & I_{yy} & I_{yz} \\ I_{xz} & I_{yz} & I_{zz} \end{bmatrix} .$$

Note that in last equations vector $\boldsymbol{\omega}$ is represented by three components ω_x, ω_y, and ω_z and vector $\boldsymbol{c}$ by three components c_x, c_y, c_z

Using the notation defined by Eqs.(6.8) and (6.9) equations of the linear identification model given by (6.4) and (6.7) can be written in the following vectorial form:

$$\begin{bmatrix} F_x \\ F_y \\ F_z \\ M_x \\ M_y \\ M_z \end{bmatrix} = \begin{bmatrix} \ddot{\boldsymbol{p}} - \boldsymbol{g} & [\dot{\boldsymbol{\omega}}\times] + [\boldsymbol{\omega}\times][\boldsymbol{\omega}\times] & 0 \\ 0 & [(\boldsymbol{g} - \ddot{\boldsymbol{p}})\times] & [\cdot\dot{\boldsymbol{\omega}}] + [\boldsymbol{\omega}\times][\cdot\boldsymbol{\omega}] \end{bmatrix} \begin{bmatrix} m \\ mc_c \\ mc_y \\ mc_z \\ I_{xx} \\ I_{xy} \\ I_{xz} \\ I_{yy} \\ I_{yz} \\ I_{zz} \end{bmatrix}, \tag{6.10}$$

or equivalently

$$\boldsymbol{w}_s = \boldsymbol{K}\, \boldsymbol{x}_l\, . \tag{6.11}$$

In Eq.(6.11) vector $\boldsymbol{w}_s$ is a six dimensional vector of the forces and torques of the reduced system, matrix $\boldsymbol{K}$ has dimensions 6×10 and depends on the configuration of the rigid body expressed in the base coordinate frame and finally $\boldsymbol{x}_l$ denotes a 10×1 dimensional vector of the dynamic parameters of the rigid body.

Note that in Eq.(6.10), ten dynamic parameters appear which characterise the load. These are: mass, m, three coordinates of the centre of mass multiplied by mass itself mc_x, mc_y, mc_z, (which are also named as elements of the first moment) and six inertia tensor elements, I_{xx}, I_{xy}, I_{xz}, I_{yy}, I_{yz}, and I_{zz}. The quantities which appear on the left side of Eq.(6.10) are the three forces and torques which are measured by the strain gauge sensor mounted between the wrist of the robot and the gripper. These values are expressed in coordinate frame with origin P. Vector $\boldsymbol{x}_l$ does not change when the body changes position and orientation in Cartesian space. Matrix $\boldsymbol{K}$ depends on a vector of linear acceleration $\ddot{\boldsymbol{p}}$ and on vectors of angular velocity and acceleration $\boldsymbol{\omega}$, $\dot{\boldsymbol{\omega}}$, respectively. Now we write explicitly all components of the matrix $\boldsymbol{K}$ which appear in Eq.(6.11) using the notation defined by Eqs.(6.8) and (6.9)

$$\boldsymbol{K} = \left[\begin{matrix} \ddot{p}_x - g_x & -\omega_y^2 - \omega_z^2 & \omega_x\omega_y - \dot{\omega}_z & \omega_x\omega_z + \dot{\omega}_y & 0 \\ \ddot{p}_y - g_y & \omega_x\omega_y + \dot{\omega}_z & -\omega_x^2 - \omega_z^2 & \omega_y\omega_z - \dot{\omega}_x & 0 \\ \ddot{p}_z - g_z & \omega_x\omega_z - \dot{\omega}_y & \omega_y\omega_z + \dot{\omega}_x & -\omega_y^2 - \omega_x^2 & 0 \\ 0 & 0 & \ddot{p}_z - g_z & g_y - \ddot{p}_y & \dot{\omega}_x \\ 0 & g_z - \ddot{p}_z & 0 & \ddot{p}_x - g_x & \omega_x\omega_z \\ 0 & \ddot{p}_y - g_y & g_x - \ddot{p}_x & 0 & -\omega_x\omega_y \end{matrix} \right. \tag{6.12}$$

$$\left. \begin{matrix} 0 & 0 & 0 & 0 & 0 \\ 0 & 0 & 0 & 0 & 0 \\ 0 & 0 & 0 & 0 & 0 \\ \dot{\omega}_y - \omega_x\omega_z & \dot{\omega}_z + \omega_x\omega_y & -\omega_y\omega_z & \omega_y^2 - \omega_z^2 & \omega_y\omega_z \\ \dot{\omega}_x + \omega_y\omega_z & \omega_z^2 - \omega_x^2 & \dot{\omega}_y & \dot{\omega}_z - \omega_x\omega_y & -\omega_x\omega_z \\ \omega_x^2 - \omega_y^2 & \dot{\omega}_x - \omega_y\omega_z & \omega_x\omega_y & \dot{\omega}_y + \omega_x\omega_z & \dot{\omega}_z \end{matrix} \right],$$

where g_x, g_y, and g_z are the components of the gravity vector $\boldsymbol{g}$ expressed in the coordinate frame associated with the strain gauge sensor.

Due to the fact that Eqs.(6.10) and (6.11) require the angular and linear accelerations to be know, which is not always the case, we integrate Eq.(6.11) in the time interval $(t, t+T)$. Note that the integral operation has to be performed in the coordinate frame associated with the force/torque sensor. As a consequence, since all quantities which appear in Eq.(6.11) are expressed in local coordinate frame associated with the sensor, we can say that the integration operation is performed in the same coordinate frame. After applying the integration operation we get:

$$\begin{bmatrix} \int_t^{t+T} \boldsymbol{f} dt \\ \int_t^{t+T} \boldsymbol{m} dt \end{bmatrix} = \left[\int_t^{t+T} \boldsymbol{K} dt \right] \boldsymbol{x}_l \, . \tag{6.13}$$

The first row of matrix $\boldsymbol{K}$ which appears in Eq.(6.10) after performing the integration operation has the following form:

$$\left[\boldsymbol{v} \,|_t^{t+T} + \int_t^{t+T} \boldsymbol{\omega} \times \boldsymbol{v} \, dt - \left(\int_t^{t+T} {}_O^P\boldsymbol{R} \, dt \right) \boldsymbol{g}, \; [(\boldsymbol{\omega} \,|_t^{t+T} \times] + \int_t^{t+T} [\boldsymbol{\omega}\times][\boldsymbol{\omega}\times] \, dt, \; \boldsymbol{0}) \right] . \tag{6.14}$$

Accordingly, the second row of the $\boldsymbol{K}$ matrix can be written as follows:

$$\left[\boldsymbol{0}, \left[\left(-\boldsymbol{v} \,|_t^{t+T} - \int_t^{t+T} \boldsymbol{\omega} \times \boldsymbol{v} \, dt + \left(\int_t^{t+T} {}_O^P\boldsymbol{R} \, dt \right) \boldsymbol{g} \right) \times \right], \; [(\cdot\, \boldsymbol{\omega} \,|_t^{t+T}] + \int_t^{t+T} [\boldsymbol{\omega}\times][\cdot\, \boldsymbol{\omega}] \, dt \right] . \tag{6.15}$$

In order to interpret Eqs.(6.14) and (6.15) notice that vector $\boldsymbol{g} = [0, 0, g_O]^T$ is the acceleration vector of the gravity which is expressed in the base coordinate frame and matrix ${}_O^P\boldsymbol{R}$ is the cosine matrix between the base coordinate frame and the coordinate frame with its origin in point P. Therefore we have to recalculate the gravity vector from the base coordinate frame to the coordinate frame in which all quantities are expressed. Derivation of the components of matrix $\boldsymbol{K}$ expressed in terms of Eqs.(6.14) and (6.15) follows from the representation of the vector and its derivative in two different coordinate frames: the base coordinate frame and the local coordinate frame associated with the force/torque sensor. Making use of Eqs.(6.14) and (6.15) we can integrate all elements which appear in matrix $\boldsymbol{K}$. Comparing Eq.(6.12) with Eqs.(6.14) and (6.15) note that in order to calculate the elements which appear in the last equations, we need to know linear and angular velocities (therefore linear and angular accelerations are not present). Linear and angular velocities $\boldsymbol{v}$ and $\boldsymbol{\omega}$ are expressed in local coordinate frame associated with the force/torque sensor. Finally, note that in Eq.(6.13) the vector of dynamic parameters of the load $\boldsymbol{x}_l$ is linear in the corresponding model which is called the implicit differential model of the load.

Now we apply the models given by Eqs.(6.11) and (6.13) to the IRp-6 robot. The kinematical structure of the robot is presented in Fig.3.6a and the kinematical parameters of the modified Denavit–Hartenberg notation are summarised in Table 3.2. The corresponding joint variables are calculated according to Eq.(3.1) and are denoted by q_1, q_2, q_3, q_4, and q_5. Note that these are joint variables associated with the links of the manipulator. Motor variables are calculated according to the expressions (3.116), (3.153), and (3.129) and Table 3.2. Here we need all these nonlinear expressions since in our system we measure the motor variables and these variables have to be transformed to the joint variables which are present in both models described by Eqs.(6.11) and (6.13).

Angular velocities are calculated from the base of the manipulator towards its tip according to the following expression [122]:

$$ {}^{j}\boldsymbol{\omega}_j = {}^{j}_{j-1}\boldsymbol{R}\ {}^{j-1}\boldsymbol{\omega}_{j-1} + \bar{\sigma}_j\, \boldsymbol{z}_0\, \dot{q}_j \, , \tag{6.16} $$

where ${}^{j}\boldsymbol{\omega}_j$ is an angular velocity of link j which is expressed in the local coordinate frame associated with this link, $\boldsymbol{z}_0 = [0, 0, 1]^T$, is an axis of notation for link j, $\bar{\sigma}_j = (1 - \sigma_j) = 1$ when link j is rotational (the IRp-6 robot has only rotational joints), $\sigma_j = 0$ when link j is translational, $\dot{q}_j$ is a joint velocity of link j, ${}^{j}_{j-1}\boldsymbol{R}$ is the cosine matrix between the coordinate frames $(j-1)$ and j. Now we make use of the recurrence relationship (6.16) in order to calculate the angular velocity of the fifth link of the kinematical structure of the IRp-6 robot and we get:

$$ {}^{5}\boldsymbol{\omega}_5 = \begin{bmatrix} -\dot{q}_1 \cos q_5 \sin(q_2 + q_3 + q_4) + \sin q_5 (\dot{q}_2 + \dot{q}_3 + \dot{q}_4) \\ \dot{q}_1 \sin q_5 \sin(q_2 + q_3 + q_4) + \cos q_5 (\dot{q}_2 + \dot{q}_3 + \dot{q}_4) \\ \dot{q}_1 \cos(q_2 + q_3 + q_4) + \dot{q}_5 \end{bmatrix} . \tag{6.17} $$

Linear velocity of the origin of coordinate frame j is calculated from the base of the manipulator towards its tip according to the following expression [122]:

$$ {}^{j}\boldsymbol{v}_j = {}^{j}_{j-1}\boldsymbol{R}\ {}^{j-1}\boldsymbol{v}_{j-1} + {}^{j}_{j-1}\boldsymbol{R}({}^{j-1}\boldsymbol{\omega}_{j-1} \times {}^{j-1}\boldsymbol{l}_j) + \sigma_j\, \boldsymbol{z}_0\, \dot{q}_j \, . \tag{6.18} $$

In the last equation vector ${}^{j-1}\boldsymbol{l}_j$ connects the origins of the coordinate frames j and $(j-1)$, respectively. Applying Eq.(6.18) we can calculate the linear velocity of the origin of the coordinate frame associated with link 5:

$$
{}^5\boldsymbol{v}_5 = \begin{bmatrix} \cos q_5[\dot{q}_2 l_2 \sin(q_3+q_4) + l_3(\dot{q}_2+\dot{q}_3)\sin q_4] + \\ + \sin q_5[-\dot{q}_1 l_2 \cos q_2 - l_3 \dot{q}_1 \cos(q_2+q_3)] \\ \\ -\sin q_5[\dot{q}_2 l_2 \sin(q_3+q_4) + l_3(\dot{q}_2+\dot{q}_3)\sin q_4] + \\ + \cos q_5[-\dot{q}_1 l_2 \cos q_2 - l_3 \dot{q}_1 \cos(q_2+q_3)] \\ \\ -\dot{q}_2 l_2 \cos(q_3+q_4) - l_3(\dot{q}_2+\dot{q}_3)\cos q_4 \end{bmatrix} . \tag{6.19}
$$

Angular accelerations are calculated according to the following expression [122]:

$$
{}^j\dot{\boldsymbol{\omega}}_j = {}^{j}_{j-1}\boldsymbol{R}\; {}^{j-1}\dot{\boldsymbol{\omega}}_{j-1} + \bar{\sigma}_j \left(\ddot{\boldsymbol{q}}_j\, \boldsymbol{z}_0 + {}^{j}_{j-1}\boldsymbol{R}\; {}^{j-1}\boldsymbol{\omega}_{j-1} \times \dot{q}_j\, \boldsymbol{z}_0\right), \tag{6.20}
$$

where $\ddot{\boldsymbol{q}}_j$ denotes the joint acceleration of link j. Similarly, the angular accelerations are calculated from the base of the manipulator towards its tip and for the kinematical structure of the IRp-6 robot we obtain:

$$
{}^5\dot{\boldsymbol{\omega}}_5 = \begin{bmatrix} \cos q_5[-\ddot{q}_1 \sin(q_2+q_3+q_4) - \dot{q}_1(\dot{q}_2+\dot{q}_3+\dot{q}_4) \\ \cos(q_2+q_3+q_4)] + (\ddot{q}_2+\ddot{q}_3+ \\ \ddot{q}_4)\sin q_5 - \dot{q}_1\dot{q}_5 \sin q_5 \sin(q_2+q_3+q_4) + \\ +\dot{q}_5(\dot{q}_2+\dot{q}_3+\dot{q}_4)\cos q_5 \\ \\ \sin q_5[\ddot{q}_1 \sin(q_2+q_3+q_4) + \dot{q}_1(\dot{q}_2+\dot{q}_3+\dot{q}_4) \\ \cos(q_2+q_3+q_4)] + (\ddot{q}_2+\ddot{q}_3+\ddot{q}_4)\cos q_5 - \\ \dot{q}_1\dot{q}_5 \cos q_5 \sin(q_2+q_3+q_4) - \dot{q}_5(\dot{q}_2+\dot{q}_3+\dot{q}_4)\sin q_5 \\ \\ \ddot{q}_1 \cos(q_2+q_3+q_4) - \dot{q}_1(\dot{q}_2+\dot{q}_3+\dot{q}_4)\sin(q_2+q_3+q_4) + \ddot{q}_5 \end{bmatrix} . \tag{6.21}
$$

Similarly, the linear acceleration of the origin of coordinate frame j is calculated from the following expression [122]:

$$
\begin{aligned} {}^j\dot{\boldsymbol{v}}_j &= {}^{j}_{j-1}\boldsymbol{R}\,(\,{}^{j-1}\dot{\boldsymbol{v}}_{j-1} + {}^{j-1}\dot{\boldsymbol{\omega}}_{j-1} \times {}^{j-1}\boldsymbol{l}_j + \\ &+ {}^{j-1}\boldsymbol{\omega}_{j-1} \times (\,{}^{j-1}\boldsymbol{\omega}_{j-1} \times {}^{j-1}\boldsymbol{l}_j)) + \\ &+ \sigma_j\,(\,2\;{}^{j-1}\boldsymbol{\omega}_{j-1} \times \dot{q}_j\, \boldsymbol{z}_0 + \ddot{q}_j\, \boldsymbol{z}_0)\,. \end{aligned} \tag{6.22}
$$

The linear accelerations are calculated from the base of the manipulator towards its tip and for the IRp-6 we get the following complicated expression:

$$
{}^5\dot{\boldsymbol{v}}_5 = \begin{bmatrix}
-l_2\cos q_5\cos q_4[(\dot{q}_1^2\cos^2 q_2+\dot{q}_2^2)\cos q_3-(\ddot{q}_2+\\
+0.5\dot{q}_1^2\sin 2q_2)\sin q_3]-l_3\dot{q}_1^2\cos q_4\cos q_5\cos^2(q_2+q_3)-\\
-l_3(\dot{q}_2+\dot{q}_3)^2\cos q_4\cos q_5+\\
+l_2\cos q_5\sin q_4[(\dot{q}_1^2\cos^2 q_2+\dot{q}_2^2)\sin q_3+(\ddot{q}_2+\\
+0.5\dot{q}_1^2\sin 2q_2)\cos q_3]+l_2(\ddot{q}_2+\ddot{q}_3)\sin q_4\cos q_5+\\
0.5l_3\dot{q}_1^2\sin q_4\cos q_5\sin 2(q_2+q_3)+\\
+\sin q_5[l_2(-\ddot{q}_1\cos q_2+2\dot{q}_1\dot{q}_2\sin q_2)-\\
-l_3(\ddot{q}_1\cos(q_2+q_3)-2\dot{q}_1(\dot{q}_2+\dot{q}_3)\sin(q_2+q_3))]\\
\\
l_2\sin q_5\cos q_4[(\dot{q}_1^2\sin^2 q_2+\dot{q}_2^2)\cos q_3-\\
-(\ddot{q}_2+0.5\dot{q}_1^2\sin 2q_2)\sin q_3]+\\
+l_3\dot{q}_1^2\cos q_4\sin q_5\cos^2(q_2+q_3)+l_3(\dot{q}_2+\\
+\dot{q}_3)^2\cos q_4\sin q_5-l_2\sin q_4\sin q_5[(\dot{q}_1^2\cos^2 q_2+\\
+\dot{q}_2^2)\sin q_3+(\ddot{q}_2+0.5\dot{q}_1^2\sin 2q_2)\cos q_3]-\\
-l_3(\ddot{q}_2+\ddot{q}_3)\sin q_4\sin q_5-0.5l_3\dot{q}_1^2\sin q_4\sin q_5\sin 2(q_2+q_3)+\\
+\cos q_5[l_2(-\ddot{q}_1\cos q_2+2\dot{q}_1\dot{q}_2\sin q_2)-l_3(\ddot{q}_1\cos(q_2+q_3)-\\
-2\dot{q}_1(\dot{q}_2+\dot{q}_3)\sin(q_2+q_3))]\\
\\
-l_2\sin q_4[(\dot{q}_1^2\cos^2 q_2+\dot{q}_2^2)\cos q_3-(\ddot{q}_2+0.5\dot{q}_1^2\sin 2q_2)\sin q_3]-\\
-l_3\dot{q}_1^2\sin q_4\cos^2(q_2+q_3)-l_3(\dot{q}_2+\dot{q}_3)^2\sin q_4-\\
-l_2\cos q_4[(\dot{q}_1^2\cos^2 q_2+\dot{q}_2^2)\sin q_3+(\ddot{q}_2+0.5\dot{q}_1^2\sin 2q_2)\cos q_3]-\\
-l_3(\ddot{q}_2+\ddot{q}_3)\cos q_4-0.5l_3\dot{q}_1^2\cos q_4\sin 2(q_2+q_3)
\end{bmatrix}. \tag{6.23}
$$

In the above calculations we did not take into account that the coordinate frame associated with the strain gauge sensor is not the same as the last coordinate frame. It is located in a distance r_5 along x_5 axis from the last coordinate frame. Its orientation is the same as the orientation of the last coordinate frame (compare Fig.3.6a and last coordinate frame $x_6y_6z_6$). Because of that the direction cosine matrix ${}^6_5\boldsymbol{R}$ is an identity matrix. Parameters of the last coordinate frame, which characterises at the same time the measuring coordinate frame, according to the modified Denavit-Hartenberg notation, are presented in the last row of Table 3.2. Based on Eqs.(6.16) and (6.20) we can state that the angular velocity and acceleration of the measuring coordinate frame are expressed by Eqs.(6.17) and (6.21). Unfortunately this does not apply to the linear velocity and acceleration of the origin of the measuring coordinate frame. Based on Eq.(6.18) one can calculate the linear velocity ${}^6v_6$ as follows:

$$
{}^6\boldsymbol{v}_6 = {}^5\boldsymbol{v}_5 + \begin{bmatrix} r_5[\dot{q}_1\sin q_5\sin(q_2+q_3+q_4)+\cos q_5(\dot{q}_2+\dot{q}_3+\dot{q}_4)] \\ -r_5[\dot{q}_1\cos q_5\sin(q_2+q_3+q_4)+\sin q_5(\dot{q}_2+\dot{q}_3+\dot{q}_4)] \\ 0 \end{bmatrix}. \tag{6.24}
$$

Now based on Eq.(6.22) we can calculate the acceleration ${}^6\dot{v}_6$ according to the following expression:

$$
{}^6\ddot{\boldsymbol{v}}_6 = {}^5\,\dot{\boldsymbol{v}}_5 + \begin{bmatrix}
\begin{array}{l}
-r_5 \sin q_5[-\ddot{q}_1 \sin(q_2+q_3+q_4) - \dot{q}_1(\dot{q}_2+\dot{q}_3+ \\
+\dot{q}_4)\cos(q_2+q_3+q_4) + \dot{q}_5(\dot{q}_2+\dot{q}_3+\dot{q}_4)]+ \\
+r_5 \cos q_5[\ddot{q}_2+\ddot{q}_3+\ddot{q}_4+\dot{q}_1\dot{q}_5 \sin(q_2+q_3+q_4)]+ \\
+r_5[-\dot{q}_1 \cos q_5 \sin(q_2+q_3+q_4) + \sin q_5(\dot{q}_2+\dot{q}_3+\dot{q}_4)] \\
[\dot{q}_1 \cos(q_2+q_3+q_4)+\dot{q}_5] \\
\\
r_5 \cos q_5[\ddot{q}_1(-\sin(q_2+q_3+q_4) - \\
-\dot{q}_1(\dot{q}_2+\dot{q}_3+\dot{q}_4)\cos(q_2+q_3+q_4) + \dot{q}_5(\dot{q}_2+\dot{q}_3+\dot{q}_4)]- \\
-r_5 \sin q_5[\ddot{q}_2+\ddot{q}_3+\ddot{q}_4+\dot{q}_1\dot{q}_5 \sin(q_2+q_3+q_4)]+ \\
+r_5[\dot{q}_1 \sin q_5 \sin(q_2+q_3+q_4)+ \\
+\cos q_5(\dot{q}_2+\dot{q}_3+\dot{q}_4)][\dot{q}_1 \cos(q_2+q_3+q_4)+\dot{q}_5] \\
\\
-r_5[-\dot{q}_1 \cos q_5 \sin(q_2+q_3+q_4) + \sin q_5(\dot{q}_2+\dot{q}_3+\dot{q}_4)]^2 - \\
-r_5[\dot{q}_1 \sin q_5 \sin(q_2+q_3+q_4) + \cos q_5(\dot{q}_2+\dot{q}_3+\dot{q}_4)]^2
\end{array}
\end{bmatrix} . \tag{6.25}
$$

Note that the coordinates of the acceleration vector $\ddot{\boldsymbol{p}}$: $\ddot{p}_x$, $\ddot{p}_y$, $\ddot{p}_z$ which appear in the matrix given in Eq.(6.12) are described in Eq.(6.25). In addition, in matrix $\boldsymbol{K}$ there are the coordinates of the implicit angular acceleration and angular velocity which are described in Eqs.(6.21) and (6.17), respectively. For the implicit differential model described by Eq.(6.13) we have to substitute the coordinates of the vector of linear and angular velocity given by Eqs.(6.24) and (6.17), respectively.

Finally, we present explicit equations for the coordinates of the acceleration gravity vector $\boldsymbol{g}$, which are expressed in coordinate frame with origin in point P (see Fig.6.1, which is the force/torque sensor coordinate frame) which result from the recalculations of the vector $\boldsymbol{g}$ to the coordinate frame with origin P

$$
\begin{aligned}
g_x &= -g_0 \cos q_5 \sin(q_2+q_3+q_4) , \\
g_y &= g_0 \sin q_5 \cos(q_2+q_3+q_4) , \\
g_z &= g_0 \cos(q_2+q_3+q_4) .
\end{aligned} \tag{6.26}
$$

6.3 Exciting trajectories for load identification

In this section we present numerical results concerning the optimisation problem of the input trajectory for the purpose of load dynamics identification. We use theoretical results presented in Section 5.2.2 where we proposed optimisation of the condition number of the input correlation matrix. Corresponding numerical calculations were carried out in C++ using object programming techniques.

Now based on Eq.(6.11) we can write the input correlation matrix in the following form:

$$
\boldsymbol{P} = \boldsymbol{K}^T \boldsymbol{K} . \tag{6.27}
$$

Note that the regression matrix $\boldsymbol{K}$ depends on joint positions, velocities, and accelerations. In the case considered here linear and angular velocities and accelerations of the centre of mass of the load depend on all joint coordinates (namely, positions, velocities, and accelerations) which complicates the situation greatly from a numerical point of view. In order to verify that compare Eqs.(6.24), (6.17), (6.25), and (6.21) which for a robot having only five degrees of freedom the corresponding expressions are quite complicated and the computation load in the case considered here is very intensive. Besides that, joint positions, velocities, and accelerations have to satisfy their constraints in the configuration space and all these constraints have to be checked during the computations. Therefore it is not surprising that acceptable results were obtained after 30 to 40 hours of the IBM compatible computer. Note also that joint accelerations are rather small values and it is difficult to accelerate the load in order to identify its dynamic parameters. Therefore it is very important to choose the exciting trajectories very carefully for successful identification. Intuition, due to the very complicated expressions, can be wrong here, as we have experienced.

In order to approach the problem in a systematic way, consider the explicit differential model described by Eq.(6.11) and the optimisation procedure written in terms of the expressions (5.31) $\div$ (5.32). As the initial trajectory we have chosen spline polynomials and sine and cosine functions of time for joint generalised coordinates. To carry out the optimisation procedure we chose calculations from two to four different rows of the regression matrix $\boldsymbol{K}$. However, reproducible and acceptable results were obtained by choosing only one row of the matrix $\boldsymbol{K}$. This row is discretised in 200 equidistant points in time. The preliminary results of the optimisation and identification procedure are presented in reference [43, 132, 134]. The best optimisation results were obtained assuming the following initial trajectories in joint space:

$$\begin{aligned}
q_1 &= 12 - 0,7\cos(0,4\pi + 1,7\alpha) + 0,8\sin(0,1\pi + 0,8\alpha)\,,\\
q_2 &= 4 + 1,8\cos(0,4\pi + 1,7\alpha) + 1,3\cos(0,1\pi + 0,3\alpha)\,,\\
q_3 &= 8 + \cos(0,45\pi + 1,4\alpha) + 0,4\sin(0,4\pi + 0,4\alpha) + \\
&\quad +0,7\sin(0,8\pi + 0,6\alpha)\,,\\
q_4 &= -13 - 1,2\sin(0,45\pi + 1,7\alpha) + 1,8\cos(0,15\pi + 0,7\alpha)\,,\\
q_5 &= -9 + 2,6\cos(0,15\pi + 2\alpha) - 1,1\sin(0,45\pi + 0,2\alpha)\,,
\end{aligned} \tag{6.28}$$

where $\alpha = 2\pi i/(K - i)$, $K = 200$, and $i = 0, 1, ..., K-1$. Choosing initial trajectories as the combination of sine and cosine functions, gave good results in the case of robot dynamic parameters identification. Therefore it was rational to choose such trajectories. Changing the frequency of the input trajectories we were able to excite the components of the input correlation matrix in such a way that peeks of these functions appear in different instants of time and excite the unknown load dynamic parameters.

Based on Eq.(6.28) we can easily calculate the corresponding velocities and acceleration signals in joint space performing the differentiation operation twice. The shape of the initial trajectories and the numerical results and some intuitive judgements are presented in reference [34].

Assuming initial trajectories given by Eq.(6.28) and their first and second derivatives, we have calculated the condition number of the input correlation matrix which was equal to 90830. After 40 iterations of the optimisation procedure its value decreased to 3000. Unfortunately applying the optimisation procedure further did not give better results. The value 3000 seems to be the minimum value of the condition number of the input correlation matrix. Remember that the resultant trajectories are suboptimal. Each trajectory was discretised in 200 points assuming the sampling time $\Delta T = 32$ms. Results of the optimisation procedure are presented in Fig.6.2. As a result we get joint accelerations which have to be integrated twice in order to obtain joint velocities and positions. A continuous line in Fig.6.2 denotes the initial trajectory (initial guess) and a dotted line is used to draw an optimal trajectory in terms of joint accelerations. Note that the optimal trajectories which appear in Fig.6.2 cannot be realised in practice due to the fact that accelerations are bigger than the real IRp-6 accelerations. The initial trajectories were not realised in practice with the exception of the trajectory for the first joint. From this example it is clear that it is very difficult to find optimal trajectories for load identification due to the constraints of the movement of the IRp-6 robot [34]. Because of these difficulties, we did not run numerical calculations of the optimisation of the condition number for the implicit differential model of load identification given by Eq.(6.13).

6.4 Static load parameters measurements

In this section we describe the static load measurements for the IRp-6 robot. Recall that by static measurement of the load we understand measurements of the mass and first order moment components. In order to realise these measurements we position the manipulator in different configurations in space and measure forces and torques. Usually we collect many measurements data, for example, between 500 and 1000. The manipulator is immobile. In order to describe this situation in mathematical terms consider Eqs.(6.1) and (6.2). Based on these equations we write the condition of the equilibrium of forces and torques as follows:

$$\boldsymbol{f} = -m\boldsymbol{g} \tag{6.29}$$

and

$$\boldsymbol{m} = -m\boldsymbol{c} \times \boldsymbol{g} \; . \tag{6.30}$$

Recall that in Eqs.(6.29) and (6.30) $\boldsymbol{f}$ denotes a force which is exerted by the force/torque sensor on the load, and $\boldsymbol{m}$ is the moment acting on the load. The gravity acceleration vector $\boldsymbol{g}$ has components g_x, g_y, and g_z. Examination of Eqs.(6.29) and (6.30) leads to the conclusion that based on the static measurements we are able to identify the mass and first order moment of the

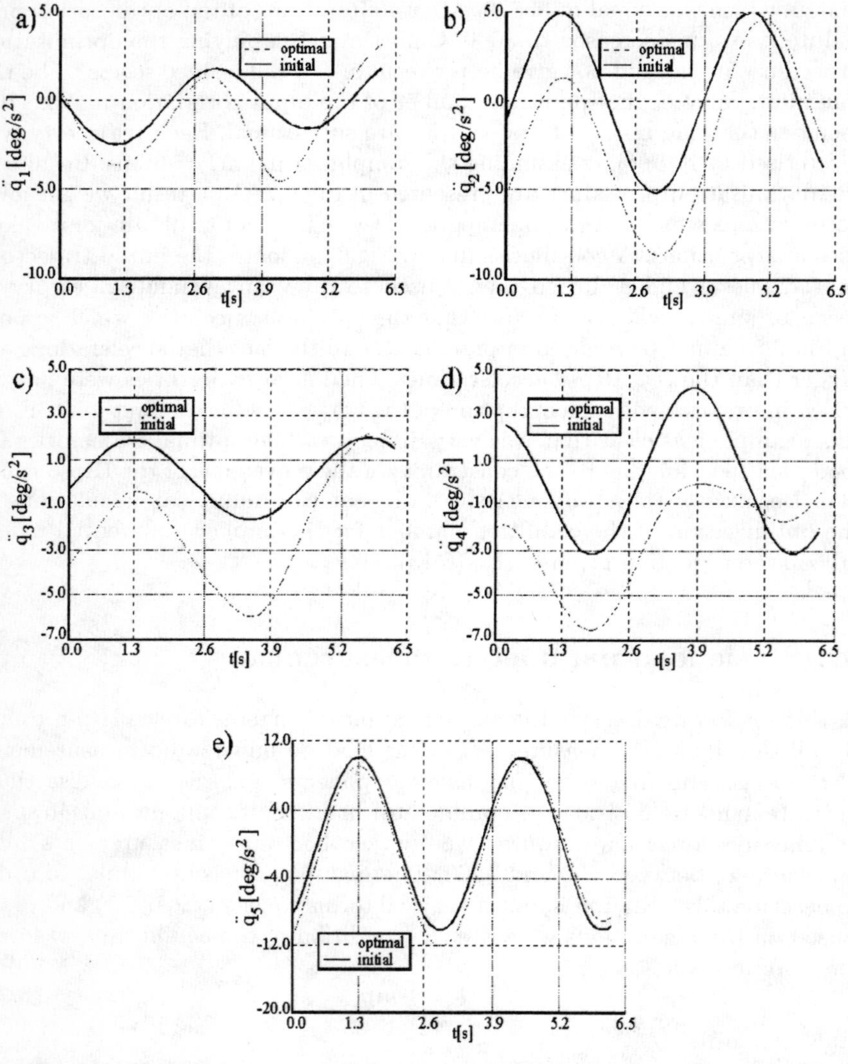

Fig. 6.2. Initial and optimal trajectories for load identification for the IRp-6 robot a) first joint, b) second joint, c) third joint, d) fourth joint, e) fifth joint

load. Vector equations (6.29) and (6.30) can be written in component form, namely we can write three scalar equations for the force components F_x, F_y, and F_z and three scalar equations for the torque components M_x, M_y, M_z. Substitution of Eq.(6.29) to Eq.(6.30) and making use of the six components results in the following three equations:

$$\begin{aligned} M_x &= F_z c_y - F_y c_z \, , \\ M_y &= F_x c_z - F_z c_x \, , \\ M_z &= F_y c_x - F_x c_y \, . \end{aligned} \tag{6.31}$$

A set of equations (6.31) has a solution when the eliminant is not equal to zero otherwise it has infinite number of solutions. In this particular case it is convenient to assume that one of the unknowns c_x, c_y, c_z is constant. This assumption has clear engineering justification. In practice the load and the gripper form one rigid body and it is quite difficult to extract the load parameters from the gripper parameters. Therefore assuming that one of the coordinates c_x, c_y, and c_z is known makes sense when we want to separate both rigid bodies. An alternative approach is to identify the mass and centre of mass of the gripper, and then to identify these parameters for a rigid body consisting of the gripper and the load. The next step is to apply the weighted average to the centre of mass (which consists of the gripper and load) and separate the centre of mass of the load from the centre of mass of the gripper assuming that masses of these two components are known. In addition we assume that grasped objects are symmetric with respect to the $x0z$ plane (associated with axes $0x$ and $0z$ of the force/torque coordinate frame with origin in point P) and its centre of mass is located at this plane. As a consequence we can state that the c_y component is constant and equals zero. Taking into account these assumptions the set of equations (6.31) can be rewritten in the following form:

$$\begin{aligned} M_x &= -F_y c_z \, , \\ M_y &= F_x c_z - F_z c_x \, , \\ M_z &= F_y c_x \, . \end{aligned} \tag{6.32}$$

We conclude that from the set of equations (6.32) the coordinates c_x and c_z can easily be calculated.

Estimation of the mass is easier because based on Eq.(6.29) we can calculate the length of the vector $m\boldsymbol{g}$ as follows:

$$|m\boldsymbol{g}| = \sqrt{F_x^2 + F_y^2 + F_z^2} \, . \tag{6.33}$$

Due to the fact that gravity forces are scaled by computer in Newtons, calculations of the load mass can be simplified according to the following expression:

$$m = \sqrt{F_x^2 + F_y^2 + F_z^2} \, . \tag{6.34}$$

In addition, in order to make the calculations of the centre of mass easier, we have assumed that the approach vector $\boldsymbol{a}$ of the gripper and the z axis of the force/sensor coordinate frame are parallel to the gravity acceleration vector expressed in the base coordinate frame. In this particular situation we have $F_x = F_y = 0$. Then based on the set of equations (6.31) we can write

$$\begin{aligned} M_x &= F_z c_y \,, \\ M_y &= -F_z c_x \,, \\ M_z &= 0 \,. \end{aligned} \tag{6.35}$$

From a set of equations (6.35) we can calculate c_x and c_y coordinates while c_z coordinate is not defined.

In order to carry out the static measurements we used two force/torque sensors a German DLR and an American JR3. Technical features of these two sensors can be found in their technical documentations [76] and [93]. Here we mention only that in the coordinate frame associated with the JR3 sensor, in which the forces and torques are measured (this coordinate frame can change its position and orientation in a manner chosen by the user), we can measure the forces in the range from -245.25N to 245.25N in x and y axes and in the range from -490.5N to 490.5N in z axis. The range of the torques which can be measured by the sensor is the same for each axis and is changes from -19.62N·m to 19.62N·m. The DLR sensor allows us to measure forces in all three directions in the range between -98.1N to 98.1N and torques along the all axes in the range -9.81N·m to 9.81N·m. The accuracy of both sensors is not the same even though they have a 12 bit analog-to-digital converter. Before real measurements we performed many measurements using the sensors not mounted at the robot. Accuracy of the DLR sensor is equal to 2% of the maximal measurement range of the sensor, namely 20g. In reality the accuracy of the sensor is equal 40g which is in contradiction to the accuracy of the 12 bit converter. Accuracy of the JR3 sensor is equal 5g which results from the accuracy of the sensor's 12 bit analog-to-digital converter. Practical application of these sensors shows us that the static measurements using the DLR sensor are worse than the same measurements carried out using the JR3 sensor. Besides that, the DLR sensor is sensitive to the parasitic signals which are present during normal work of the IRp-6 robot. Implementation of the shielding of the DLR sensor did not drastically improve the situation. The JR3 sensor is more resistant to the noise generated by the controller of the IRp-6 robot. These preliminary results and observations convinced us to use the JR3 sensor for the static measurements. The DLR sensor was not used in any further experiments. In addition we have to mention that measurement data using the DLR sensor can be collected with frequency 30Hz and using JR3 sensor with frequency 900Hz.

Preliminary results concerning the use of German sensor can be found in reference [34]. The measurements were taken in different positions of the

IRp-6 robot in configuration space. It is important to calibrate the sensor in each configuration of the manipulator due to the fact that in any configuration the gripper influence (in the sense of reaction forces and torques) is different and has to be carefully precompensated. Each measurement point consists of ten measurement points and based on these measurements each point is averaged. In total 100 points were collected (which is equal to 1000 measurement data). From these measurements we calculated the mean value and standard deviation of the measurement data. Coordinates of the centre of mass were calculated according to Eqs.(6.32) and (6.35) and mass was calculated by making use of Eq.(6.34). Calculations of the coordinates of the mass centre based on Eq.(6.32) were not satisfactory.

As an example consider mass and centre of mass measurements of the load (which is essentially a weight) which has nominal mass equal to 3kg. The three coordinates of the mass centre have the following values $c_x = -10$mm, $c_y = 0$, and $c_z = 320$mm. These geometrical values were obtained based on the dimensions of the weight. Applying the estimation procedure outlined in the previous section, we obtained the mass estimate as 2.66kg and the estimates of the centre of mass as $\hat{c}_x = -0.6$mm, $\hat{c}_y = 0$ (according to the assumption), and $\hat{c}_z = 24$mm. The standard deviation of the corresponding estimates are $\delta c_x = 3.4$mm, $\delta c_y = 0$mm, $\delta c_z = 14.6$mm, and $\delta m. = 0.012$kg. By making use of Eq.(6.35) we get $\hat{c}_x = -11.2$mm, $\hat{c}_y = 2.06$mm, and $\hat{c}_z$ is not calculated. Accordingly, we get the following standard deviations $\delta c_x =$ 0.2mm, $\delta c_y = 0.04$mm, and δc_z cannot be calculated. Mass estimates are the same as obtained before. Several measurement tests were carried out with different objects and all results were similarly unsatisfactory. Summarising, we can state that calculations of the mass centre estimates based on Eq.(6.32) are far from real values obtained from geometrical measurements of the load. An error associated with the estimates of the c_z coordinate can be equal up to 40 cm and for the c_x coordinate up to 6 cm. Of course, these results are not acceptable.

This situation can be explained as follows. The force and torque measurements are very noisy signals due to the noise generated by the IRp-6 controller. The proposed measurement method which requires calculation of the quotient is prone to measurement errors. The measurement deviations of the forces and torques from their nominal values are very high and, in general, not acceptable. Based on the calculated statistics the estimates are reliable for only 10% of the measurements when the load has a mass of 1kg and 30% of the measurements when the load has a mass of 3 kg (this applies to the centre of mass estimates which are calculated based on Eq.(6.32). It seems that the results of the estimates would be much better if we considered loads having a mass bigger than 10 kg. This results from the accuracy of the DLR sensor which is, according to the technical documentation, 2% of the maximum range of the measurement range of the sensor. When we apply Eq.(6.35) for calculations of the mass centre coordinates, the estimates differ from the real values by about 2 to 4mm. Unfortunately, when we apply Eq.(6.35) we are not able to calculate the coordinate c_z of the centre of mass.

During experimental work we investigated the DLR sensor. We noticed that one of the reasons for noisy measurements was a problem of inaccurate compensation of the sensor. Therefore, it is necessary to use statistical methods to process the measurement data from the very beginning to obtain data which are within an acceptable confidence interval. The second drawback of the DLR sensor is the measurement error which appears when the temperature increases. This is probably because of wrong selection of the operational amplifier which has the temperature drift. We tried to compensate the temperature drift but we only marginally improved our results. Besides that, the sensor is very sensitive to the noise generated by the IRp-6 controller. This is mainly due to the pulse feeder designed for the controller. We used shielding but it did not help significantly. Due to these difficulties we have decided not to use the DLR sensor for further experiments.

Now we compare the experimental results using the JR3 sensor [93] both for static and dynamic measurements. This sensor does not have the drawbacks mentioned above, is more resistent to measurement noise and can be easily and better precompensated. For the experiments we used a 50cm aluminium bar with a mass equal to 400g. This mass was weighed by a very accurate scale. Also it was possible to calculate accurately the coordinates of the centre of mass of the bar. The estimates of the mass and centre of mass coordinates were calculated based on Eqs.(6.34) and (6.35). The results of the estimates are presented in Table 6.1.

Table 6.1. Mass and centre of mass estimates for the aluminium bar

y [cm]	$\hat{m}$ [kg]	δm [kg]	L_m	$\hat{c}_x$ [cm]	δc_x [cm]	L_x	$\hat{c}_y$ [cm]	δc_y [cm]	L_y
0.0	0.388	0.001	62	1.011	0.040	100	0.268	0.010	32
7.5	0.391	0.001	89	0.899	0.005	77	7.515	0.013	89
12.5	0.401	0.000	89	1.133	0.005	82	12.127	0.015	89
17.5	0.386	0.010	60	0.826	0.005	100	17.676	0.027	60
22.5	0.388	0.000	100	0.788	0.011	23	22.391	0.016	100
27.5	0.392	0.000	59	0.904	0.008	42	26.945	0.030	100
32.5	0.381	0.000	100	1.356	0.007	60	32.932	0.039	100
37.5	0.391	0.010	35	1.596	0.005	66	37.005	0.088	37
42.5	0.360	0.001	79	1.599	0.010	28	44.943	0.115	79
47.5	0.414	0.000	93	1.718	0.005	100	44.408	0.048	100

The bar was grasped by the gripper in different distances from its centre of mass. The current distances in which the bar was grasped are indicated in the first column of Table 6.1. In the second column of Table 6.1 the estimates of the bar are collected while in the third the standard deviation of the mass is depicted. The next column denotes the number of measurement data which were accepted for the processing. These data were qualified based on the statistical test. Different tests were used but eventually we decided to use the t-Student statistical test with $n-1$ degrees of freedom. The next two columns contain data which are the estimates of the centre of mass coordinates c_x and c_y with corresponding number of samples and errors. From Table 6.1 we deduct that the accuracy of the mass estimate is 10g which corresponds to the accuracy of the analog-to-digital converter of the sensor. Accordingly, the estimates of the centre of mass along y axis is about 5mm and in x-axis is from 10 to 15mm. The accuracy of the centre of mass estimates depends also on the position of the aluminium bar in the gripper. Each position was very carefully tuned but we realise that there were still inacurracies in the positioning of the bar. Nevertheless note that the estimates are very stable and the results are definitely better in comparison to those obtained by the DLR sensor. We have implemented Eq.(6.32) in order to calculate the estimates of the centre of mass. The results were of similar nature.

The static measurement using JR3 sensor are quite promissing and this sensor was used for further dynamic measurements described in the next section.

6.5 Dynamic load parameters measurements

In this section we present dynamic parameters measurements of the load grasped by the IRp-6 robot, when it moves along a predefined trajectory. This kind of measurement is known as the dynamic measurement in contrary to the static measurement described in the previous section. Some preliminary results can be found in references [38, 41, 43, 44, 46, 50] set-up was described in Chapter 2 and it was used to carry out the experiments. Note that the experimental set-up is universal and allows us to perform different experiments concerning the IRp-6 robot.

Due to the arguments discussed in the previous section we carried out dynamic experiments using only the JR3 sensor. The differential model of load dynamic parameters is described by Eq.(6.11)and is a fundamental equation for further considerations. We suggest using the least squares method for the purpose of identification. Its recurrence form is the most prefarable with a combination of the Agee–Turner factorisation [15]. In order to avoid a difficult problem with load and gripper seperation due to the complicated shape of the gripper, we decided to separate it from the load. Because of the shape and construction of the sensor, it is an easy matter to connect the load directly to it. Actually, if it is not the case, we have to carry out two experiments first one with the gripper and load and second with the gripper only. Making use of the Steiner theorem [81] we can separate the inertia parameters of the load from the gripper. Note that, from a control point of view, it is interesting what dynamic properties the load which is held by the gripper has or equivalently, which forces and torques are exerted on the gripper by the load.

Therefore, it is not necessary to separate the parameters of the load from the gripper. One can imagine that the load which is fixed to the force/torque sensor plays the role of the gripper and load together as one rigid body. Many different objects were used to symbolise this situation and many experiments were carried out using the set-up described in Chapter 2. Here we present the estimates of the dynamic parameters of the load which is a rectangular object made of steel. It is fairly easy to calculate all the dynamic parameters of this object. These dynamic parameters are expressed in load coordinate frame associated with the force/torque sensor. The dynamic parameters are collected in the second column of Table 6.2.

Table 6.2. Load dynamic parameters and their estimates

Parameters	CAD values	Estimates	Standard deviation	Confidence interval
m [kg]	1.910	1.916	0.024	0.011
mc_x [kg·m]	0.0573	0.0568	0.0028	0.0013
mc_y [kg·m]	–0.0210	–0.0207	0.0088	0.0042
mc_z [kg·m]	0.1394	0.1396	0.0091	0.0044
I_{xx} [kg·m^2]	0.01050	0.01180	0.0051	0.0024
I_{xy} [kg·m^2]	0.00094	–0.00083	0.0014	0.00067
I_{xz} [kg·m^2]	–0.00450	–0.00751	0.0017	0.00084
I_{yy} [kg·m^2]	0.01635	0.01567	0.0071	0.0034
I_{yz} [kg·m^2]	0.00169	0.00001	0.00006	0.00029
I_{zz} [kg·m^2]	0.00520	0.00449	0.00037	0.00018

During experiments we assumed that the joint velocity signals are measured by the rate generators and these measurements are not very accurate due to the transient state of the rate generators. The positional signal are measured very accurately. We differentiated the position signals and compared them with the measured velocity signals. Prior to numerical differentiation we filtered the positional signal. We implemented the differentiation by taking samples on both sides of the point in which the derivative is calculated. The discretisation time is $\Delta T = 32$ms and we took two, four, six, and eight samples in order to calculate the time derivative at a point which is located at the centre of the samples. It is surprising, but the results of the numerical differentiation were better than the measurement data from the rate generators. Velocity signals when differentiated, result in acceleretion signals which are worse than twice differentiated positions signals.

In our experiment we assumed that the joint trajectories are fifth order polynomials in joint space which were designed such that the joint velocities achieve their maximum values. The range of the movement for each joint is as follows:

$$-40^\circ \leq q_1 \leq 30^\circ,$$

$$-20^\circ \le q_2 \le 10^\circ,$$
$$-30^\circ \le q_3 \le 40^\circ,$$
$$-10^\circ \le q_4 \le 30^\circ,$$
$$-150^\circ \le q_5 \le -10^\circ.$$

The total time of the movement was the same for each joint and equal to 1.83s. The position signals were collected using the interface with direct access to the resources of the IRp-6 controller. In the position boundaries above we did not include offsets which result from Table 3.2 for the second, third, and fourth joints and we used degrees instead of radians.

As an example, consider the joint position, velocity, and acceleration signals for the third joint of the IRp-6 robot. These three signals are illustrated in Figs.6.3a, b, c, respectively. The effect of twice differentiation operation can easily be seen in Figs.6.3b and 6.3c. Observe the noise which is contaminated in the acceleration signal. This noise was filtered using additional filters. The input signals for the other joints can be found in reference [36].

In Fig.6.4 we can observe the estimates of the mass (Fig.6.4a) and three components of the first moment (Figs.6.4b÷6.4d). Note that the estimates are very stable and usually stabilise after 30 iterations, namely after half of the total observation time. Estimates of the inertia tensor elements can be found in reference [34].

Note that each measurement data is processed by the set of equations (6.11) in about half of the observation time which is equivalent to 180 iterations. Now comparing the results of the estimates with the real values calculated based on the dimension of the rectangular load measurements (see second column of the Table 6.2) we can state that the mass and the three coordinates of the first order moment are estimated very accurately. Similarly, the inertia moments on the diagonal are estimated very accurately. The deviation inertia elements (off-diagonal elements), which have smaller values, are not so accurate, but nevertheless their numerical values are still close to the real values. The results presented here are comparable to the results obtained by An and co-authors [1]. In order to evaluate the estimates for each of them based on 30 measurement data we calculated the standard deviation and confidence interval which was assumed to be 99%. The confidence interval was calculated from the following expression

$$u = s \cdot \mathrm{const} \cdot \sqrt{m_u}\ .$$

In the last equation s denotes the standard deviation, const equals 2.576, and $m_u = 30$. Based on these assumptions we calculated the statistical features and collected them in the last two columns of Table 6.2. The calculated standard deviation and the confidence interval show that the estimates have good statistical characteristics.

In Fig.6.5 we can observe two force curves. The first one denoted by a continuous line, shows the measured forces by the sensor and the second,

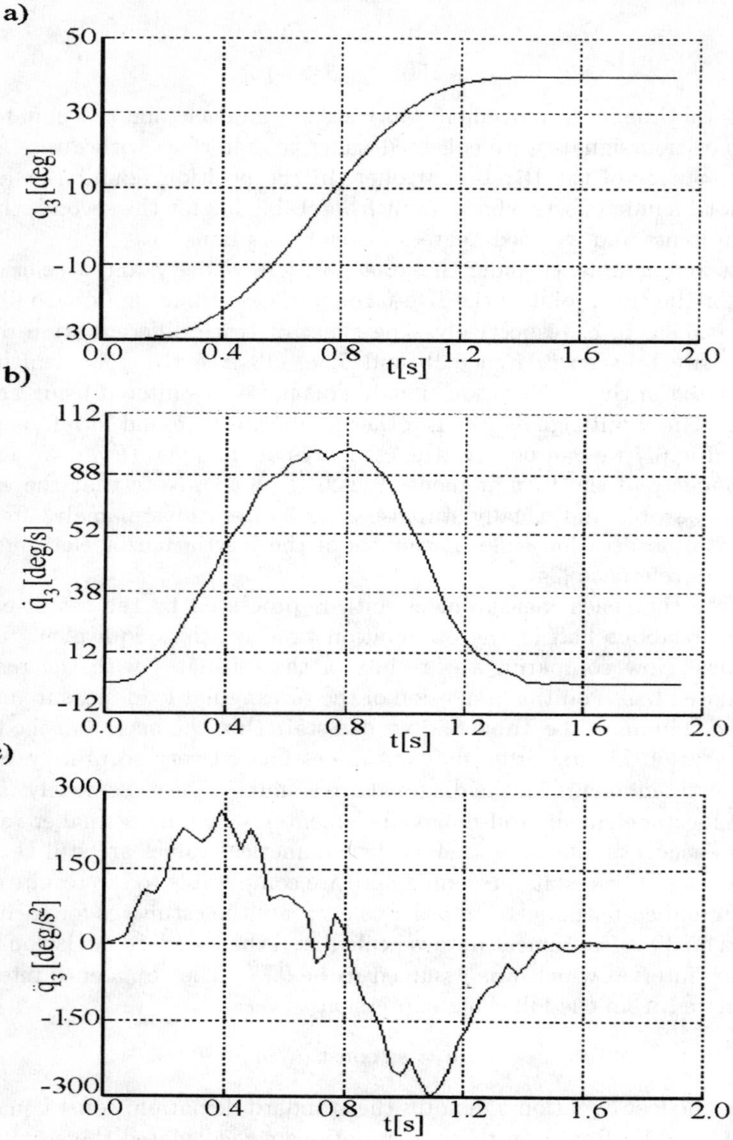

Fig. 6.3. The IRp-6 robot signals for the purpose of load identification a) position, b) velocity, and c) acceleration

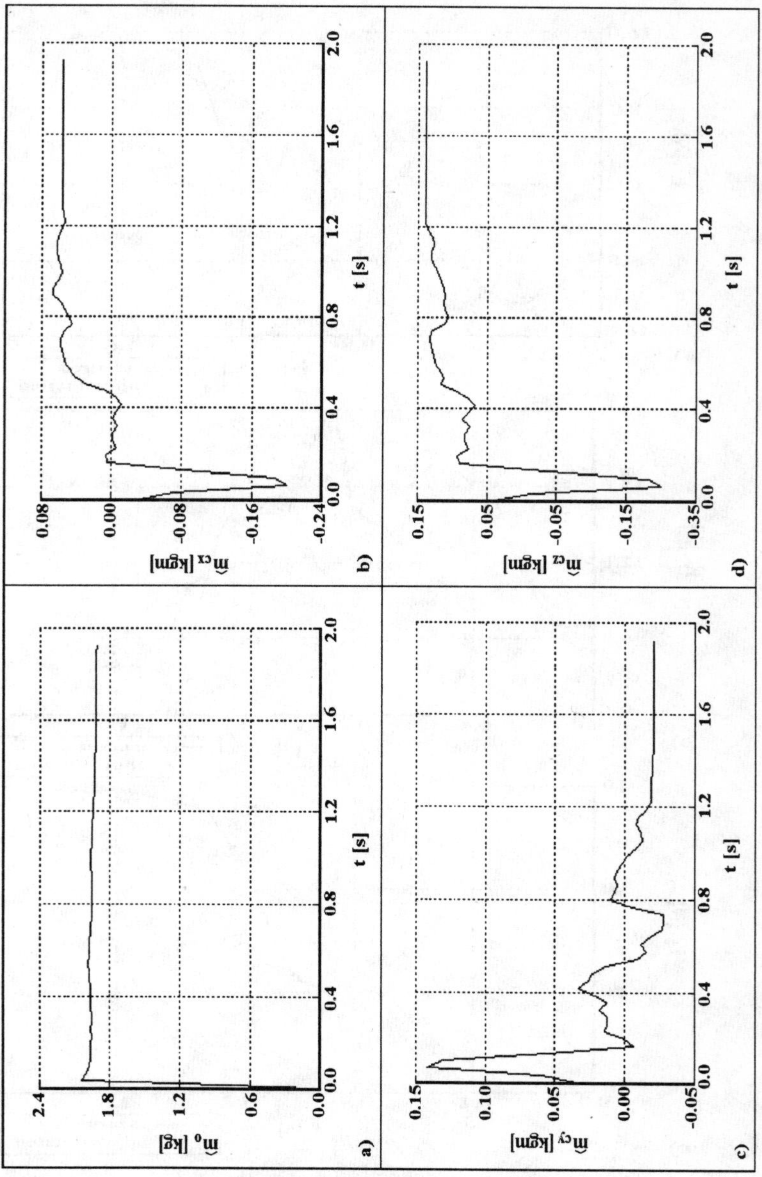

Fig. 6.4. Estimates of the mass and centre of mass of the load

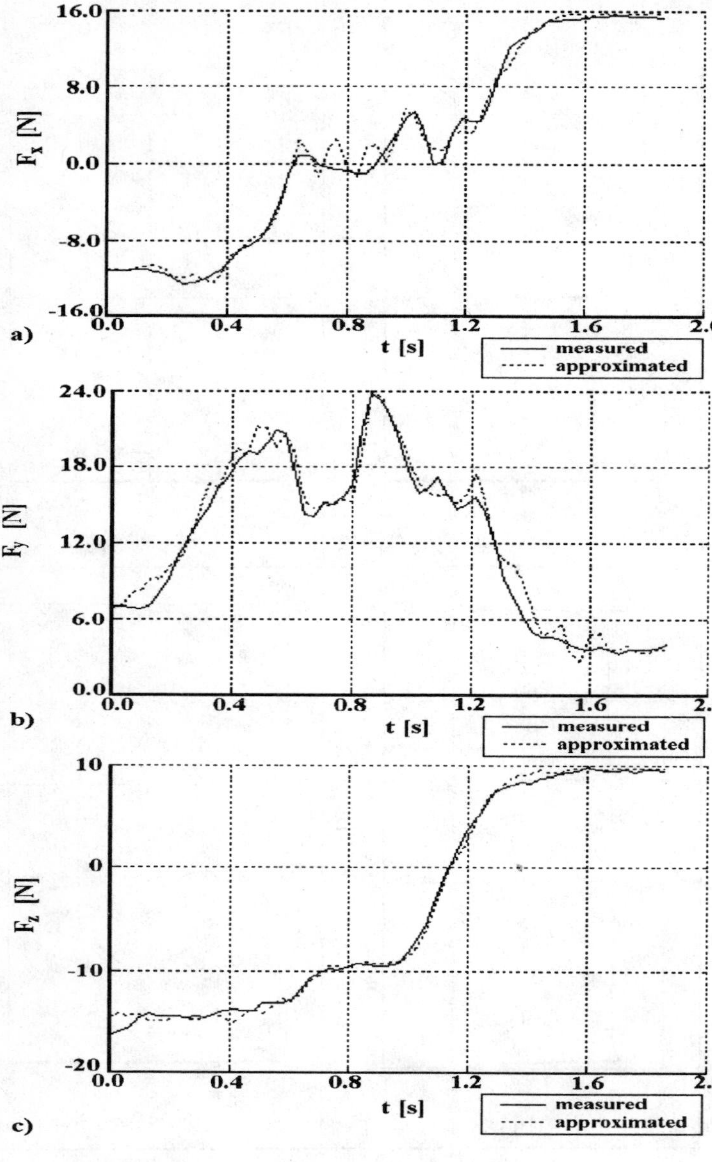

Fig. 6.5. Measured and approximated forces

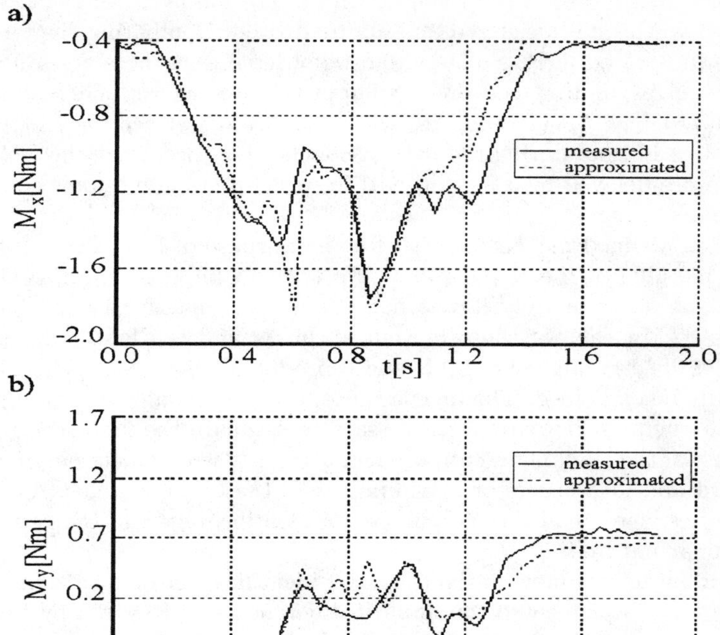

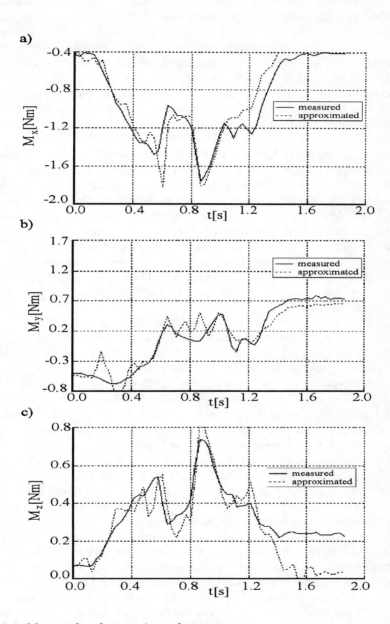

Fig. 6.6. Measured and approximated torques

indicated by dotted line, plots the forces calculated based on estimated parameters and measured joint positions, velocity, and accelerations (namely calculated based on the right side of Eq.(6.11)). Similarly, in Fig.6.6 we can observe two torques measured and approximated. Comparing the forces and torques illustrated in Figs.6.5 and 6.6 we observe that there are some differences in corresponding measured and approximated curves. These differences are not too big. All of them are within the range of accuracy of the JR3 sensor. In addition, during the measurements we noticed that the JR3 sensor signals were delayed by 3×32ms with respect to measured joint generalised coordinates.

The predefined movement used for the purpose of identification was the fastest possible of the IRp-6 robot. We tried to implement different movements, but unfortunately it was not possible to identify the inertia tensor parameters. In each of these movements mass and the first order moment were identified giving very stable and reproducible estimates. This situation can easily be explained. The inertia tensor element should be excited by very fast movements which are not necessary to estimate the first order moment and mass of the load. Different polynomials were used, namely polynomials of the third and fifth orders or combinations of these polynomials. All of these trajectories were good for the purpose of identification and the results were of a similar nature.

In order to get more insights of the identification of load dynamic parameters, we implemented the implicit differential model given by Eq.(6.13). In general we obtained estimates which had worse statistical characteristics than the estimates from the differential model. Implementation of the integration operation cuases the noise signals which have low frequency to be accumulated during the experiment. This observation was also noticed by authors of references [26, 154]. In order to decrease this error we chose different equations from the set of equations (6.13) which resulted in better estimates.

In Section 6.3 we have described the optimal trajectories for the purpose of load estimation. Unfortunatelly due to the velocity, and acceleration limits we were not able to use the optimal trajectories for the purpose of identification. Obtained accelerations cannot be realised in practice; see Fig.6.2. The estimation results and the results of the optimisation procedure evidently show that the inertia tensor identification requires fast trajectories. It is interesting to note that for robot dynamics identification this requirement was not crucial and applying slower trajectories we were generally able to identify the dynamic parameters of the canonical model of the IRp-6 robot. One possible explanation is that the inertial parameter of the load has smaller values (recall that maximum load of the IRp-6 robot is 5kg) in comparison to the inertia parameters of IRp-6 robot. This is substantiated by the numerical values listed in Tables 6.1 and 6.2. As a consequence, if we want to identify these parameters we have to move them very fast. The fastest trajectories which we were able to appy gave good reproducible identification results, which for a class of geared robots, are difficult to obtain.

At the end of this chapter we discuss the possibilities of applying the least squares method for identification of load dynamic parameters when the recursive algorithm is implemented on–line. In order to find out which instructions are the most computationally intensive we have used the standard Turbo Profiler Version 2.2 programme Copyright (c) 1990, 92 Borland International. The total experimental time was equal to 3.9722s when executed on an IBM compatible computer (exactly IBM PC 486 DX 33MHz) which when calculated per one iteration step results in a time equal to 66.215ms. The discretisation time is 32ms. Based on the iteration procedure described in terms of Eqs.(4.10)-(4.11) we observed, that the calculations of the input correlation matrix $\boldsymbol{P} = \boldsymbol{U}\boldsymbol{D}\boldsymbol{U}^T$ are the most time consuming. Matrix multiplications take about 55% of the total computation time in one iteration step. Taking into account the particular structure of the matrices $\boldsymbol{U}$, $\boldsymbol{D}$ ($\boldsymbol{U}$ – is an upper triangular matrix, $\boldsymbol{D}$ – diagonal matrix) it was possible to reduce the computational time by half of the total time. Besides that, we moved the equation which updates the vector of the estimates (compare Eq.(4.10)) to the procedure which performs the Agee–Turner algorithm [15]. As a result, we have, a programme which calculates the estimates on an IBM compatible computer in 0.5211s which results in an iteration time of 8.7ms.

Taking into account the results of the optimisation of the programme, it is now possible to carry out the identification procedure on–line. Compare that the discretisation time is equal to 32ms. In order to calculate the total time per one iteration we have to take into account the computational time which is necessary to perform the calculations from the motor coordinates to the joint coordinates. We also have to add the calculations of the joint velocities and accelerations which result in a total time per one iteration of 24ms. This computational time allows us to carry out the identification of load parameters on–line. Approximately the same time is necessary to carry out identification on–line for robot dynamics. Concluding, presented algorithms are suitable for off– and on–line identification.

CHAPTER 7
HYBRID CONTROL OF THE IRp-6 ROBOT

7.1 Introduction

In this chapter we present some issues on the hybrid control of robot manipulators taking as an example the IRp-6 robot. The considerations discussed in this chapter are restricted to the robots which are equipped with position controllers. The IRp-6 robot belongs to this type of robots. In general the experimental set-up described in Chapter 2 allows us to examine this kind of control. We developed the necessary tools such as a straight line trajectory planning in Cartesian space and data acquisition system for collecting and analysing data from the force/torque sensor JR3 to realise the hybrid control of the IRp-6 robot. Examples presented here show how an industrial robot with all its limitations and restrictions can be used for the purposes of advanced control.

Hybrid control plays very important role in industrial applications. A notion of hybrid control can be explained in the following way. When a manipulator moves in its working space and does not interact with an external environment, its position controller follows a described trajectory. When the manipulator is in contact with a stiff surface and we use only the position controller it is possible that the interaction forces between the environment and the robot can reach high values and the manipulator can either destroy the surface or itself. In this particular case the controller has to take into account the interaction forces, namely we need a separate force controller to handle this situation. Briefly this is the idea of hybrid control. It is important to note that hybrid control plays a very important role in the integration of the robot with the environment in which it is working. Different sensors, for example, vision, force, touch, and others can be used to collect data from the external environment which surrounds the robot. This is known as sensory feedback in robotics. Force and position control have many practical applications, for example, in assembly, welding, grinding surfaces, manipulation of the objects by two or more robots etc.

As can be seen from the above, there are many potential applications of hybrid control. On the other hand, this kind of control requires expensive force/torque sensors to be used which limits the applications. In order to overcome this difficulty we can use inexpensive sensors which are able to measure force or torque only in one direction. Besides that we need a controller with an open architecture in which we have direct access to the servo loop, namely to the torques. Therefore many industrial robots are not suitable to execute these tasks due to their inherent limitations. The approach presented in this chapter tries to solve this problem assuming that the original structure of the position controller remains unaltered.

The standard hybrid position/force controller was presented by Raibert and Craig in reference [187] and Mason [157]. Khatib [107] developed the results presented by Raibert and Craig introducing the operational space formulation. Both approaches assume that in the robot controller both position and current loops can be controlled. This is not the case for the IRp-6 robot. This robot has only a position loop and the user does not have access to the servo loop. The problem of hybrid control was considered in the following books [8, 31] and [235]. A very good survey of hybrid control can be found in references [223, 224]. In Polish literature there are also several references concerning this subject, see for example, references [32, 42, 49, 77, 78, 138]. Finally, Goldenberg and Song [71] proposed a novel approach to this problem.

Other researchers have faced the same problem and have presented various solutions. For example Degoulange and co-authors [33] proposed a control scheme for the PUMA 560 robot. They assumed, that measured forces and torques can be transformed to the position and orientation variations in Cartesian space. They proposed selecting the directions in task space in which the measured forces and torques can be transformed to equivalent changes in the position and orientation of the robot. Note that in order to completely specify the position and orientation of the robot in Cartesian space we need six independent coordinates. By making use of these coordinates we can specify the position and orientation of a rigid body in Cartesian coordinates. A contact between the force/torque sensor and an environment is modelled as a spring and a mass. The desired and measured forces and torques are subtracted from each other and the difference is used in the force controller in the directions in which the desired force and torque are exerted. Next, the corrections of position in all directions are added to the standard position controller and final corrections are calculated through inverse kinematics. A similar approach was used by Grodecki and Gosiewski in papers [77] and [78], respectively. They designed a hybrid controller for the IRp-6 robot. Note that the algorithm presented in reference [33] cannot be applied to the IRp-6 robot due to the fact that this robot has only five degrees of freedom. Therefore Grodecki and Gosiewski proposed defining a tool which is symmetric along the axis, along which there is no torque exerted by the environment. Taking into account these limitations it is possible to control the robot in Cartesian space in a similar fashion to that described above.

In this monograph we extend the results presented by Grodecki and Gosiewski in papers [77] and [78]. First we have calculated, based on ideas presented in reference [8], the local stiffness of each joint of the IRp-6 robot. The stiffness in Cartesian space changes due to the changes of the robot Jacobian [8, 56, 57]. Therefore, for each task the stiffness in Cartesian space has to be recalculated. We suggest measuring the stiffness in Cartesian space in an experimental way.

The global compliance is also calculated by means of the local stiffness of the manipulator itself [8] and is compared with the results of the above method. These two approaches are calculated against each other and compared to the IRp-6 robot.

These results are very useful for trajectory design with contact force and torque measurements. Several experiments were run and illustrate how force and torque measurements can be used for the IRp-6 robot to follow a prescribed trajectory with desired forces and torque acting along the trajectory. Force and torque sensor measurements are incorporated with other sensor which measure positions, velocities, and currents for each joint of the IRp-6 robot.

The chapter is organised as follows. In Section 7.2 we introduce two different approaches to construct the hybrid controller for robots having only position controllers. These two algorithms are based on references [33] and [77, 78]. In Section 7.3 we propose local and global stiffness measurements for the IRp-6 robot which were developed in our laboratory. We believe that this method is new and very useful in combination with the hybrid controller for the IRp-6 robot. Finally, in the last section we illustrate the theoretical consideration presented in the previous sections showing experimental results carried out at the Robot Control Laboratory on the experimental set-up around the IRp-6 robot.

7.2 Different control algorithms for robots with position controllers

7.2.1 A hybrid controller for the PUMA 560 robot

In this section we present a hybrid control algorithm developed at the University of Montpellier, France by Dégoulange and co-authors [33]. Conceptually the hybrid control scheme is presented in Fig.7.1. The above scheme uses the standard position controller and therefore there are no hardware changes necessary in order to implement the hybrid control scheme. In addition we have to measure three dimensional forces and torques using the force/torque sensor mounted between the wrist of the manipulator and the gripper. Now suppose that we assume a vector of desired forces and torques to be known which is expressed in the coordinate frame ${}^{H}\boldsymbol{w}_d$ (it is sometimes called a hybrid frame, and this will be discussed in more detail in the next section) in the following form:

$$ {}^{H}\boldsymbol{w}_d = \left[{}^{H}\boldsymbol{f}_d^T ,\ {}^{H}\boldsymbol{m}_d^T \right]^T . \tag{7.1}$$

In the last equation ${}^{H}\boldsymbol{f}_d$ is a vector of desired forces, while ${}^{H}\boldsymbol{m}_d$ is a vector of desired torques and T denotes a transpose operation. Now assume that the

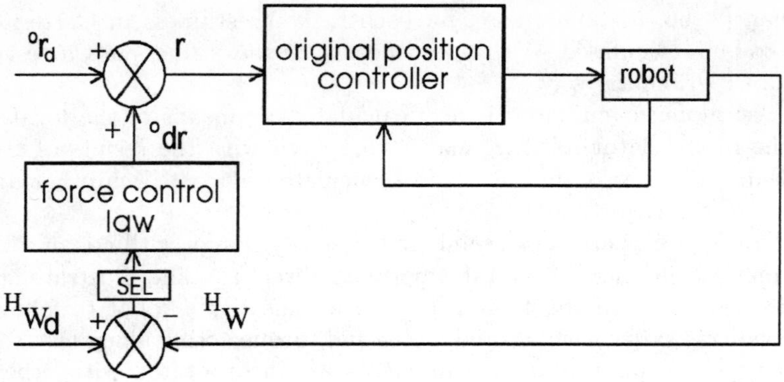

Fig. 7.1. Simplified force control scheme

vector of the forces and torques measured by the sensor is transformed to the hybrid coordinate frame H (recall that, in general, the coordinate frame which is associated with the sensor is not the same as the hybrid coordinate frame) and is as follows:

$$^{H}\boldsymbol{w} = \left[{}^{H}\boldsymbol{f}^{T},\ {}^{H}\boldsymbol{m}^{T}\right]^{T}. \tag{7.2}$$

In the above equation $^{H}\boldsymbol{f}$ and $^{H}\boldsymbol{m}$ are the measured forces and torques, respectively transformed to the hybrid coordinate frame. We also introduce a vector of displacement:

$$^{0}\boldsymbol{dr} = \left[{}^{0}\boldsymbol{dp}^{T},\ {}^{0}\boldsymbol{d\nu}^{T}\right]^{T}, \tag{7.3}$$

where $^{0}\boldsymbol{dp}$ is a vector of position increments, and $^{0}\boldsymbol{d\nu}$ is a vector of orientation angles increments which are expressed in the base coordinate frame. The vector given by Eq.(7.3) results from the force control law (see Fig.7.1). We can say that an error between the desired and measured forces causes the increment $^{0}\boldsymbol{dr}$ which is the result of applying the force control law. The vector $^{0}\boldsymbol{dr}$ is added to the vector of the desired position and orientation $^{0}\boldsymbol{r}_{d}$ which is written as:

$$^{0}\boldsymbol{r}_{d} = \left[{}^{0}\boldsymbol{p}_{d}^{T},\ {}^{0}\boldsymbol{\nu}_{d}^{T}\right]^{T}. \tag{7.4}$$

In this equation we assume that the orientation vector is expressed in terms of the Euler angles [195, 198]. Mathematically $^{0}\boldsymbol{r}_{d}$ is not a vector since the typical operations characterising a vector space cannot be defined on a set of Euler angles. Therefore this vector is recognised as a pseudovector. To be more precise we can always calculate the orientation error by changing the Euler angles to the equivalent rotation matrix both for the desired orientation and the measured orientation. Then both these orientations are expressed by the orthonormal vectors $\boldsymbol{n}_{d}$, $\boldsymbol{o}_{d}$, $\boldsymbol{a}_{d}$, and $\boldsymbol{n}$, $\boldsymbol{o}$, $\boldsymbol{a}$ and the orientation error is calculated in terms of the cross product of these vectors $\frac{1}{2}(\boldsymbol{n} \times \boldsymbol{n}_{d} + \boldsymbol{a} \times$

$\boldsymbol{a}_d + \boldsymbol{o} \times \boldsymbol{o}_d$) (see references [195, 198]). It seems that for the user it is easier to express the orientation in terms of the Euler angles, nevertheless we have to remember that the orientation error has to be calculated by the above cross product. For understanding how the control scheme works we use the definitions given by Eqs.(7.3) and (7.4).

Now we comment on the control scheme presented in Fig.7.1. When an external force acts on the manipulator, as a result of this force, the manipulator will change its position (here by position we mean position and orientation). As a consequence we can calculate the position error which is calculated in the external loop of the Fig.7.1 and this error is used to compensate for the external forces and torques. If there are no external forces and torques the robot uses its original position controller and changes its position to follow a predefined trajectory. Conceptually the idea of the hybrid controller from Fig.7.1 is very simple, it contains two loops; position loop and force/torque loop. The last one is scaled in equivalent position changes which are used as input values to the standard position controller. A logical block SEL which is placed in the force control loop activates the last function of the system. A diagonal matrix SEL (with ones and zeros) is known as a selection matrix. It selects the directions which must be force controlled and the directions which must be position controlled. Usually it is assumed that the force control and position control are orthogonal actions.

Now we explain how to calculate the equivalent increments of the position and orientation as the result of the forces and torques errors. In order to do so we assume that the contact between the robot and the environment is treated as a model of generalised spring. Therefore we can assume that the position increment is proportional to the force error (later on in this chapter we discuss this assumption in more detail). Therefore we can write the following equations:

$$ {}^{H}\boldsymbol{e}_f = {}^{H}\boldsymbol{f}_d - {}^{H}\boldsymbol{f}, \tag{7.5} $$

and

$$ {}^{0}\boldsymbol{dp} = {}^{0}\dot{\boldsymbol{p}}dt = \boldsymbol{C}_{f_3}\left({}^{H}\boldsymbol{f}_d - {}^{H}\boldsymbol{f}\right). \tag{7.6} $$

In the latter equation, $\boldsymbol{C}_{f3}$ is a 3×3 matrix diagonal consisting of the proportional coefficients (these coefficients characterise the compliance of the forces in the hybrid frame). Note that ${}^{0}\boldsymbol{dp}$ is formally a differential of the position and at the same time an infinitesimal position error.

Similarly, we proceed with the orientation angle error and the torques. We can write:

$$ {}^{H}\boldsymbol{e}_m = {}^{H}\boldsymbol{m}_d - {}^{H}\boldsymbol{m} = \boldsymbol{\omega} dt, \tag{7.7} $$

$$ {}^{0}\boldsymbol{d\nu} = {}^{0}\dot{\boldsymbol{\nu}}dt, \tag{7.8} $$

where $\boldsymbol{\omega}$ and $\dot{\boldsymbol{\nu}}$ denote angular velocity and the time derivative of the orientation of the hybrid coordinate frame with respect to the base coordinate frame (recall the task is defined in the hybrid frame). The angular velocity $\boldsymbol{\omega}$ in terms of the time derivative $\dot{\boldsymbol{\nu}}$ can be written as:

$$\boldsymbol{\omega} = \boldsymbol{B}_0(\boldsymbol{\nu})\, {}^0\dot{\boldsymbol{\nu}}, \tag{7.9}$$

where $\boldsymbol{B}$ represents the orientation matrix expressed in terms of the Euler angles [33]. The inverse of the orientation matrix can be written as:

$$\boldsymbol{B}_0^{-1}(\boldsymbol{\nu}) = \begin{bmatrix} \sin O \cdot \tan A & -\cos O \cdot \tan A & 1 \\ \cos O & \sin A & 0 \\ \sin O / \cos A & -\cos O / \cos A & 0 \end{bmatrix}. \tag{7.10}$$

Finally, taking into account Eqs.(7.9) and (7.10) we can write:

$${}^0 d\boldsymbol{\nu} = \boldsymbol{B}_0^{-1}(\boldsymbol{\nu})\boldsymbol{\omega} dt = \boldsymbol{B}_0^{-1}(\boldsymbol{\nu})\boldsymbol{C}_{m_3}\left({}^H\boldsymbol{m}_d - {}^H\boldsymbol{m}\right). \tag{7.11}$$

In Eq.(7.11) $\boldsymbol{C}_{m3}$ is a 3×3 matrix diagonal which consists of the proportional coefficients associated with the torques (these coefficients characterise the compliance of the torques in the hybrid frame). Both matrices $\boldsymbol{C}_{f3}$ and $\boldsymbol{C}_{m3}$ which appear in Eqs.(7.6) and (7.11) are designed by an engineer and can be changed according to the executed task in the hybrid frame. Based on Eqs.(7.6) and (7.11) we are able to calculate the position increments ${}^0 d\boldsymbol{p}$ and the orientation angle increments ${}^0 d\boldsymbol{\nu}$ which result from the desired position ${}^0\boldsymbol{r}_d$ (here again position and orientation) and the desired force ${}^H\boldsymbol{w}_d$ (here spatial force consisting of the vector of forces and torques). At the same time the vector of forces and torques is a result of measured voltage $\boldsymbol{v}_{vol}$ and the calibration matrix $\boldsymbol{K}_a$ (which are associated with the force/torque sensor JR3 [93]) according to the following formula:

$$\left[{}^S\boldsymbol{f}^T, \; {}^S\boldsymbol{m}^T\right] = {}^S\boldsymbol{w} = \boldsymbol{K}_a \boldsymbol{v}_{vol}. \tag{7.12}$$

The superscript which appears in the vector ${}^H\boldsymbol{w}_d$ denotes the coordinate frame associated with the force/torque sensor (see description of the JR3 sensor in Chapter 2). Equation (7.12) presents a practical expression from which the spatial vector force can be calculated based on the measurements. The calibration matrix $\boldsymbol{K}_a$ is known to the user.

Due to the fact that the spatial forces have to be transformed to the coordinate frame associated with the gripper we have to do the following calculations:

$${}^H\boldsymbol{f} = {}^H_S\boldsymbol{R}\, {}^S\boldsymbol{f}, \tag{7.13}$$

$${}^H\boldsymbol{m} = {}^H_S\boldsymbol{R}\left(\left({}^S\boldsymbol{f} \times {}^S\boldsymbol{l}_{NA}\right) + {}^S\boldsymbol{m}\right), \tag{7.14}$$

In Eqs.(7.13) and (7.14) ${}^H_S\boldsymbol{R}$ denotes the direction cosine matrix between the hybrid coordinate frame and the sensor coordinate frame and ${}^S\boldsymbol{l}_{NA}$ denotes the vector which describes the position of the tip of the tool (the coordinate frame which is associated with the tool is denoted by NA) with respect to the coordinate frame S. This will be discussed in more detail in the next section. This scheme cannot be applied to the IRp-6 robot due to the fact that it has five degrees of freedom.

At the end of this section we make one comment which is associated with Eq.(7.9) in which it is necessary to calculate the time derivative of the Euler angles. This operation is straightforward and from a practical point of view it is easier to imagine the time derivative of the Euler angles. From a mathematical point of view, as we did above with the formal representation concerning the Euler angles (in terms of the rotation matrix associated with them), we can calculate the angular velocity vector based on the time derivative of the Euler angles and the Jacobian matrix 3×3 which is associated with the analytical Jacobian (compare the considerations carried out by Sciavicco and Siciliano [195] concerning this matter). Based on the angular velocity we are able to calculate formally the time derivative of the orientation error (as before compare the corresponding lengthily expression in reference [195]) and formally eliminate the pseudovector consisting of the Euler angles and its time derivative. This updates the considerations discussed by Dégoulange and co-authors [33]. In defining orientation it is also very convenient to use its quaternion representation which is another alternative for solving the orientation representation and its error and time derivative (compare considerations presented by Angeles [2]).

7.2.2 A hybrid controller for the IRp-6 robot

As mentioned above, the hybrid control scheme presented in the previous section cannot be directly applied to the IRp-6 robot which has five degrees of freedom. We cannot apply the transformation equations (7.6) and (7.11) due to the limited number of degrees of freedom. Compare this concept with the idea of the sensor ball described in Chapter 2 where the exerted forces and torques are transformed to the joint generalised coordinates in the case of the robot having six degrees of freedom. Here we present considerations based on references [77, 78]. As before, we assume that the force control and position control are orthogonal actions in task space with the restriction discussed by Goldenberg and Song [71]. As a result due to the error of the spatial forces, correction position and orientation errors are created in the force direction which are summarised with the position vector due to the trajectory reconstruction. Both errors are compensated for by the joint controllers. Due to the fact that correction in the force direction does not create movement as a result we obtain forces in this direction. A block diagram of the hybrid controller is presented in Fig.7.2. Because the IRp-6 robot has only five degrees of freedom an arbitrary orientation in its working space cannot be achieved. Therefore, in order to obtain an inverse kinematics solution (Invkin in Fig.7.2) we introduce a notion of the symmetric tool reducing at the same time the generalised space to five dimensions. As a consequence we cannot exert a moment of a force along the axis of rotation of the tool. Therefore we eliminate one of the axes of the coordinate frame associated with the tool. This simplified assumption and careful trajectory planning allows the

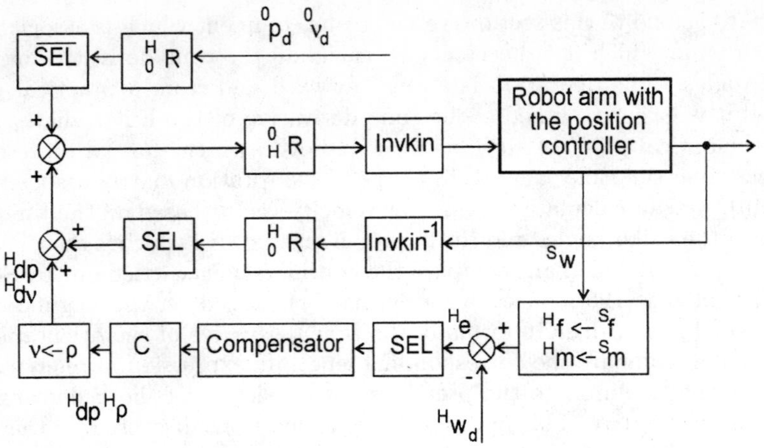

Fig. 7.2. A hybrid controller for the IRp-6 robot

construction of a hybrid controller along the lines presented in the previous section.

The input trajectory ${}^0\boldsymbol{r}_d$ is described in terms of Eq.(7.4). As before we assume that the trajectory is defined in Cartesian space. Components of the vector ${}^0\boldsymbol{p}_d$ describe the position of the tool tip:

$$ {}^0\boldsymbol{p}_d(t) = \left[{}^0p_{dx}(t),\ {}^0p_{dy}(t),\ {}^0p_{dz}(t)\right]^T , \tag{7.15}$$

Three components of the orientation of the tool which is symmetric along the axis of rotation are:

$$ {}^0\boldsymbol{\nu}_d(t) = \left[{}^0\nu_{dx}(t),\ {}^0\nu_{dy}(t),\ {}^0\nu_{dz}(t)\right]^T . \tag{7.16}$$

The actual trajectory differs from the desired trajectory which results in the correction vector ${}^0\boldsymbol{dr}$ described by Eq.(7.3) which is responsible for the force control. The correction vector is associated with the directions in which the forces are exerted and therefore at any time has to be orthogonal to the desired velocity vector, namely we can write

$$ {}^0\dot{\boldsymbol{r}}_d(t)^T \cdot {}^0\boldsymbol{dr} = 0. \tag{7.17}$$

A resultant position vector has the following form:

$$ {}^0\boldsymbol{r}_d(t) + {}^0\boldsymbol{dr} . \tag{7.18}$$

This vector is an input data to the module which solves the inverse kinematics problem, which supplies the joint generalised positions to the joint controllers [78]. The correction vector is created as follows. Recall that the three components of the forces and torques are measured in the coordinate frame associated with the sensor, which is denoted by S. Actually we are interested in reaction forces which are exerted at the tip of the tool because these forces and torques are important for the realisation of the task. Therefore, we have to find an equivalent vector of forces and torques which when applied to the tip of the manipulator creates an equivalent vector to the measured vector with its origin in the coordinate frame S. An equivalent spatial vector of forces and torques is described in the hybrid coordinate frame, H, due to the fact that in this coordinate frame we selected the forces and torques in which the force controller is active. We can write (see Eq.(7.13))

$$^{H}F_x = \boldsymbol{n} \cdot \, ^{S}\boldsymbol{f}, \tag{7.19}$$

$$^{H}F_y = \boldsymbol{o} \cdot \, ^{S}\boldsymbol{f}, \tag{7.20}$$

$$^{H}F_z = \boldsymbol{a} \cdot \, ^{S}\boldsymbol{f}, \tag{7.21}$$

where $^{S}\boldsymbol{f}$ is the force vector measured by the sensor and $\boldsymbol{n}$, $\boldsymbol{o}$, and $\boldsymbol{a}$ are normal, orientation, and approach vectors of the rotation matrix which can be written as follows (here $\cdot$ denotes a scalar operation of two vectors)

$$^{H}_{S}\boldsymbol{R} = \begin{bmatrix} n_x & o_x & a_x \\ n_y & o_y & a_y \\ n_z & o_z & a_z \end{bmatrix}. \tag{7.22}$$

The inverse of the above rotation matrix can be calculated from the following sequences of multiplications

$$^{S}_{H}\boldsymbol{R} = \, ^{S}_{6}\boldsymbol{R} \cdot \, ^{0}_{6}\boldsymbol{R}^{-1} \cdot \, ^{0}_{L}\boldsymbol{R} \cdot \, ^{L}_{H}\boldsymbol{R}. \tag{7.23}$$

The rotation matrices which appear in Eqs.(7.22) and (7.23) have the following meaning: $^{S}_{6}\boldsymbol{R}$ denotes a rotation matrix between coordinate frames: the tip of the manipulator and frame S, $^{0}_{6}\boldsymbol{R}$ is a rotation matrix between the tip of the manipulator and the base coordinate frame, $^{0}_{L}\boldsymbol{R}$ is the rotation matrix from the local coordinate frame in which the task is described and the tip of the manipulator and finally, $^{L}_{H}\boldsymbol{R}$ is a rotation matrix between frames H and L. In the last coordinate frame the task is described and sometimes this frame is exactly the same as the H frame. The hybrid coordinate frame is chosen because of restrictions imposed on the torques. It is assumed here that this coordinate frame has the same origin as the tool frame NA. If it is not the case we can find an appropriate rotation between frames NA and H. This assumption is valid since the external forces and torques are exerted on frame NA. The components of the three torques in the hybrid frame are calculated as follows (compare also Eq.(7.14)):

$$\begin{aligned} ^{H}M_x &= \boldsymbol{n} \cdot \left(\left(\, ^{S}\boldsymbol{f} \times \, ^{S}\boldsymbol{l}_{NA} \right) + \, ^{S}\boldsymbol{m} \right), \\ ^{H}M_y &= \boldsymbol{o} \cdot \left(\left(\, ^{S}\boldsymbol{f} \times \, ^{S}\boldsymbol{l}_{NA} \right) + \, ^{S}\boldsymbol{m} \right), \\ ^{H}M_z &= \boldsymbol{a} \cdot \left(\left(\, ^{S}\boldsymbol{f} \times \, ^{S}\boldsymbol{l}_{NA} \right) + \, ^{S}\boldsymbol{m} \right), \end{aligned} \tag{7.24}$$

where ${}^{S}\boldsymbol{m}$ is the vector of torques measured in the sensor frame and vector ${}^{S}\boldsymbol{l}_{NA}$ describes the position of the tip of the tool in frame S. This vector is calculated as

$$ {}^{S}\boldsymbol{l}_{NA} = {}^{S}\boldsymbol{l}_6 + {}^{S}_{6}\boldsymbol{R} \cdot {}^{6}\boldsymbol{l}_N. \tag{7.25}$$

Equations (7.19) to (7.21) and (7.24) recalculate the measured forces and torques to the hybrid coordinate frame H. We assume that the desired forces and torques ${}^{H}\boldsymbol{w}_d$ are also expressed in the hybrid coordinate frame. If they are expressed in the local coordinate frame L associated with the task they have to be recalculated to the frame H which allows us handle torque limitations imposed on the tool. Next, we calculate the force error in all directions of the hybrid coordinate frame as:

$$ {}^{H}\boldsymbol{e} = {}^{H}\boldsymbol{w}_d - {}^{H}\boldsymbol{w} = \left[{}^{H}\boldsymbol{e}_f^T, \, {}^{H}\boldsymbol{e}_n^T\right]^T . \tag{7.26}$$

Now we can select the direction of the forces and at the same time directions of the force controllers using the selection matrix SEL ($\overline{\text{SEL}} = \mathcal{I} - \text{SEL}$ and SEL are complementary matrices, see Fig.7.2). Therefore the force controller generates a hypothetical generalised force which, assuming that the stiffness between the robot and environment of the task is described by matrix $\boldsymbol{C}$,

$$ \boldsymbol{C} = \begin{bmatrix} C_x & & \dots & & 0 \\ & C_y & & & \\ \dots & & C_z & & \dots \\ & & & C_{\rho x} & \\ 0 & & \dots & & C_{\rho y} \end{bmatrix} \tag{7.27}$$

causes the position increment ${}^{H}\boldsymbol{dp}$ and orientation angle increment ${}^{H}\boldsymbol{dr}$. As before we comment on a problem with orientation representation. In general, orientation should be easily physically interpreted and independent of the sequence of transformations imposed on it. It is known from section 7.2.1 that any set of Euler angles does not result in a vector from a mathematical point of view. Therefore it is necessary to define an orientation in a non-standard way. In order to do so, observe that changes of the orientation are the result of torque changes therefore the coordinate frame H changes its orientation along a resultant vector $\boldsymbol{m}$. The angle of rotation along vector $\boldsymbol{m}$ is proportional to the angular mechanical susceptibility. In the case when the stiffness is not uniform in all directions we define the angles of rotation independently (proportionally to the mechanical susceptibility in each direction), and next we calculate a resultant vector of rotation. The angle of rotation of the coordinate frame is proportional to the length of the resultant vector according to the rule of thumb. Now recalling that the tool is symmetric (which means that we cannot change the orientation of the tool along the z axis and at the same time we cannot exert the torque along this axis), the resultant vector $\boldsymbol{\rho}$, which is a kind of orientation representation, has the following form:

$$ \boldsymbol{\rho} = \left[C_{\rho x} m_x, \, C_{\rho y} m_y, 0\right]^T . \tag{7.28}$$

This orientation, assuming that a particular mechanical stiffness is knows, causes the generation of the corresponding force.

The new orientation vector, which is obtained by rotating the coordinate frame and taking into account geometrical considerations, has the following form:

$$\boldsymbol{\nu} = \left[\frac{\rho_y}{\| \boldsymbol{\rho} \|} \sin \| \boldsymbol{\rho} \|, \; -\frac{\rho_x}{\| \boldsymbol{\rho} \|} \sin \| \boldsymbol{\rho} \|, \; \cos \| \boldsymbol{\rho} \|\right]^T . \tag{7.29}$$

Assuming that the initial orientation vector is of the form $\boldsymbol{\nu} = [0, 0, 1]^T$ increment of the orientation vector, expressed in the coordinate frame H, can be written as:

$${}^{H}d\boldsymbol{\nu} = \boldsymbol{\nu}' - \boldsymbol{\nu} = \left[\frac{{}^{H}\rho_y}{\| {}^{H}\boldsymbol{\rho} \|} \sin \| {}^{H}\boldsymbol{\rho} \|, -\frac{{}^{H}\rho_x}{\| {}^{H}\boldsymbol{\rho} \|} \sin \| {}^{H}\boldsymbol{\rho} \|, \cos \| {}^{H}\boldsymbol{\rho} \| -1\right]^T . \tag{7.30}$$

Increment of the orientation corresponds to a model of a generalised spring in which the rotation angle is proportional to the torque. The increments of the orientation vector ${}^{H}d\boldsymbol{\nu}$ and the increment of the position vector ${}^{H}d\boldsymbol{p}$ are transformed to the base coordinate frame using the following rules:

$$\begin{aligned} {}^{0}d\boldsymbol{p} &= {}^{0}_{H}\boldsymbol{R} \cdot {}^{H}d\boldsymbol{p}, \\ {}^{0}d\boldsymbol{\nu} &= {}^{0}_{H}\boldsymbol{R} \cdot {}^{H}d\boldsymbol{\nu}. \end{aligned} \tag{7.31}$$

Vector ${}^{0}d\boldsymbol{r}$ has two components (compare Eq.(7.3)) and these are added to the vector of the desired position (here position and orientation), ${}^{0}\boldsymbol{r}_d(t)$, and this result constitutes an input to the block which solves the inverse kinematics problem, which calculates a sequence of new joint generalised positions. This method is actually an admittance method [224] in which the original position controller of the robot has an additional feedback from the force controller. The control law is realised in the hybrid coordinate frame in which the desired trajectory and spatial forces are defined. In this coordinate frame we select the directions in which the force controller is active according to the artificial constraints (see, for example, Raibert and Craig [187]) concerning the spatial forces. At the same time the path control is realised in a standard way using the original position controllers (recall that the trajectory in terms of the positions and orientations is specified in the hybrid coordinate frame). The desired trajectory is modified only in those directions in which the forces cannot be exerted according to the general philosophy of the decomposition of the forces and positions. Due to the fact that the IRp-6 robot has only five degrees of freedom it is possible to control only five coordinates of the spatial vector of forces and five joint generalised coordinates. Note that only special tasks which do not require the torque acting along the axis z of the tool can be executed by the proposed hybrid controller. Observe also that the considerations presented in the section are the modifications of the algorithm described is Section 7.2.1 with the assumption that the robot has five degrees of freedom.

7.3 Local and global stiffness measurement of the IRp-6 robot

7.3.1 A method description

In this section we present local stiffness and global compliance measurements for the IRp-6 industrial robot. Local stiffness measurements are based on ideas presented in reference [8]. One can imagine that each joint is represented by a spring. This seems to be a reasonable assumption due to the fact that there is a gear mechanism associated with each joint which can be treated as an elastic element. Notice that over a wide range of tasks the stiffness of each link is a constant value for the IRp-6 robot. The results of these measurements are given in references [42, 49] and when applied to the hybrid controller, were not satisfactory. Therefore we propose using a similar approach to that given in references [56] and [57].

It is assumed that the end effector of the robot is constrained by an infinitely stiff environment and therefore we can neglect constant friction. The reaction force is directed along the normal to the constraint surface. According to this method we look for directions in which gear ratio matrix multiplied by the Jacobian of the manipulator and aforementioned normal vector to the surface are zero. This approach is very systematic and finds the configurations of the manipulator in which only one entry of the product mentioned above is zero. Next we introduce a small perturbation motor position force i and we measure the corresponding variation of the reaction force amplitude by making use of the force/torque sensor. Then stiffness along prescribed direction can be calculated.

In mathematical terms this algorithm can be summarised in the following steps.

1. Choose a configuration $\boldsymbol{q}$ (in terms of joint coordinates) of the manipulator and an orientation $\bar{\boldsymbol{n}}$ of the surface such that only entry i of vector $\boldsymbol{P}\boldsymbol{J}_p^T(\boldsymbol{q})\bar{\boldsymbol{n}}$ is different from zero:

$$(\boldsymbol{P}\boldsymbol{J}_p^T(\boldsymbol{q})\bar{\boldsymbol{n}})_j = 0, \quad \forall j \neq i. \tag{7.32}$$

   In the above equation $\boldsymbol{P}$ denotes a matrix of gear ratios of the manipulator, $\boldsymbol{J}_p(\boldsymbol{q})$ is the Jacobian of the Cartesian position of the end effector (first three rows of the complete Jacobian $\boldsymbol{J}(\boldsymbol{q})$) and $\bar{\boldsymbol{n}}$ is the three dimensional unit vector normal to the constraint surface.
2. Given a small perturbation, a step function for example, to motor controller i, calculate corresponding joint torque perturbation according to the following formula:

$$\Delta\boldsymbol{\tau} = \boldsymbol{P}\boldsymbol{J}_p^T(\boldsymbol{q})\bar{\boldsymbol{n}}\Delta\boldsymbol{w}, \tag{7.33}$$

   where $\Delta\boldsymbol{w}$ is a vector of the reaction force variation measured by the force/torque sensor mounted at the wrist of the manipulator.

3. Finally, compute the local stiffness coefficient as follows:

$$C_i = \frac{\Delta \tau}{\Delta q}. \tag{7.34}$$

From the algorithm described above it is clear that we need joint positions and torques measurements as well as the reaction force measurements.

In the above algorithm the most important part is to check the conditions of Eq.(7.32). Note that in order to calculate the stiffness we need the Jacobian of the manipulator. In the case of the IRp-6 robot, recall its kinematical structure given in Fig.3.6a and its parameters defined according to the modified Denavit-Hartenberg notation which is summarised in Table 3.2.

The Jacobian of the IRp-6 robot has the following form:

$$\boldsymbol{J} = \begin{bmatrix} a_1 & a_2 & a_3 & a_4 & 0 \\ b_1 & b_2 & b_3 & b_4 & 0 \\ 0 & g_2 & g_3 & 0 & 0 \\ d_1 & d_2 & d_3 & d_4 & 0 \\ e_1 & e_2 & e_3 & e_4 & 0 \\ f_1 & 0 & 0 & 0 & 1 \end{bmatrix}, \tag{7.35}$$

where non-zero elements are functions of joint variables and have the form:

$$\begin{aligned}
a_1 &= \sin q_5[-r_5 \cos(q_4 - q_2 - q_3) - l_2 \cos q_2 - l_3 \cos(q_2 + q_3)], \\
a_2 &= \cos q_5(r_5 + l_3 \cos q_4 + l_2 \cos(q_4 - q_3)), \\
a_3 &= \cos q_5(r_5 + l_3 \cos q_4), \\
a_4 &= r_5 \cos q_5, \\
b_1 &= \cos q_5[-r_5 \cos(q_4 - q_2 - q_3) - l_2 \cos q_2 - l_3 \cos(q_2 + q_3)], \\
b_2 &= -\sin q_5[r_5 + l_3 \cos q_4 + l_2 \cos(q_4 - q_3)], \\
b_3 &= -\sin q_5(r_5 + l_3 \cos q_4), \\
b_4 &= -r_5 \sin q_5, \\
g_2 &= -l_3 \sin q_4 - l_2 \sin(q_4 - q_3), \\
g_3 &= -l_3 \sin q_4, \\
d_1 &= \cos q_5 \cos(q_4 - q_2 - q_3), \\
d_2 &= \sin q_5, \\
d_3 &= \sin q_5, \\
d_4 &= \sin q_5, \\
e_1 &= -\sin q_5 \cos(q_4 - q_2 - q_3), \\
e_2 &= \cos q_5, \\
e_3 &= \cos q_5, \\
e_4 &= \cos q_5, \\
f_1 &= -\sin(q_4 - q_2 - q_3).
\end{aligned} \tag{7.36}$$

Now taking into account only three columns of the Jacobian of the IRp-6 robot Eq.(7.32) can be rewritten as follows:

$$\boldsymbol{PJ}_p^T(\boldsymbol{q})\bar{\boldsymbol{n}} = \begin{bmatrix} p_1 & 0 & 0 & 0 & 0 \\ 0 & p_2 & p_3 & 0 & 0 \\ 0 & 0 & p_4 & p_5 & 0 \\ 0 & 0 & 0 & p_6 & p_7 \\ 0 & 0 & 0 & p_8 & p_9 \end{bmatrix} \begin{bmatrix} a_1 & b_1 & 0 \\ a_2 & b_2 & g_2 \\ a_3 & b_3 & g_3 \\ a_4 & b_4 & 0 \\ 0 & 0 & 0 \end{bmatrix} \begin{bmatrix} \cos_6^5(x) \\ \cos_6^5(y) \\ \cos_6^5(z) \end{bmatrix} =$$

$$= \begin{bmatrix} p_1 a_1 \cos_6^5(x) + p_1 b_1 \cos_6^5(y) \\ (p_2 a_2 + p_3 a_3)\cos_6^5(x) + (p_2 b_2 + p_3 b_3)\cos_6^5(y) + \\ +(p_2 g_2 + p_3 g_3)\cos_6^5(z) \\ (p_4 a_3 + p_5 a_4)\cos_6^5(x) + (p_4 b_3 + p_5 b_4)\cos_6^5(y) + p_4 g_3 \cos_6^5(z) \\ p_6 a_4 \cos_6^5(x) + p_6 b_4 \cos_6^5(y) \\ p_8 a_4 \cos_6^5(x) + p_8 b_9 \cos_6^5(y) \end{bmatrix}, \tag{7.37}$$

where vector $\bar{\boldsymbol{n}}$ has components $cos_6^5(x)$, $cos_6^5(y)$, and $cos_6^5(z)$. Notice from Eq.(7.37) that as a result we get a vector with 5 components.

In order to follow the procedure outlined in points 1 to 3, the manipulator has to be poised in these positions in Cartesian space for which only one of the components in non-zero. It means that contact forces between the robot and its environment are exerted on one joint of the robot. Note that analytical expressions for these components are complicated nonlinear functions of joint variables. Here we propose an alternative solution. It is easier to consider a kinematical structure of the IRp-6 robot and a situation under which the contact forces projected on joint axes result in zero components. In addition, we use the properties of the product of vectors and projection of forces and torques on the desired direction which simplifies the situation greatly. The results of this approach are published in references [42, 49] as the results of independent research. The method outlined in references [56, 57] is applicable only to the measurement of the stiffness of the first three degrees of freedom of the manipulator. This method does not include the singularities of the kinematical manipulator structure. This proposition was successfully applied to all five axes of the IRp-6 robot. Note that it is essentially equivalent to the procedure outlined in points 1 to 3, but comes from physical reasoning.

In order to illustrate it consider the first joint of the IRp-6 robot. This situation is depicted in Fig.7.3. From Fig.7.3 it is clear that in order to follow the proposition outlined above the axis of the contact forces should be parallel to axes 2, 3, and 4 of the IRp-6 robot because projection of the moment on these axes results in zero components (note that the torque is perpendicular to the axis of the contact forces).

As a consequence the force and torque sensor measures forces and torques on xy plane which can be calculated as follows:

$$F = \sqrt{F_x^2 + F_y^2}. \tag{7.38}$$

The forces which are exerted on the stiff surface are within the range of 1 to 98.1 N. This range was experimentally verified and is limited by the accuracy

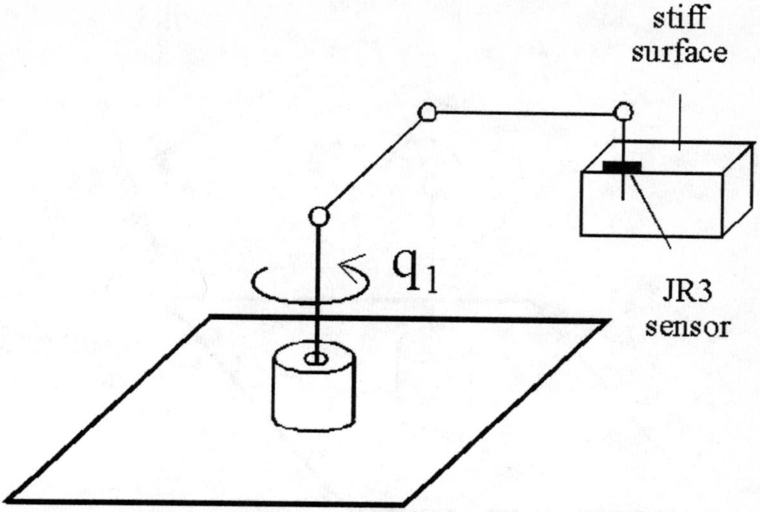

Fig. 7.3. The IRp-6 configuration and surface orientation for the first joint

of the JR3 sensor. A distance between a contact point and the first joint of the robot is equal to 0.807m and results from the forward kinematics solution.

Table 7.1 collects the measurement results for the case considered. From

Table 7.1. Measurement and computation results for the first joint of the IRp-6 robot

No.	$\frac{\Delta F}{\Delta q_1}$ [N/deg]	C_1 [N·m/deg]
1	2.9037	2.3457
2	2.9933	2.4180
3	2.9958	2.4200
4	3.0736	2.4829
5	3.2435	2.6201
6	3.1192	2.5197
7	3.0764	2.4852
8	3.1625	2.5547
9	3.1782	2.5674
10	3.3119	2.6754

the measurements results presented in Table 7.1, we calculate the stiffness of the first joint which is $C_1 = 2.51$N·m/deg and consequently its compliance $k_1 = \frac{1}{C_1} = 0.3985$ deg/N·m.

Figure 7.4 illustrates orientation of the infinitely stiff surface seen in the end-effector coordinate frame. Note that the vector described by Eq.(7.37) in this particular situation has the following form:

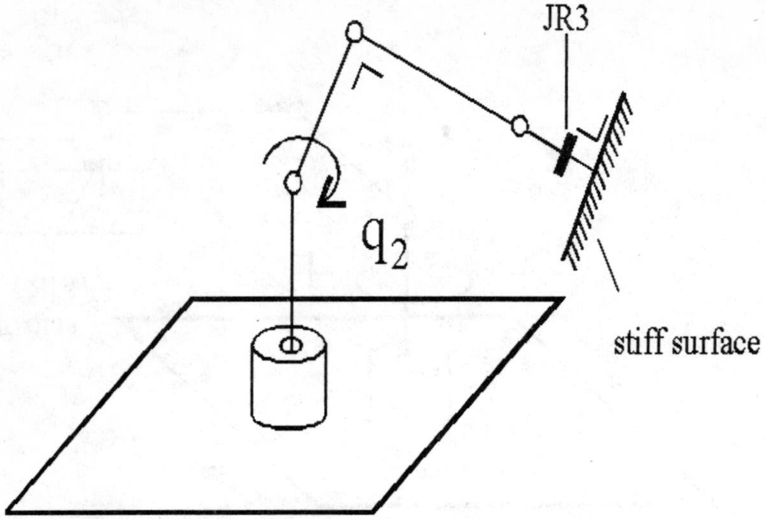

Fig. 7.4. The IRp-6 configuration and surface orientation for the second joint

$$\begin{bmatrix} -0,807p_1 \\ 0 \\ 0 \\ 0 \\ 0 \end{bmatrix}. \tag{7.39}$$

Components 2, 3, 4, and 5 of this vector are equal to zero due to the fact that generalised coordinate $q_5 = 0$.

For the second axis of the IRp-6 robot we carry out a similar procedure. In order to measure the stiffness of the second joint of the manipulator we use its configuration and orientation of the infinitely stiff surface presented in Fig.7.4. We measure the F_z component of the vector of force exerted on the sensor, therefore we have $F = F_z$. Measurements of the F_z component are summarised in Table 7.2. The forces were measured in the range of 1 to 196.2N and arm of a force is equal to 0.45m.

Finally, the stiffness for the second joint is $C_2 = 2.04$N·m/deg and consequently the compliance $k_2 = \frac{1}{C_2} = 0.4906$ deg/N·m. In order to verify these calculations we check the vector given by Eq.(7.37) which in this particular situation has the following form:

$$\begin{bmatrix} 0 \\ -0,45p_2 \\ 0 \\ 0 \\ 0 \end{bmatrix}. \tag{7.40}$$

Table 7.2. Measurements and computational results for the second joint of the IRp-6 robot

No.	$\frac{\Delta F}{\Delta q_2}$ [N/deg]	C_2 [N·m/deg]
1	4,71559	2,12201
2	4,43950	1,99777
3	4,51265	2,03069
4	4,32950	1,94827
5	4,59019	2,06558
6	4,58798	2,06459
7	4,44629	2,00083
8	4,65134	2,09310
9	4,59473	2,06763
10	4,41940	1,98873

Now it is of interest to show how to measure a local stiffness for the third joint of the IRp-6 robot. It happens that the condition described by equation (7.37) is never satisfied due to the fact that the second and third axes are parallel. In order to measure the local stiffness of the third joint the second link has to be kept idle and therefore it does not take part in the reaction due to an exerted torque. The configuration of the IRp-6 robot in this particular case is depicted in Fig.7.5. Measurement results for the third joint are summarised in Table 7.3. The arm of a force was 0.67m and

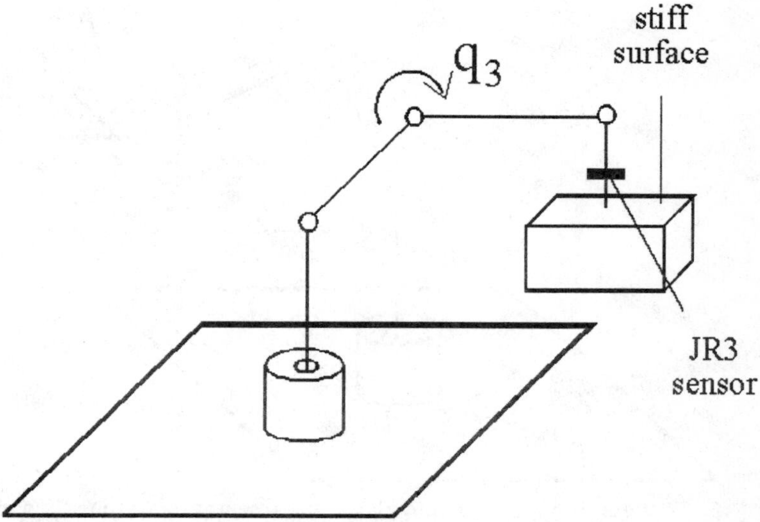

Fig. 7.5. The IRp-6 configuration and surface orientation for the third joint

measured forces were in the range of 1 to 98.1N.

Table 7.3. Measurement and computational results for the third joint of the IRp-6 robot

No.	$\frac{\Delta F}{\Delta q_3}$ [N/deg]	C_3 [N·m/deg]
1	3.34399	2.21161
2	3.45352	2.28405
3	3.68165	2.43493
4	3.47544	2.29855
5	3.65031	2.41420
6	3.77313	2.49543
7	3.64455	2.41040
8	3.73400	2.46956
9	3.62389	2.39673
10	3.56613	2.35853

Based on measurement results collected in Table 7.3 the local stiffness for the third joint was calculated as $C_3 = 2.37$N·m/deg which results in a global compliance of $k_3 = \frac{1}{C_3} = 0.4206$deg/N·m.

Configuration of the manipulator and orientation of the stiff surface for measurements of the stiffness of the fourth axis of the IRp-6 robot is depicted in Fig. 7.6. The manipulator is supported under the fourth axis in order to eliminate the influence of the drive systems for joints 1,2, and 3. Now we

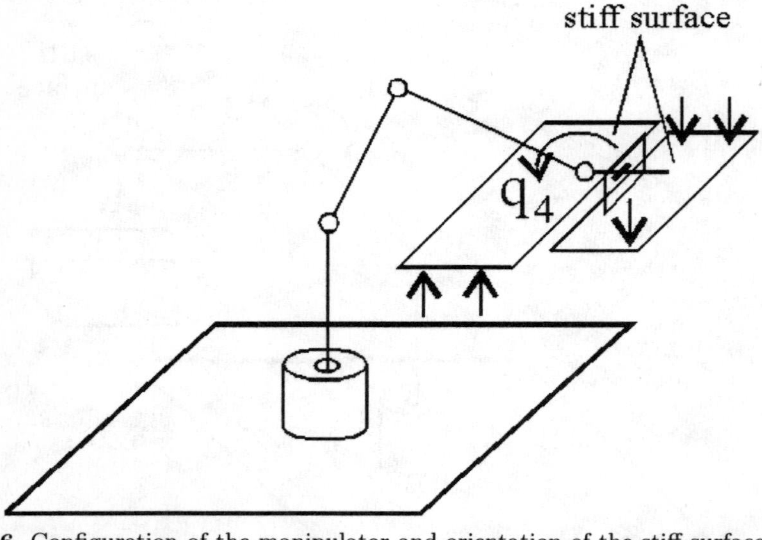

Fig. 7.6. Configuration of the manipulator and orientation of the stiff surface for the fourth axis of the IRp-6 robot

move the tip of the manipulator in the direction opposite to the gravity and at the same time we press on the stiff measurement surface. Measurement and computational results for the fourth axis of the IRp-6 robot are shown in Table 7.4. The forces were measured in the range of 1 to 98.1N and the

Table 7.4. Measurement and computational results for the fourth axis of the IRp-6 robot

No.	$\frac{\Delta F}{\Delta q_4}$ [N/deg]	C_4 [N·m/deg]
1	0,86314	0,10918
2	0,87844	0,11112
3	0,86643	0,10960
4	0,86413	0,10931
5	0,85437	0,10807
6	0,82879	0,10484
7	0,83248	0,10530
8	0,77910	0,09855
9	0,78187	0,09890
10	0,82669	0,10457

arm of the force was 0.125m.

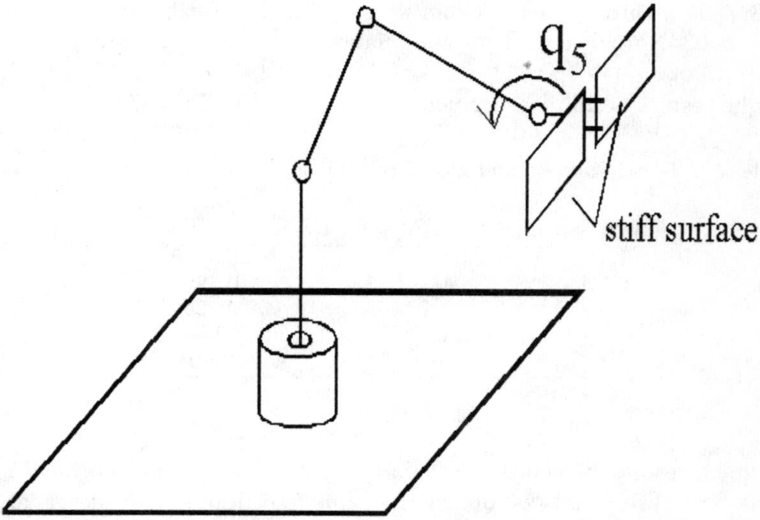

Fig. 7.7. Configuration of the manipulator and orientation of the stiff surface for the fifth axis of the IRp-6 robot

Finally, we obtain the following constants based on the measurements from Table 7.4: $C_4 = 0.1059$N·m/deg and $k_4 = \frac{1}{C_4} = 9,43$ deg/(N·m) which are the constant stiffness and its compliance.

In order to measure the stiffness of the fifth joint of the IRp-6 manipulator, consider Fig.7.7 in which the configuration and orientation of the stiff surface are presented. Measured torques M_z in the coordinate frame associated with the force/torque sensor recalculated to the increment generalised angles. The

Table 7.5. Computational results of the stiffness of the fifth joint robot

No.	C_5 [deg/N·m]
1	0,03398
2	0,03304
3	0,03523
4	0,03408
5	0,03594
6	0,03477
7	0,03559
8	0,03459
9	0,03407
10	0,03219

measurement range of the torques is 1 to 2.94N·m. The stiffness calculations for the fifth joint are summarised in Table 7.5.

Based on these measurements we calculated the stiffness of the fifth joint $C_5 = 0,034$N·m/deg, and its compliance $k_5 = \frac{1}{C5} = 29,08$deg/N·m of the IRp-6 robot.

The results of all measurements are collected in Table 7.6. From the mea-

Table 7.6. Local stiffness and global compliance of the IRp-6 robot

Axis	Stiffness [Nm/deg]	Compliance [deg/N·m]
1	2.50	0.3986
2	2.04	0.4907
3	2.38	0.4207
4	0.106	9.44
5	0.034	29.08

surement results presented in Table 7.6, it is clear that the local stiffness of the first three joints is almost the same and differs by a factor 20 for the other joints. This situation was verified by several experiments using a hybrid controller.

7.3.2 Local and global stiffness calculations

In this section, based on reference [8], we describe a method for calculating the local stiffness measurements to obtain the global stiffness of the manipulator.

In the previous section we calculated the local stiffness of each joint of the IRp-6 based on the experimental method. First we observe the relationship between the torque increment and the corresponding joint increment, namely

$$\Delta\tau_i = C_i \Delta q_i, \tag{7.41}$$

where C_i is the local stiffness (compare also Eq.(7.34). Therefore, for the manipulator we can write

$$\boldsymbol{\Delta\tau} = \boldsymbol{C}\boldsymbol{\Delta q}. \tag{7.42}$$

From the last equation we can calculate Δq as

$$\boldsymbol{\Delta q} = \boldsymbol{C}^{-1}\boldsymbol{\Delta\tau}, \tag{7.43}$$

where

$$\boldsymbol{\Delta\tau} = \begin{bmatrix} \Delta\tau_1 \\ \cdots \\ \Delta\tau_n \end{bmatrix}, \ \boldsymbol{\Delta q} = \begin{bmatrix} \Delta q_1 \\ \cdots \\ \Delta q_n \end{bmatrix}, \ \boldsymbol{C} = \begin{bmatrix} C_1 & 0 & 0 & . & 0 \\ 0 & C_2 & 0 & . & 0 \\ 0 & 0 & C_3 & . & 0 \\ . & . & . & . & . \\ 0 & 0 & 0 & 0 & C_n \end{bmatrix}. \tag{7.44}$$

Now we write the differential relationship between the joint generalised coordinates and the coordinate in Cartesian space as follows:

$$d\boldsymbol{r} = \boldsymbol{J}d\boldsymbol{q}. \tag{7.45}$$

Substituting Eq.(7.43) into the last equation, and assuming that the appropriate differential are substituted by the increments, yields

$$\boldsymbol{\Delta r} = \boldsymbol{J}\boldsymbol{C}^{-1}\boldsymbol{\Delta\tau}. \tag{7.46}$$

Now we assume that the friction and gravitational torques are compensated for by the manipulator (the friction torques are precomputed and subtracted from the generalised forces and the gravitational forces and the torques are compensated for by a system of counterbalances; both assumptions were verified in practice) we can write the relationship between the external forces acting at the tip of the manipulator and the generalised forces [8] as:

$$\boldsymbol{\Delta\tau} = \boldsymbol{J}^T\boldsymbol{\Delta w}. \tag{7.47}$$

Substitution of Eq.(7.46) to the last equation yields the following equation:

$$\boldsymbol{\Delta r} = \boldsymbol{J}\boldsymbol{C}^{-1}\boldsymbol{J}^T\boldsymbol{\Delta w}. \tag{7.48}$$

Having obtained the local stiffness of the manipulator we can calculate a global compliance of the manipulator

$$\boldsymbol{K}_p = \boldsymbol{J}\boldsymbol{C}^{-1}\boldsymbol{J}^T \tag{7.49}$$

and rewrite Eq.(7.48) as
$$\Delta r = K_p \Delta w \ . \tag{7.50}$$

In the case considered the global compliance of the IRp-6 has the following form (which has to be recalculated to the hybrid frame):

$$K_p = \begin{bmatrix} \sum_{i=1}^{4} k_i a_i^2 & \sum_{i=1}^{4} k_i a_i b_i & \sum_{i=2}^{3} k_i a_i g_i & \sum_{i=1}^{4} k_i a_i d_i & \sum_{i=1}^{4} k_i a_i e_i & k_1 a_1 f_1 \\ \sum_{i=1}^{4} k_i a_i b_i & \sum_{i=1}^{4} k_i b_i^2 & \sum_{i=2}^{3} k_i b_i g_i & \sum_{i=1}^{4} k_i b_i d_i & \sum_{i=1}^{4} k_i b_i e_i & k_1 b_1 f_1 \\ \sum_{i=2}^{3} k_i a_i g_i & \sum_{i=2}^{3} k_i b_i g_i & \sum_{i=2}^{3} k_i g_i^2 & \sum_{i=2}^{3} k_i d_i g_i & \sum_{i=2}^{3} k_i e_i g_i & 0 \\ \sum_{i=1}^{4} k_i a_i d_i & \sum_{i=1}^{4} k_i b_i d_i & \sum_{i=2}^{3} k_i d_i g_i & \sum_{i=1}^{4} k_i d_i^2 & \sum_{i=1}^{4} k_i d_i e_i & k_1 d_1 f_1 \\ \sum_{i=1}^{4} k_i a_i e_i & \sum_{i=1}^{4} k_i b_i e_i & \sum_{i=2}^{3} k_i e_i g_i & \sum_{i=1}^{4} k_i d_i e_i & \sum_{i=1}^{4} k_i e_i^2 & k_1 e_1 f_1 \\ k_1 a_1 f_1 & k_1 b_1 f_1 & 0 & k_1 d_1 f_1 & k_1 e_1 f_1 & k_1 f_1^2 + k_5 \end{bmatrix} . \tag{7.51}$$

At the end of this section we make one comment. Notice that it is not possible to measure the local stiffness for the fourth and fifth joints separately due to the kinematical structure of the IRp-6 robot. The fifth axis is connected to two motors working simultaneously, by string. The situation is the same for the fourth joint. Therefore it is interesting to notice that the measurement results characterise the same drive system which for the last two joints, works in a different manner.

7.4 Experimental results

In this section we present some experimental results carried out on the experimental set-up described in Chapter 2. They take into account the global compliance measurements described in the previous section. Based on these measurements we can tune coefficients of the PID position controller according to the compliance measurements which were changed on-line during the experiment.

All results presented in this section show the possible application of the position controller which is used in the hybrid control of robot manipulators. All experiments presented here were carried out on the IRp-6 robot. First we tested the algorithm which is based on the generalised spring concept which we have updated with the calculations of the global compliance matrix. The results of these experiments are presented in references [42, 49]. Local stiffness coefficients remain constant for a wide range of applications while the global compliance matrix changes only when the manipulator configuration changes (see Eq.(7.49)). Then we implement the method cited in the previous section to the global compliance matrix calculations such that the above algorithm is not sensitive to the manipulator configuration changes. It is one of the drawbacks of the algorithm presented in Section 7.2.2. The experimental results which we obtained using the updated compliance calculations were satisfactory. We checked the control algorithm on many examples in task space and we obtained reproducible results.

As an example, consider the following case. A stiff pointer is attached to the gripper of the robot. This pointer moves along a straight line on an infinitely stiff metal plate which is bent to form a triangle. A view of the straight line in a coordinate frame is presented in Fig.7.8.

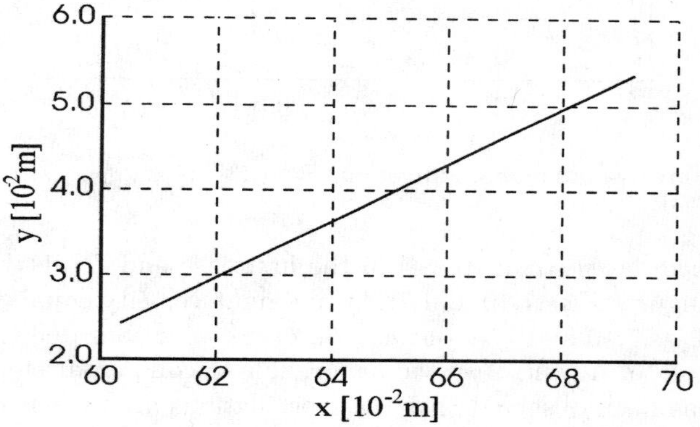

Fig. 7.8. A view of a straight line in xy coordinates

In addition the value of this force which is parallel to the z axis of the base coordinate system is 13N. The control problem considered here can be decomposed into two stages. The position controller is constrained to velocities which have to be parallel to the z axis of the base coordinate system. The force controller has constraints in the plane which has to be parallel to xOy plane. In Fig.7.9 the force values along z direction are presented. The complete time of the movement was 12s. Oscillations which are presented in Fig.7.9 are caused by roughness of the metal plate. In the middle of the movement oscillations are bigger crossing the edge of the plate. The compliance coefficient in the force direction was changed in their absolute values from $1 \cdot 10^{-4}$ to $5 \cdot 10^{-5}$.

Now consider a second example of the hybrid control. On a horizontal table there is a plate which is fixed to the table. This plate has a variable profile. The task considered here is to draw a straight line by a scriber mounted at the wrist of the robot. As in the first example a projection of the line on an xOy surface is a straight line. When the input value of a force along z axis (a

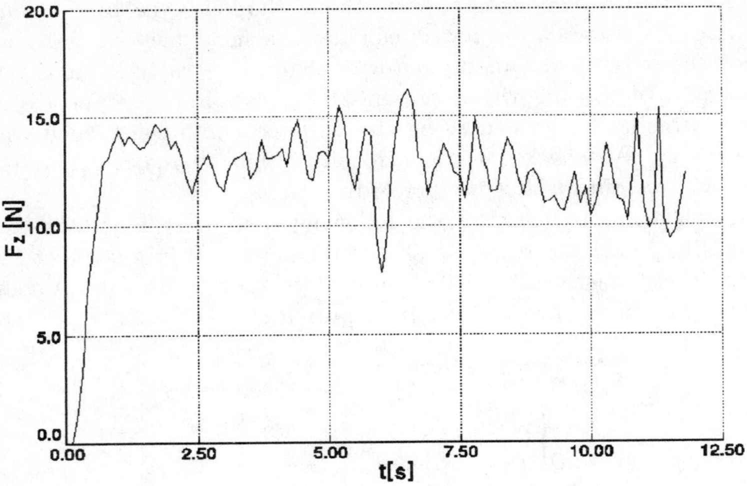

Fig. 7.9. Force values along z direction, $F_z = 13\text{N}$

holding down face) equals to 8N in the first case and 4N in the second case (compare figures Figs.7.10 and 7.11) the input velocity equals 10^{-3}m/s.

From Figs.7.10 and 7.11 one can observe oscillations due to roughness of the metal plate. A control system is not able to compensate for these oscillations. It has been observed that these oscillations are higher in amplitude if the input velocity is faster. It means that a force control system cannot execute these tasks. When the scriber is idle these oscillations are much smaller in amplitude. This situation can be explained as follows. There is a relatively long time delay in the measurement feedback loop. Besides that, discretisation of the force executed on the environment causes oscillations too. Delay in the measurement system is due to data acquisition system of the JR3 sensor. Notice that the 3 forces and torques are measured with high frequency (up to 800Hz), but these results are input to the system with a constant time delay which equals 0.1s. The IRp-6 robot moves its tip in discrete time values defined by position sensors mounted at the joint of the manipulator. It has been observed that the higher the stiffness of a contact robot-environment, the bigger values of the forces.

In Fig.7.10 we can notice a dynamical behaviour of the force control loop. After 3 seconds a contact between the scriber and the metal plate is activated. The input force is reached after 4 seconds and next the oscillations are observed around 8N. From Fig.7.11 is clear that the force control loop results in higher oscillations, around 4N. It was observed that higher oscillations are associated with smaller input force values. Note that the results presented in reference [33] are of a similar nature.

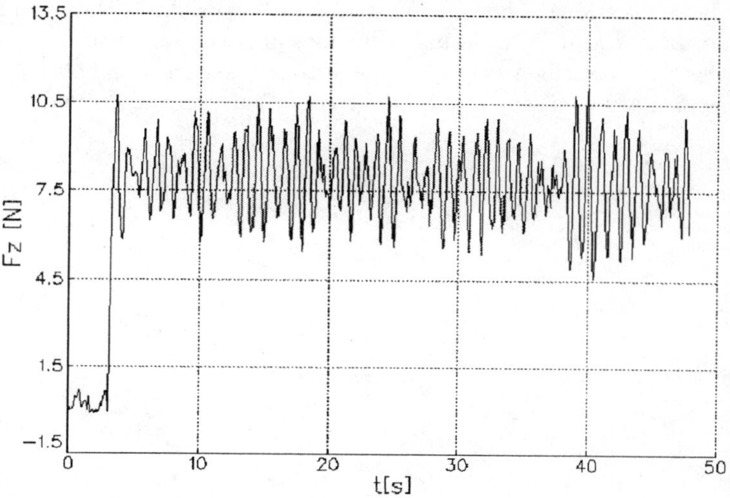

Fig. 7.10. Force values along z direction, $F_z = 8\text{N}$

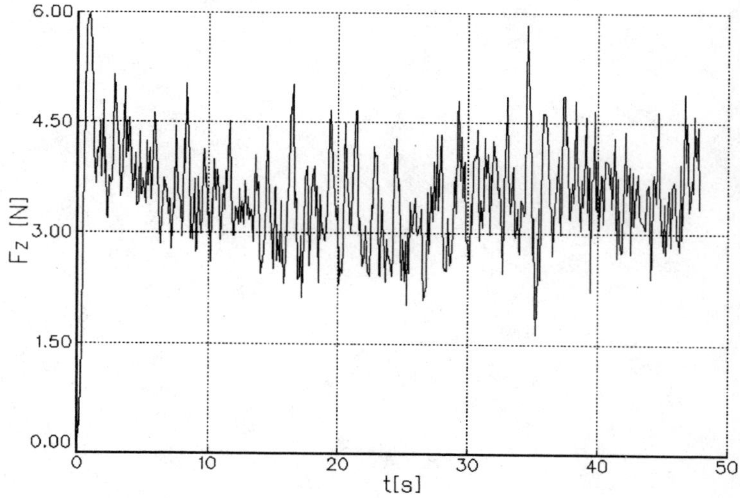

Fig. 7.11. Force values along z direction, $F_z = 4\text{N}$

We have tried different movements in order to test the controller and measurement method of the global compliance of the manipulator (for example peg in hole insertion). The presented approach shows that the controller has good performance in completing tasks in which the forces and torques acting along the prescribed trajectory have to be taken into account as well as on-line calculations of the manipulator compliance in the desired directions.

CHAPTER 8
CONCLUDING REMARKS

In this monograph we present the experimental results concerning the identification of dynamic parameters of the robot and load. Two types of models, in which dynamic parameters appear, were used: the differential and the integral. In general the differential model uses the knowledge of joint generalised positions, velocities, accelerations and torques, while the integral model requires all mentioned signals to be known except the joint accelerations. In both types of models dynamic parameters appear in a linear form. For the purposes of identification we defined canonical models which are expressed in terms of the minimum set of dynamic parameters. In the canonical model, functions which are associated with the dynamic parameters are linearly independent. The parameters which appear in the canonical model are combinations of the link individual dynamic parameters and do not have a clear physical interpretation. Nevertheless existance of the canonical model is a necessary condition for successful identification. A formal proof, which shows that both models are expressed in terms of the same set of parameters, is given in the monograph.

All experimental work was performed on three experimental stands. The first one was built around the IRp-6 robot. The experimental set-up is equipped with several sensors and a data acquisition system which collects data from the different sensors. We developed a high level robot programming language which was implemented for various sophisticated robot programming tasks. Our aim was to keep the programming language as general as possible, however, some parts of the software are dedicated to the IRp-6 robot only. To write the software system took us about two and a half years of work of a group of 8 students studying Robotics and Automation. This part of the work is not seen by the user, but if we want to carry out complicated experiments we need a lot of useful tools which constitute only engineering work and are not purely scientific. The second experimental set–up was designed and built for research purposes and is essentially a one link geared robot. This system has an open architecture and can also be used for control experiments. The third type of robot used in our experiments was the EDDA robot, which belongs to the direct drive class.

From the experimental results presented here, it follows that a kind of trade off exists between engineering intuition and a rigorous approach to the identification problems in robotics. It happens that engineering intuition cannot be neglected in the phase of planning the experiment and in the design of the optimal trajectories for the purpose of robot and load dynamics identification. Obviously, intuition can be wrong, but fortunately very seldom. For example, in optimal trajectory design the initial trajectory for the purpose of identification can be selected intuitively. From this point we can start a rigorous optimisation procedure which usually leads to very good results. Intuition is also very helpful in the planning stage of the identification experiment when we use knowledge about the kinematical structure in order to design the movements of one or two links. By careful designing of the movement we are able to judge which functions associated with dynamic parameters will be excited. We should point out that there are no available results of the optimisation procedure for the integral model trajectory design. This monograph shows that the optimisation of the input trajectory is particularly useful for the integral model where we have less information available. At the same time we have shown that the integral model can also be very useful in identification experiments. Usually researchers claim that the differential model is superior to the integral model. We prove that it is not true.

The next difficult problem solved in the monograph is the friction characteristic construction for the integral model. We propose a special procedure applicable to integral models only, which makes them even more attractive for the purposes of identification. Friction modelling is particularly important for geared robots and this was verified experimentally. The procedure developed here incorporates full Tustin friction modelling and also includes temperature dependence which is seldom considered for robot friction modelling. We also show that by using optimal trajectories it is possible to identify the friction coefficients of the simple friction model which is described in terms of the Coulomb and viscous friction coefficients. Friction modelling was also considered for direct drive robots and all friction phenomena was observed too. As a first approximation we can neglect the friction effects for the direct drive robots. Identification of friction coefficients is a difficult problem, but we have shown that it can be solved either by a special optimal trajectory design or by using the procedure introduced in this monograph. The accuracy of the identified friction coefficients is satisfactory. We compare the coefficients identified from the friction characteristics with the coefficient identified using the optimal trajectories. The accuracy of the identified dynamic parameters of robot and load is high. The identified model was verified by calculating the joint generalised accelerations and next by integrating them and finding the generalised velocities and positions and comparing them with the input trajectories. All results discussed in the monograph are compared to the results obtained in robotics literature using different approaches and different models.

From our experimental work we can deduce that the results of the identification for the integral model are better when we use the short time integral. The long time integral is prone to accumulate errors and in general is not recommended. It is very difficult to identify the friction coefficients for the long time integral. Using the differential model we were able to identify load dynamic parameters for robots which move rather slowly. Unfortunately, the integral model was not useful for the purposes of load identification for the IRp-6 robot. The differential models are more suitable for identification when the robot moves very fast and the friction effects have to be included without necessarily using optimal trajectories. For the integral models it is better to use optimal trajectories particularly when we want to identify the friction coefficients. The considerations presented in this monograph are applicable to geared and non–geared robots.

One of the results of this work is the dynamic model of the first three degrees of freedom of the IRp-6 robot which includes friction modelling. The results presented here are more accurate and more general than those presented in reference [216]. Due to the fact that a few scientific institutions in Poland have this robot is seems that it was worth putting effort into identifying its dynamic model and friction characteristics.

One example of how the experimental set-up can be used is the hybrid control scheme discussed in Chapter 7. We proposed also a method to measure local and global stiffness coefficients of the IRp-6 robot. These measurements have not been carried out before for the IRp-6 robot. Note that we have also implemented the hybrid control scheme for a robot which has only positional controllers.

One of the drawbacks of the set-up discussed here is the lack of the torque control which is required when we want to implement advanced model based control schemes and this prevented us from working on control schemes. Therefore we decided to build the experimental set–up which is also described in the monograph and is the one link geared robot. The set–up has as open architecture and allows control of the manipulator from the current side which is proportional to the torque. The one link geared robot was identified and the results are presented here. Some preliminary results concerning control of the system which is based on the model are presented in the following references [135, 137, 141, 142]. The work is still under progress.

INDEX

REFERENCES

1. An Ch. H., Atkeson Ch. G., Hollerbach J. M. (1988): *Model-based control of a robot manipulator.* The MIT Press, Cambridge
2. Angeles J. (1997): *Fundamentals of robotic mechanical systems theory, methods and algorithms.* Springer-Verlag, New York
3. Armstrong B., Khatib O., Burdick J.(1986): The explicit dynamic model and inertial parameters of the PUMA 560 Arm. *Proceedings of the IEEE International Conference on Robotics and Automation*, 510–518
4. Armstrong B.(1988): Dynamics for robot control, friction modelling and ensuring excitation during parameter identification. Ph. D. Thesis, Stanford University, Dept. of Electrical Engineering
5. Armstrong B. (1989): On finding exciting trajectories for identification experiments involving systems with nonlinear dynamics. *The International Journal of Robotics Research* **8**, 28–48
6. Armstrong-Hélouvry B. (1991): *Control of machines with friction.* Kluwer Academic Publishers, Boston
7. Arnold V.I. (1978): *Mathematical methods of classical mechanics.* Springer-Verlag, New York
8. Asada H., Slotine J. J. (1986): *Robot analysis and control.* J.Wiley, New York
9. Asada H., Youcef-Toumi K. (1987): *Direct drive robots theory and practice.* The MIT Press, Cambridge
10. Atkeson Ch. G., An Ch. H., Hollerbach J. M. (1985): Rigid body load identification for manipulators. *Proceedings of the 24th Conference on Decision and Control.* Ft. Lauderdale, 996–1002
11. Atkeson Ch. G., An Ch. H, Hollerbach J. M. (1986): Estimation of inertial parameters of manipulator loads and links. *The International Journal of Robotics Research.* **5**, 101–119
12. A.Bartosiewicz (1993): Adaptive control of robot manipulators using their simplified dynamical models. Ph.D. Thesis, Lodz University of Technology (in Polish)
13. Bejczy A.K. (1974): Robot arm dynamics and control. Technical Memorandum, 33-669, Jet Propulsion Laboratory
14. Biennis F., Khalil W. (1990): Minimum inertial parameters of robots with parallelogram closed-loops. *Proceedings of the 1990 International Conference on Robotics and Automation.* Cincinnati, 1026–1031
15. Bierman G. J. (1977): *Factorization methods for discrete sequential estimation.* Academic Press, New York
16. Blume C., Jakob W. (1986): *PASRO. Pascal for Robots.* Springer-Verlag, Berlin
17. Brady M., Hollerbach J.M., Johnson T.L., Lozano-Perez T., Mason M.T. (1983): *Robot motion: planning and control.* The MIT Press, Cambridge

18. Bridges M.M., Dawson D.M., Martindale S.C. (1993): An experimental study of flexible joint robots with harmonic drive gearing. *Proceedings of the 2-nd Conference on Control Applications,* Vancouver, 499-504
19. Bryson A. E. and Ho Y. (1975): *Applied optimal control.* Hemisphere Publishing Co., New York
20. Burdick J.W. (1986): An algorithm for generation of efficient manipulator dynamic equations. *Proceedings of the 1986 IEEE International Conference on Robotics and Automation,* San Francisco, 212-218
21. Caccavale F., Chiacchio P. (1994): Energy based identification of dynamic parameters for a conventional industrial manipulator. *Preprints of the Fourth IFAC Symposium on Robot Control.* Capri, 619–624
22. Caccavale F., Chiacchio P. (1994): Identification of dynamic parameters for a conventional industrial manipulator. *Proceedings of the 10th IFAC Symposium on System Identification.* Kopenhagen, 583–588
23. Canudas de Wit C., Aström R. J., Braun K. (1987): Adaptive friction compensation in DC-motor drives. *IEEE Journal of Robotics and Automation.* RA-3, 681–685
24. Canudas de Wit C., Carrillo J. (1990): A modified EN-RLS algorithm for systems with bounded disturbances. *Automatica, Vol* **26**, No. **3**, 599-606
25. Canudas de Wit C., Noël P., Aubin A., Brogliato B. (1991): Adaptive friction compensation in robot manipulators: low velocities. *The International Journal of Robotics Research.* **10**, 189–199
26. Canudas de Wit C., Aubin A. (1991): Robot parameter identification via sequential hybrid algorithm. *Proceedings of the IEEE International Conference on Robotics and Automation.* Sacramento, 952–957
27. Canudas de Wit C., Olsson H., Åström K.J., Lischinsky A. (1995): A new method for control of systems with friction. *IEEE Transactions on Automatic Control,* **40**, 419-425
28. Canudas de Wit C., Siciliano B., Bastin G. (1996): *Theory of robots.* Springer-Verlag London Limited, Berlin, Heidelberg, New York
29. Chedmail P., Gautier M., Khalil W. (1986): Automatic dynamic modelling of robot including parameters of links and actuators. *Proc. of the IFAC/IFIP/IMACS International Symposium on Theory of Robots.* Vienna, 195–199
30. Craig J. J. (1988): *Adaptive control of mechanical manipulators.* Addison-Wesley Publishing Company, Reading, Massachusetts
31. Craig J. J. (1989): *Introduction to Robotics. Mechanics and control.* Addison-Wesley Publishing Company, Reading, Massachusetts
32. Czajka A. (1994): Hybrid control of the IRp-6 robot. M.S. Thesis, Poznań University of Technology (in Polish)
33. Dégoulange E., Dauchez P., Pierrot F. (1993): Force control of an industrial Puma 560 robot under environmental constraints: Implementation issues and experimental results. *Proceedings of the IEEE International Conference on Robotics and Automation,* Atlanta, 213–218
34. Dutkiewicz P., Kozłowski K., Królikowski A., Wróblewski W. (1992): Programming system of the IRp-6 ASEA robot, its model and force and torque sensor feedback. Technical report, No. 1096/55/92/02 for the State Committee for Scientific Research, Warsaw (in Polish)
35. Dutkiewicz P., Kozłowski K., Wróblewski W. (1992): Repetitive control of the IRp-6 robot. Technical report, Poznań University of Technology (in Polish)
36. Dutkiewicz P., Kozłowski K., Wróblewski W. (1993): Experimental identification of robot and load dynamic parameters. *Proceedings of the 2-nd IEEE Conference on Control Applications,* Vancouver, 767–776

37. Dutkiewicz P., Kozłowski K., Wróblewski W. (1993): Experimental identification of dynamic parameters of the IRp-6 robot. *Scientific Papers of the Institute of Technical Cybernetics of the Technical University of Wroclaw, Fourth National Conference on Robotics,* **93**: 77–84 (in Polish)
38. Dutkiewicz P., Kozłowski K., Wróblewski W. (1993): Experimental identification of load dynamics. *Scientific Papers of the Institute of Technical Cybernetics of the Technical University of Wroclaw, Fourth National Conference on Robotics,* **93**: 246–253 (in Polish)
39. Dutkiewicz P., Kozłowski K., Wróblewski W. (1993): Robot programming system for research purposes. *Scientific Papers of the Institute of Technical Cybernetics of the Technical University of Wroclaw, Fourth National Conference on Robotics,* **94**, 335–342 (in Polish)
40. Dutkiewicz P., Kozłowski K., Wróblewski W. (1993): Robot programming for research purposes. *Proceedings of the COMPEURO'93.* Paris, 94–101
41. Dutkiewicz P., Kozłowski K., Wróblewski W. (1993): Experimental identification of load parameters. *Proceedings of the IEEE International Symposium on Industrial Electronics,* Budapest, 361–366.
42. Dutkiewicz P., Kozłowski K. (1994): A force controller for manipulators with closed positioning control system. *Proceedings of the First International Symposium on Mathematical Models in Automation and Robotics,* Międzyzdroje, Poland, 398–403
43. Dutkiewicz P., Kozłowski K. (1994): Identification of robot dynamics with optimal exciting trajectories. *Proceedings of the First International Symposium on Mathematical Models in Automation and Robotics,* Międzyzdroje, Poland, 368–372
44. Dutkiewicz P., Kozłowski K., Wróblewski W. (1994): Experimental results of robot and load dynamic parameters identification. *Automation and Informatics Studies,* **19**, 5–26
45. Dutkiewicz P. (1994): Friction: Experimental determination based on an integral model of the IRp-6 robot.*Proceedings of the First International Symposium on Mathematical Models in Automation and Robotics,* Międzyzdroje, Poland, 416–421
46. Dutkiewicz P., Kozłowski K., Królikowski A., Wróblewski W. (1994): Programming system of the IRp-6 ASEA robot, its model and force and torque sensor feedback. *Bulletin of the Industrial Research Institute for Automation and Measurements,* Warsaw, 51–63 (in Polish)
47. Dutkiewicz P., Kozłowski K., Wróblewski W. (1994): Programming system for the IRp-6 robot. *Archives of Control Sciences.* **3**, 1/2, 51–68
48. Dutkiewicz P., Kozłowski K. (1995): Exciting trajectories for the identification of robot dynamics. *Proceedings of Fourth International Symposium on Measurement and Control,* Smolenice, Slovakia, 371–376
49. Dutkiewicz P., Kozłowski K. (1995): A hybrid controller for the IRp-6 robot. *Proceedings of the IXth Conference of the Electrical Engineering Faculty of the Poznań University of Technology,* Poznań, 201–204 (in Polish)
50. Dutkiewicz P. (1995): Robot and load dynamics identification of the IRp-6 robot with hybrid control and programming elements. Ph.D. Thesis, Poznań University of Technology (in Polish)
51. Dutkiewicz P., Kozłowski K. (1996): Optimal exciting trajectories for robot dynamics identification. *Zeitschrift für Angewandte Mathematik und Mechanik,* **76**: 415–416
52. Dutkiewicz T. (1992): Learning control scheme for the IRp-6 robot. M.S. Thesis, Poznań University of Technology (in Polish)

53. Egeland O. (1986): Adaptive control of a manipulator when the mass of the load is unknown. *Proceedings of the IFAC/IFIP/IMACS International Symposium on Theory of Robots.* Vienna, 251–255
54. Elsefari K., Khalil W. (1991): Energy based indirect adaptive control of robots. *Proceedings of the IEEE International Conference on Robotics and Automation,* Sacramento, 2142–2147
55. Featherstone R. (1987): *Robot dynamics algorithms.* Kluwer Academic Publishers
56. Ferreti G., Maffezzoni C., Magnani G., and Rocco P. (1994): Joint stiffness estimation based on force sensor measurement in industrial manipulators. *Journal of Dynamic Systems, Measurement and Control,* **116**, 163–167
57. Ferreti G., Magnani G., and Rocco P. (1994): Estimation of resonant transfer functions in the joints of an industrial robot. *Proceedings of the 10-th IFAC Symposium on System Identification,* Kopenhagen, 583–588
58. Fu K. S., Gonzalez R. C., Lee C. S. G. (1987): *Robotics control, vision and intelligence.* McGraw-Hill Book Company, Industrial Engineering Series, New York
59. Gautier M. (1986): Identification of robot dynamics. *Proceedings of the IFAC/ IFIP/ IMACS International Symposium on Theory of Robots,* Vienna, 351–356
60. Gautier M., Khalil W. (1988): On the identification of the inertial parameters of robots. *Proceedings of the 27th Conference on Decision and Control,* Austin, 2264–2269
61. Gautier M., Khalil W. (1988): A direct determination of minimum inertial parameters of robots. *Proceedings of the IEEE International Conference on Robotics and Automation,* Philadelphia, 1682–1686
62. Gautier M., Khalil W. (1989): Identification of the minimum inertial parameters of robots. *Proceedings of the IEEE International Conference on Robotics and Automation,* Scottsdale, 1529–1534
63. Gautier M. (1990): Numerical calculation of the base inertial parameters of robots. *Proceeding of the 1990 IEEE International Conference on Robotics and Automation,* Cincinnati, 1020–1025
64. Gautier M., Presse C. (1991): Sequential identification of base parameters of robots. *Proceedings of the Fifth International Conference on Advanced Robotics,* Pisa, 1105–1110
65. Gautier M., Khalil W. (1992): Exciting trajectories for the identification of base inertial parameters of robots. *The International Journal of Robotics Research,* **11**, 362–375
66. Gautier M (1992): Optimal motion planning for robots inertial parameters identification. *Proceedings of the 31st IEEE Conference on Decision and Control,* Tucson, 70–73
67. Gautier M., Khalil W., Presse C., Restrepo P.P. (1994): Experimental identification of dynamic parameters of robot. *Preprints of the Fourth IFAC Symposium on Robot Control,* Capri, 625–630
68. Gautier M., Khalil W., Restrepo P.P. (1995): Identification of the dynamic parameters of a closed loop robot. *Proceedings of the 1995 IEEE International Conference on Robotics and Automation,* Nagoya, 3045–3050
69. Gautier M. (1996): A comparison of filtered models for dynamic identification of robots. *Proceedings of the 35th Conference on Decision and Control,* Kobe, 875-880
70. Goldenberg A.A., Xiaogeng He., Ananthanarayanan S.P. (1992): Identification of inertial parameters of a manipulator with closed kinematic chains. *IEEE Transactions on SMC,* **22**, 799-805

71. Goldenberg A., Song P. (1997): Principles for design of position and force controllers for robot manipulators. To appear in *Robotics and Automation Systems*
72. Golub G. H. and Van Loan C. F. (1989): *Matrix computation.* John Hopkins University Press, Baltimore and London
73. Gosiewski A., Kurman K., Masłowski P., Wieczorek W., Olbrot A. (1984): Investigation over dynamics of the IRp-6 and IRp-50 robots and design of tracking control systems. Research Report of the Institute of Automatic Control of Warsaw University of Technology, Project 06.6, PR-01.03.04, Part I (in Polish)
74. Gosiewski A. (1986): Dynamic intersections in 3-main-axes robot and their influance on CP control. *Proceedings of the IFAC/IFIP/IMACS International Symposium on Theory of Robots,* Vienna, 385-390
75. Graham J. H. (1989): Special computer architectures for robotics: tutorial and survey. *IEEE Transactions on Robotics and Automation,* **5**, 543–554
76. Griff Kräfte und Momente in die Erfassen und Steuern Zukunft (1988). Mess- und Regeltechnik GmbH (technical documentation, in German)
77. Grodecki A., and Gosiewski A. (1993): Hybrid controller for the IRp-6 robot. Part I: Kinematical Description. *Scientific Papers of the Institute of Technical Cybernetics of the Technical University of Wroclaw, Fourth National Conference on Robotics,* **93**, 126–133 (in Polish)
78. Grodecki A. (1993): Hybrid controller for the IRp-6 robot. Part II: Structure of controller and experimental results. *Scientific papers of the Institute of Technical Cybernetics of the Technical University of Wroclaw, Fourth National Conference on Robotics,* **93**, 134–141 (in Polish)
79. Grossman D.D., Taylor R.H. (1978): Interactive Generation of object models with a manipulator, *IEEE Transactions on SMC,* **8**, 667–679
80. Gruver W. A., Soroka B. I., Craig J. J., Turner T. L. (1984): Industrial robot programming languages: A comparative evaluation. *IEEE Transactions on SMC,* **14**, 560–565
81. Gutowski R. (1971): *Analytical mechanics.* Polish Scientific Publishers, Warsaw (in Polish)
82. Ha I., Ko M., Kwon S. K. (1989): An efficient estimation algorithm for the model parameters of robotic manipulators. *IEEE Transactions on Robotics and Automation,* **5**, **3**, 386–394
83. Hayward V., Paul R. P. (1986): Robot manipulator control under unix RCCL: A robot "C" library. *The International Journal of Robotics Research,* **5**, 94–111
84. Hirzinger G. (1989): Issues on low-dimensional sensing and feedback. *IEEE Transactions on SMC,* **19**, **4**, 832–839
85. Hirzinger G., Dietrich J. (1986): Multisensory robots and sensor-based path generation. *Proceedings of the IEEE International Conference on Robotics and Automation,* San Francisco, 1992–2001
86. Hollerbach J.M., Lokhorst D.M. (1995): Closed–loop kinematic calibration of the RSI-6DOF hand controller. *IEEE Transactions on Robotics and Automation,* Vol **11**, No. **6**, 352-359
87. Hsu P., Bodsen M., Sastry S., Paden B. (1987): Adaptive identification and control of manipulators without using joint accelerations. *Proceedings of the 1987 IEEE International Conference on Robotics and Automation,* 1210-1215
88. Ignasiak S. (1997): Verification of the new adaptive control algorithms for a one degree of freedom geared manipulator with joint elasticity. *Automation and Informatics Studies,* **22**, 7-28 (in Polish)
89. Industrial robots IRp-6/60 (1988). Technical documentation. Ostrów Wlkp, Poland (in Polish)
90. ISRA RCT B Benutzerhandbuch (1988): Technical documentation, (in German)

91. Jain A., Rodriguez G. (1992): Recursive dynamics for manipulators with joint flexibility. Unpublished document
92. Jezierski E. (1993): Experiments with algorithms estimation the parameters of a robot drive system. *Proceedings of the Conference on Modern Control for Power Electronics and Electrical Drives,* Vol.1, Dobieszkow, 200–213 (in Polish)
93. JR3 Universal Force-Torque Sensor System (1991): Technical documentation
94. Juang J-N. (1994): *Applied system identification.* PTR Prentice Hall Englewood Cliffs, New Jersey
95. Kallenbach R. (1987): *Kovariantzmethoden zur Parameteridentification Zeitkontinuierlicher Systeme.* VDI Verlag, Stuttgart
96. Karaśkiewicz E. (1976): *Theory of vectors and tensors.* Polish Scientific Publishers, Warsaw (in Polish).
97. Kasiński A., Kozłowski K., Wróblewski W. (1988): Optimisation methods for nonlinear robot control. Scientific report for Polish Academy of Sciences, Institute of Fundamental Problems of Technology. Part III, 46–72 (in Polish)
98. Kasiński A., Kozłowski K., Pińczak M. (1989): Optimisation methods for nonlinear robot control. Scientific report for Polish Academy of Sciences. Institute of Fundamental Problems of Technology, Part IV, 21–94 (in Polish)
99. Khalil W, Kleinfinger J. F., Gautier M. (1986): Reducing the computational burden of the dynamic model of robots. *Proceedings of the 1986 IEEE International Conference on Robotics and Automation,* San Francisco, 525–531
100. Khalil W., Kleinfinger J.F. (1986): A new geometric notation for open and closed-loop robots. *Proceedings of the IEEE International Conference on Robotics and Automation,* San Francisco, 1174–1179
101. Khalil W., Gautier M. (1986): Identification of geometric parameters of robots. *Proceedings of the IFAC/IFIP/IMACS International Symposium on Theory of Robots,* Vienna, 191–194
102. Khalil W., Gautier M., Kleinfinger J.F. (1986): Automatic generation of identification models of robots. *International Journal of Robotics and Automation,* **1**, 2–6
103. Khalil W., Kleinfinger J. F. (1987): Minimum operations and minimum parameters of the dynamic models of tree structure robots. *IEEE Journal of Robotics and Automation,* RA-3, 517–526
104. Khalil W., Bennis F., Chevallereau C. and Kleinfinger J. F. (1989): SYMORO: A software package for the symbolic modelling of robots. *Proceedings of the 20th ISIR,* Tokyo
105. Khalil W., Bennis F., Gautier M. (1990): The use of the generalised links to determine the minimum inertial parameters of robots. *Journal of Robotic Systems,* **7**, **2**, 225–242
106. Khalil W. (1994): A system for generating the symbolic models of robots. *Preprints of the Fourth IFAC Symposium on Robot Control,* Capri, 527–634
107. Khatib O. (1987): A unified approach for motion and force control of robot manipulators, the operational space formulation. *IEEE Journal of Robotics and Automation,* RA-3, 43–52
108. Khosla P. K. (1986): Real-time control and identification of direct-drive manipulators. Ph.D. Thesis, Cornegie Mellon University, Department of Electrical and Computer Engineering, The Robotics Institute
109. Khosla P.K. (1989): Categorization of parameters in the dynamic robot model. *IEEE Transactions on Robotics and Automation,* **5**, 261–268
110. Klema V. C., Laub A. J. (1980): The singular value decomposition: its computation and some applications. *IEEE Transactions on Automatic Control,* **25**, 164–176

111. Kominek J. (1993): Model-based control of the IRp-6 robot. M.S. Thesis, Poznań University of Technology (in Polish)
112. Kozłowski K., Kasiński A., Wróblewski W. (1987): Optimisation methods for nonlinear robot control. Scientific report for Polish Academy of Sciences, Institute of Fundamental Problems of Technology, Part II (in Polish)
113. Kozłowski K. (1988): Identification problems in robotics. *Scientific Papers of the Institute of Technical Cybernetics of the Technical University of Wroclaw, Second National Conference on Robotics,* **76**, **2**, 151–158 (in Polish)
114. Kozłowski K. (1988): A Pascal–based robot programming language. *Proceedings of the X International Symposium, Computer at the University,* Dubrownik, 3.18.1–3.18.7
115. Kozłowski K (1988): A Pascal–based robot programming language. *Scientific Papers of the Institute of Technical Cybernetics of the Technical University of Wroclaw,* **76**, 159–166 (in Polish)
116. Kozłowski K., Romanowski K., Warczyński J. (1989): Industrial robots. Scientific report for the Industrial Research Institute for Automation and Measurements (in Polish)
117. Kozłowski K. (1989): Dynamics of robot manipulators. *Automation and Informatics Studies,* **15**, 47–79 (in Polish)
118. Kozłowski K. (1989): POLROB - A manipulator level programming language. *IFAC Proceedings of the Symposium on Robot Control (SYROCO'88),* Karlsruhe, 453–458
119. Kozłowski K., Wróblewski W. (1990): Optimisation method for nonlinear robot control. Scientific report for the Polish Academy of Sciences, Institute of Fundamental Problems of Technology, Part V (in Polish)
120. Kozłowski K. (1991): POLROB - A manipulator level programming language based on Pascal. *Journal of Microcomputer Applications,* **14**, 49–60
121. Kozłowski K. (1991): POLROB – A programming system for robot manipulators. *Automation and Informatics Studies,* **16–17**, 31–47 (in Polish)
122. Kozłowski K. (1992): Mathematical dynamic robot models and identification of their parameters. Poznań University of Technology Press, Habilitation Thesis, Poznań (in Polish)
123. Kozłowski K. (1992): Identification of model parameters of robotic manipulators. *Systems Science,* **18**, **2**, 165–187
124. Kozłowski K., Prüfer M. (1992): Parameter identification of an experimental 2-link direct drive arm. *Proceedings of the IASTED International Conference, Control and Robotics,* Vancouver, 313–316
125. Kozłowski K. (1993): Computational requirements for a discrete Kalman filter in robot dynamics. *Robotica,* **11**, 27–36
126. Kozłowski K., Wróblewski W., Dutkiewicz P. (1994): Development and experimental validation of real-time inverse and forward dynamics algorithms for an industrial robot. Reports on research and developments assisted in fiscal year 1991, Oshinomura, Foundation for Promotion of Advanced Automation Technology, Japan, 29–32
127. Kozłowski K. at al. (1994): Implementation of a force sensor in hybrid controller of the IRp-6 industrial robot. Scientific report No.94/45–025 (in Polish)
128. Kozłowski K., Dutkiewicz P. (1995): An experimental study on a one link geared robot. *Proceedings of the Second International Symposium on Methods and Models in Automation and Robotics,* Międzyzdroje, Poland, 585–590
129. Kozłowski K. (1995): Modelling of the geared robot dynamics. *Proceedings of the First National Conference of MATLAB Users,* Kracow, 68–75 (in Polish)

130. Kozłowski K., Dutkiewicz P., Dawidowska J., Mularczyk A., Kamoś P., Woźniak K. (1995): Optimal trajectory design for the robot dynamics identification. *Proceedings of the IXth Conference on Electrical Engineering Faculty of the Poznań University of Technology,* Poznań, 197–200 (in Polish)
131. Kozłowski K., Dutkiewicz P. (1995): Identification of friction and robot dynamics for a class of geared robots. *Proceedings of the Fourth International Symposium on Measurement and Control in Robotics,* Smolenice, Slovakia, 301–306
132. Kozłowski K., Dutkiewicz P. (1995): Optimal trajectory design for the identification of robot and load dynamics. *Applied Mathematics and Computer Science,* **5**, **4**, 101–117
133. Kozłowski K, Królikowski A., Wróblewski W., Dutkiewicz P. (1996): Acquisition, knowledge processing and parallel computations of the optimal trajectory for robots. Scientific report for the State Committee for Scientific Research, Warsaw (in Polish)
134. Kozłowski K., Dutkiewicz P. (1996): Experimental identification of robot and load dynamics. *13-th IFAC World Congress, San Francisco,* Vol. A, 397–402
135. Kozłowski K., Dutkiewicz P. (1996): Adaptive control of a one link geared robot. *Proceedings of the IEEE International Symposium on Industrial Electronics,* Warsaw, 990–995
136. Kozłowski K., Dutkiewicz P. (1996): Experimental identification of dynamic parameters for a class of geared robots. *Robotica,* **14**, 561–574
137. Kozłowski K., Dutkiewicz P. (1996): Robust control of a one link geared robot. *Third International Symposium on Methods and Models in Automation and Robotics,* Międzyzdroje, Poland, 947-954
138. Kozłowski K., Dutkiewicz P. (1996): Robot trajectory planning with force and torque sensor measurements. *Proceedings of the IEEE-SMC, CESA'96 IMACS Multiconference,* Lille, 676–681 (invited paper)
139. Kozłowski K., Dutkiewicz P. (1996): *Modelling and identification in robotics.* Poznań University of Technology Press, Poznań (in Polish)
140. Kozłowski K. (1996): Optimal trajectory design for identification purposes in robotics. *Proceedings of the Third International Symposium on Methods and Models in Automation and Robotics,* Międzyzdroje, Poland, 871–877 (invited paper)
141. Kozłowski K. (1997): Adaptive control of robots with gears. *Proceedings of the 30th International Symposium on Automotive Technology and Automation,* Florence, 403-410 (invited paper)
142. Kozłowski K. (1997): Experimental verification of control algorithms for a one link geared robot. *Proceedings of the 1997 IEEE International Conference on Robotics and Automation,* Albuquerque, 2240-2245
143. Kręglewska U., Kruszyński H, Szynkiewicz W. (1988): An experimental set-up for the IRb-6 robot. *Scientific Papers of the Institute of Technical Cybernetics of the Technical University of Wroclaw, Second National Conference on Robotics,* **35**, 189–192 (in Polish)
144. Kręglewska U. (1990): Direct access interface between IBM AT and the IRp-6 robot. *Scientific Papers of the Institute of Technical Cybernetics of the Technical University of Wroclaw, Third National Conference on Robotics,* **37**, 90–95 (in Polish)
145. Kuncewicz M., Foremniak P. (1995): Design and investigations of a one link geared robot. M.S. Thesis, Poznań University of Technology (in Polish)
146. Lee C. S. G. (1982): Robot arm kinematics, dynamics and control. *Computer,* **15**, 62–80
147. Lee C. S. G., Gonzalez R. C. (1986): *Tutorial on robotics.* IEEE Computer Society

148. Lee C. S. G., Lee H.B. (1987): Development of generalized d'Alembert equations of motion for robot manipulators. *IEEE Transactions on SMC,* **17**, 311–325
149. Leyko J. (1978): *Mechanics.* Polish Scientific Publishers, Warsaw (in Polish)
150. Lin S. K. (1994): Identification of a class of nonlinear deterministic systems with application to manipulators. *IEEE Transactions on Automatic Control,* **39**, 1886–1893
151. Lin S. K. (1992): An identification method for estimating the inertia parameters of a manipulator. *Journal of Robotic Systems,* **4**, 505–528
152. Ljung L., Söderström (1983): *Theory and practice of recursive identification.* The MIT Press, Cambridge, Massachusetts
153. Lozano-Perez T. (1983): Robot programming. *Proceedings of the IEEE,* **71**, 821–841
154. Lu Z., Shimoga K. B., Goldenberg A. A. (1993): Experimental determination of dynamic parameters of robotic arms. *Journal of Robotic Systems,* **10**, 1009–1029
155. Małecki J. (1987): On some practical aspects of numerical generation of equations of the dynamics of the manufacturing robots, *Proceedings of the Robcon,* Sofia, **4**, 227–235
156. Mareels I. M. Y., Bitmaed R. R., Gevers M. et al. (1987): How exciting can a signal really be. *System and Control Letters,* **8**, 197–204
157. Mason M.T. (1981): Compliance and force control for computer controlled manipulators. *IEEE Transactions of SMC,* **11**, **6**, 418–432
158. Mayeda H., Osuka K., Kangawa A. (1984): A new identification method for serial manipulator arms. *Preprints of the IFAC 9th World Congress,* Budapest, **4**, 74–79
159. Mayeda H., Yoshida K., Osuka K. (1990): Base parameters of manipulator dynamic models. *IEEE Transactions on Robotics and Automation,* **6**, 312–321
160. McInnis B. C., Chen-Kang F. L. (1986): Kinematics and dynamics in robotics: A tutorial based upon classical concepts of vectorial mechanics. *IEEE Journal of Robotics and Automation,* RA-2, **4**, 181–187
161. Middleton R.H., Goodwin G.C. (1988): Adaptive computed torque control for rigid link manipulators. *Systems and Control Letters,* **10**, 9-16
162. Moder K. (1993): Compilator output of the IRL Language in terms of the VRS-1 instructions. M.S. Thesis, Poznań University of Technology (in Polish)
163. Morecki A. (1987): Fundamental problems of technology. *Vol XXV, Robotics, Polish Scientific Publishers,* Warsaw (in Polish)
164. Mujtaba M. S., Goldman R., Bindford T. (1982): The AL robot programming language. *Computers in engineering,* G. D. Gupta (ed), 77–86
165. Murphy S.H., Wen J.T., Saridis G.N. (1990): Recursive calculation of geared robot manipulator dynamics. Department of Electrical, Computer and Systems Engineering, CIRSSE Document #49
166. Murphy S. H., Wen J. T., Saridis G. N. (1990): Simulation and analysis of flexible joined manipulators. Center for Intelligent Robotic System for Space Exploration Document #56, Rensselaer Polytechnic Institute
167. Murphy S.H., Wen J.T., Saridis G.N. (1990): Recursive calculation of geared robot manipulator dynamics. *Proceedings of the 1990 IEEE International Conference on Robotics and Automation,* Cincinnati, 839-844
168. Müller L. (1972): *Gear transmissions: strength calculations.* Scientific Technical Publishers, Warsaw (in Polish)
169. Narasimhan S., Siegel D. M., Hollerbach J. M. (1989): CONDOR: An architecture for controlling the Utah-MIT dexterous hand. *IEEE Transactions on Robotics and Automation,* **5**, 616–627

170. Standard ISO/DP 10562–1–ICR–Intermediate Code for Robots (1990)
171. *Standard DIN 66 312 Industrieroboter Programiersprachen Industrial Robot Language (IRL)* (1992). Berlin, Verlag GmbH
172. Nowicki K. (1989): Identification of robot dynamic parameters. M.S. Thesis, Poznań University of Technology (in Polish)
173. Olsen H. B., Bekey G. A. (1986): Identification of robot dynamics. *Proceedings of the 1986 IEEE International Conference on Robotics and Automation,* San Francisco, 1004–1010
174. Otter M., Türk S. (1988): The DFVLR models 1 and 2 of the r3 robot. *Technischer Bericht 88–13, DFVLR, Oberpfaffenhofen,* Institut für Dynamic der Flugsysteme
175. Papoulis A. (1965): *Probability, random variables, and stochastic processes.* Mc Graw–Hill, Inc.
176. Paul R. P. (1981): *Robot Manipulators: mathematics, programming and control,* The MIT Press, Cambridge
177. Pfeiffer F., Hölz J. (1995): Parameter identification for industrial robots. *Proceedings of the 1995 IEEE International Conference on Robotics and Automation,* Nagoya, 1469–1476
178. Piotrowski L. (1988): Interactive programming system for robots. M.S. Thesis, Poznań University of Technology (in Polish)
179. Presse C., Gautier M. (1993): New criteria of exciting trajectories for robot identification. *Proceedings of the IEEE International Conference on Robotics and Automation,* Atlanta, 907–912
180. Presse C., Gautier M. (1991): Identification of robot parameters via sequential energy method. *Proceedings of the Third IFAC Symposium on Robot Control,* Vienna, 117–122
181. Prüfer M., Kozłowski K. (1992): Identification of inertial and friction parameters of an experimental 2-link direct drive arm. Technischer Bericht 2-92-1, Institut für Robotik und Prozessinformatik, Technische Universität Braunschweig
182. Prüfer M., Wahl F. (1993): Analyse und Vorkompensation von Reibungseffecten bei Industrierobotrn mit Getrieben, *VDI Berichte,* **1094**, 597–605
183. Prüfer M., Schmidt C., Wahl F. (1994): Identification of robot dynamics with differential and integral models: A comparison. *Proceedings of the IEEE International Conference on Robotics and Automation,* San Diego, 340–345
184. Prüfer M., Wahl F. (1994): Friction analysis and modelling for geared robots. *Preprints of the Fourth IFAC Symposium on Robot Control,* Capri, 551–556
185. Prüfer M. (1995): Reibungsanalyse und Identifikation von Dynamikparametern von direktangetriebenen und getriebebehafteten Robotern. Ph.D. Thesis, Verlag Shaker, Aachen
186. Qu Z., Dawson D.M. (1996): *Robust tracking of robot manipulators.* IEEE Press, New York
187. Raibert H. M., Craig J. J. (1981): Hybrid position force control of manipulators. *Journal of Dynamic Systems, Measurement, and Control,* **102**, 126–133
188. Ranky P. G., Ho C. Y. (1985): *Robot modelling, control and applications with software.* Springer–Verlag, Berlin
189. Rembold U., Dillmann R. (1986): *Computer-aided design and manufacturing,* Springer–Verlag, Berlin
190. Rodriguez G. (1987): Kalman filtering, smoothing and recursive robot arm forward and inverse dynamics. IEEE *Journal of Robotics and Automation,* **RA-3**, 624-639
191. Schaefers J., Xu S. J., Darouach M. (1994): On parameter estimation of industrial robots without using acceleration signal, *Proceedings of the 10th IFAC Symposium on System Identification,* Kopenhagen, 589–593

192. Schiehlen W. O., Kallenbach R. G. (1986): Identification of robot system parameters. *IFAC/IFIP/IMACS Symposium on Theory of Robots,* Vienna, 339–444
193. Schmidt C., Prüfer M. (1997): Parameter identification in robot control. *Applied Mathematics and Computer Science,* **7**, 377-399
194. Sciavicco L., Siciliano B., Villani L. (1994): On dynamic modelling of gear–driven rigid body manipulators. *Preprints of the Fourth IFAC Symposium on Robot Control,* Capri, 543-549
195. Sciavicco L., Siciliano B. (1996): *Modeling and control of robot manipulators.* The McGraw-Hill Companies, Inc. New York
196. Seeger G., Leonhard W. (1989): Estimation of rigid body models for a six-axis manipulator with geared electric drives. *Proceedings of the IEEE International Conference on Robotics and Automation,* Scottsdale, 1690–1695
197. Seeger S. (1991): Self-tuning control of a commercial manipulator based on an inverse dynamic model. *Proceedings of the Symposium on Robot Control SYROCO'91,* Vienna, 453–458
198. Selig J.M. (1996): *Geometrical methods in robotics.* Springer-Verlag, New York
199. Seyfferth W., Maghzal A.J., Angelas J. (1995): Nonlinear modelling and parameter identification of harmonic drive robotic transmissions. *Proceedings of the 1995 IEEE International Conference on Robotic and Automation,* Nagoya, 3027-3032
200. Shapiro R. (1978): Direct Linear Transformation Method for Three–Dimensional Cinematography. *The Research Quartely, Vol 49,* No. 2, 197-205
201. Shaumma J.S., Whitney D.E. (1987): A method for inverse robot calibration. *ASME Journal of Dynamic Systems, Measurement, and Control, Vol* **109**, 36-43
202. Shih-Ying Sheu., Walker M.M. (1989): Estimating the essential parameter space of the robot manipulator dynamics. *Proceedings of the 28th Conference on Decision and Control,* Tampa, Florida, 2135–2140
203. Shih-Ying Shue, Walker M. W. (1989): Basic sets for manipulator inertial parameters. *Proceedings of the IEEE International Conference on Robotics and Automation,* Scottsdale, 1517–1522
204. Shih-Ying Sheu, Walker M.W. (1990): Dynamic modeling and analysis of a geared robot manipulator. *Proceedings of the 29th Conference on Decission and Control*
205. Shich-Ying Sheu, Walker M. W. (1993): Identifying the independent inertial parameter space of robot manipulators. *The International Journal of Robotics Research,* **10**, **6**, 668–683
206. Silver W. M. (1982): On the equivalence of Lagrangian and Newton-Euler dynamics for manipulators. *The International Journal of Robotics Research,* **1**, **2**, 60–70
207. Slotine J-J. E., Li W. (1991): *Applied nonlinear control.* Prentice Hall, Englewood Cliffs, New Jersey
208. Specht R., Isermann R. (1989): On-line identification of inertia, friction and gravitational forces applied to an industrial robot. *IFAC Proceedings Series SYROCO'88, Pergamon Press,* Oxford, 219–224
209. Spong M. W. (1987): Modelling and control of elastic joint robots. *Journal of Dynamic Systems, Measurement, and Control,* **109**, 310–319
210. Spong M. W., Vidyasagar M. (1989): *Robot dynamics and control.* John Wiley and Sons, New York
211. Swevers J., Ganseman C., Bilgin Tükel D., De Schutter J., Van Brussel H.: Optimal robot excitation and identification. *IEEE Transactions on robotics and Automation, Vol. 13,* No. 5, 1997, 730 - 740

212. Symon K.R. (1971): *Mechanics.* Addison-Wesley Publishing Company, Reading, Massachusetts
213. Szkodny T. (1990): *Mathematical models of robot manipulators.* Silesian Technical University Press, Gliwice (in Polish)
214. Türk S. (1988): Dynamische Robotermodelle am Beispiel des Manutec r3. Technischer Bericht 88-16, DFVLR
215. Türk S. (1990): Zur Modellierung der Dynamik von Robotern mit rotatorischan Gelenken. *Fortschritt–Berichte VDI,* Reihe 8, Band 211, VDI Verlag
216. Szynkiewicz W., Gosiewski A., Janecki D. (1990): Experimental verification of dynamics model of the IRp-6 robot. *Scientific Papers of the Institute of Technical Cybernetics of the Technical University of Wroclaw, Third National Conference on Robotics,* **37**, 255–260 (in Polish)
217. Takase T., Paul R. P., Berg E. J. (1981): A structured approach to robot programming and teaching. *IEEE Transactions on SMC,* **11**, 514–529
218. Tarn T. J., Bejczy A. K., Shuotiao Han, Xiaoping Yun (1995): Inertial parameters of PUMA 560 robot arm. Department of Systems Science and Mathematics, Washington University, Robotics Laboratory Report, SSM-RL-85-01
219. Tuttle T. D. (1992): Understanding and modeling the behaviour of a harmonic drive gear transmissions. M.S. Thesis MIT Artificial Intelligence Laboratory
220. Uhl T., Lisowski W., Bojko T., Ród J. (1995): Identification of robotic joints using model adjustment technics. *Proceedings of the Second International Symposium on Method and Models in Automation and Robotics,* Międzyzdroje, Poland, 659–664
221. Van Der Linden G. W., Van Der Weiden A. J. J. (1994): Practical rigid body parameter estimation. *Preprints of the Fourth IFAC Symposium on Robot Control,* Capri, 631–636
222. Vandanjon P. O., Gautier M., Desbats P. (1995): Identification of robot inertial parameters by means of spectrum analysis. *Proceedings of the 1995 IEEE International Conference on Robotics and Automation,* Nagoya, 3033–3038
223. Volpe R., Khosla P. (1993): Theoretical and experimental investigation of explicit force control strategies for manipulators. *IEEE Transactions on Automatic Control,* **38**, **11**, 1634–1650
224. Vukobratovic M., Tuneski A. (1994): Contact control concepts in manipulation robotics-an overview. *IEEE Transactions on Industrial Electronics,* **41**, 12–23
225. Vukobratovic M., Potkonjak V. (1982): *Dynamics of manipulation robots.* Springer Verlag, Berlin
226. Voltz R. A. (1988): Report of the programming language working group: NATO workshop on robot programming languages. *IEEE Journal of Robotics and Automation,* **4**, 86–90
227. Wampler C.W., Hollerbach J.M., Arai T. (1995): An implicit loop method for kinematics calibration and its application to closed–chain mechanisms. *IEEE Transactions on Robotics and Automation,* Vol **11**, No. **5**, 710-724
228. Wang L.T., Ravani B. (1985): Recursive computations of kinematic and dynamic equations for mechanical manipulators. *IEEE Journal of Robotics and Automation,* **RA-1**, 124-131
229. Wittenburg J. (1977): *Dynamics of systems of rigid bodies.* B. G. Teuber, Stuttgart
230. Wolfram S. (1992): *Mathematica, a system for doing mathematics.* Addison-Wesley Publishing Company, Reading, Massachusetts
231. Wolski G. (1991): Identification of dynamic parameters of the IRp-6 integral model. M.S. Thesis Poznań University of Technology (in Polish)

232. Woźniak E. (1985): Robot programming languages. *Scientific Papers of the Institute of Technical Cybernetics of the Technical University of Wroclaw, First National Conference on Robotics,* 251–263 (in Polish)
233. Wróblewski W. (1992): Robot dynamics models for the purpose of control. Ph.D. Thesis, Poznań University of Technology (in Polish)
234. Yang D.C.H., Tzeng S.W. (1996): Simplification and linearization of manipulator dynamics by the design of inertia distribution. *The International Journal of Robotics Research,* **5**, 120-128
235. Yoshikawa T. (1990): *Fundamentals of robotics. Analysis and control.* The MIT Press, Cambridge
236. Zieliński C., Grodecki A., Kreglewska U., Śluzek A., Zielińska T. (1990): The concept of robot controllers designed for control purposes. *Scientific Papers of the Institute of Technical Cybernetics of the Technical University of Wroclaw, Third National Conference on Robotics,* **37**, 336–341 (in Polish)
237. Zieliński C., Śluzek A. (1985): Kinematical aspects of control of robots with 5 degrees of freedom. *Scientific Papers of the Institute of Technical Cybernetics of the Technical University of Wroclaw, First National Conference on Robotics,* **66**, 121–129 (in Polish)
238. Zieliński C. (1990): Description of semantics of robot programming instructions. *Archives of Control Engineering and Telemechanics.* **35**, 15–45
239. Zieliński C. (1995): *Robot programming methods.* Warsaw University of Technology Press, Warsaw
240. Zimmermann U., Wunderlich H., Rake H. and Bruns M. (1989): Identification of time varying parameters of the robot dynamics. *IFAC Proceedings Series 1989, SYROCO'88,* Pergamon Press, Oxford, 225–229